Interfacial Properties of Petroleum Products

CHEMICAL INDUSTRIES

A Series of Reference Books and Textbooks

Founding Editor

HEINZ HEINEMANN
Berkeley, California

Series Editor

JAMES G. SPEIGHT
Laramie, Wyoming

Interfacial Properties of Petroleum Products

Lilianna Z. Pillon
Petroleum Products Consulting
Sarnia, Ontario, Canada

CRC Press
Taylor & Francis Group
Boca Raton   London   New York

CRC Press is an imprint of the
Taylor & Francis Group, an **Informa** business

CRC Press
Taylor & Francis Group
6000 Broken Sound Parkway NW, Suite 300
Boca Raton, FL 33487-2742

First issued in paperback 2020

© 2008 by Taylor & Francis Group, LLC
CRC Press is an imprint of Taylor & Francis Group, an Informa business

No claim to original U.S. Government works

ISBN-13: 978-0-367-57761-2 (pbk)
ISBN-13: 978-1-4200-5100-1 (hbk)

Library of Congress Cataloging-in-Publication Data

Pillon, Lilianna Z.
  Interfacial properties of petroleum products / Lilianna Z. Pillon.
    p. cm. -- (Chemical industries ; 120)
  Includes bibliographical references and index.
  ISBN 978-1-4200-5100-1 (hardback : alk. paper)
  1. Petroleum products--Refining. 2. Petroleum products--Stability. 3. Petroleum--Refining. 4. Petroleum--Stability. 5. Gas-liquid interfaces. I. Title. II. Series.

TP690.P54 2007
665.5'38--dc22   2007022604

Visit the Taylor & Francis Web site at
http://www.taylorandfrancis.com

and the CRC Press Web site at
http://www.crcpress.com

Dedication

To
David, Sylvia, Monica, and Samantha
for their encouragement and patience

Contents

Preface

Crude oils are complex mixtures of different molecules and their phase stability is dependent on many factors. The various issues related to crude oil stability, such as asphaltene and wax deposition, were studied and many predictive models were developed. With an increase in exploitation of many oil fields, the properties of crude oils are changing. Many high value crude oils have an increased viscosity, metal, salt, and acid contents leading to deterioration in their interfacial properties. Poor properties of crude oils at their oil/air, oil/water, and oil/metal interfaces lead to increased foaming tendencies, stable emulsions, rust, and corrosion. Crude oils are the feedstocks used to manufacture different petroleum products, such as fuels, lube oil base stocks, wax, and asphalts, and their transportation and storage are becoming more difficult.

Refining of crude oils is based on removing unwanted molecules or modifying their chemistry to increase the yields and improve properties of petroleum products. After desalting and distillation, many petroleum oils have poor stability leading to sediment formation, rust, and corrosion. After additional processing, such as solvent extraction, dewaxing, and finishing, the lube oil base stocks foam when mixed with air, form emulsions when mixed with water, and cause rusting and corrosion. After hydroprocessing, catalyst deactivation and poor stability of some petroleum products increase the processing cost. Hundreds of additives, including surface active molecules, are being developed and used to improve the interfacial properties of petroleum products; however, while their use is cost effective, their effect is limited. As the viscosity of petroleum products increases, from fuels to lube oil base stocks, their interfacial properties change which should be of interest to refining of heavy oils from oil sands.

Acknowledgments

I would like to acknowledge the help of Dr. James G. Speight, the editor of the *Petroleum Science and Technology* journal, in publishing this book. The data used in this book are based on many literature searches and I would like to acknowledge the help provided by many libraries, such as the University of Windsor, the University of Western Ontario, McMaster University, and Sarnia Public Library in Canada. The many tables with the experimental data, included in this book, are the result of obtaining copyright permissions from many publishers around the world and I would like to thank the many authors for the use of their technical data.

Author

Lilianna Z. Pillon grew up in Poland where she received the MSc degree in chemistry from the University of Lodz. She came to Canada to obtain a doctorate degree in chemistry and graduated with a PhD degree from the University of Windsor in Ontario. She was awarded a National Research Council Canada Postdoctoral Fellowship and studied polymer blends at the NRC Industrial Materials Research Institute in Boucherville, Quebec. She joined Polysar Ltd., Latex R&D Division, where she developed an interest in polymer emulsions and was awarded the first prize for the best technical presentation on "Crosslinking Systems for Latex Materials" at the Polysar Global Technology Conference. She also worked for Imperial Oil, Research Department, where she was awarded patents related to interfacial properties of petroleum products and the use of surface active additives.

1 Crude Oils

1.1 STABILITY

1.1.1 VARIATION IN COMPOSITION

Petroleum and the equivalent term "crude oil" is a mixture of gaseous, liquid, and solid molecules that occur in rock deposits found in different parts of the world. Oil recovery from porous sedimentary rocks depends on the efficiency with which oil is displaced by some other fluids (Morrow, 1991). The literature reported on enhanced oil recovery (EOR) with microemulsions (Miller and Qutubuddin, 1987). The flow behavior of crude oil, gas, and brine in the porous rock medium of petroleum reservoirs is controlled largely by the interactions occurring at the interfaces within the various fluids. Natural gas is found in many petroleum reservoirs and, in some cases, is the only occupant. The main component of natural gas is methane but other hydrocarbons, such as ethane, propane, and butane, are also present. The sulfur, oxygen, and nitrogen contents of crude oils vary. Early literature reported that the oxygen and nitrogen contents of crude oils are small but the sulfur content can be as high as 2–3 wt% (Roberts, 1977). Crude oils contain different hydrocarbon and heteroatom containing molecules and the comprehensive review of the composition and the chemistry of crude oils was published (Speight, 2006). The content and the chemistry of typical heteroatom containing molecules found in crude oils are shown in Table 1.1.

The carbon content of crude oils was reported to vary in a range of 83–87 wt% and the hydrogen content in a range of 10–14 wt%. There are basically two types of hydrocarbons, saturated and aromatics. According to the literature, the olefinic hydrocarbons are usually not present in crude oils (Speight, 2006). The sulfur content can vary in a range of 0–6 wt% and the sulfur compounds found in crude oils include thiols, sulfides, and thiophenes. The oxygen content of crude oils is usually less than 2 wt% and the typical oxygen compounds are alcohols, ethers, acids, and phenols. The presence of ketones, esters, and anhydrides was also reported (Speight, 2006). The nitrogen content of crude oils is low, in the range of 0.1–0.9 wt%, and the typical nitrogen molecules are pyridines, quinolines, pyrroles, and indoles. The proportion of saturated hydrocarbons varies with the type of crude oil and can be as high as 75 wt% but it decreases with an increase in their molecular weight (Speight, 2006). The typical chemistry of hydrocarbons found in crude oils is shown in Table 1.2.

TABLE 1.1

Content and Chemistry of Heteroatom Molecules Found in Crude Oils

Content	Range, wt%	Typical Chemistry
Carbon	83–87	Hydrocarbons
Hydrogen	10–14	Hydrocarbons
Sulfur	0–6	Thiols, sulfides, thiophenes
Oxygen	0–2	Alcohols, ethers, acids, phenols
Nitrogen	0–0.9	Pyridines, quinolines, pyrroles, indoles.

Source: From Speight, J.G., *The Chemistry and Technology of Petroleum*, 4th ed., CRC Press, Boca Raton, 2006.

Around 1940, it was discovered that n-paraffins formed solid adducts with urea in methyl alcohol solution. The n-paraffin-urea adducts were used to separate n-paraffins, C_6 to C_{20}, from hydrocarbon mixtures by filtration. Thiourea was found to form adducts with highly branched chain paraffins and cyclic compounds (Goldstein, 1958). The branched chain paraffins, having a similar molecular weight, were found not to form solid adducts with urea which allowed separation of the n-paraffins from branched chain paraffins known as iso-paraffins. The separation depended on the fact that there was sufficient room inside the crystal molecules of urea for n-paraffins but branched chain paraffins were too big. The carbon number of paraffins (C_1–C_{35}), iso-paraffins (C_4–C_{23}), and naphthenes (C_5–C_{12}) was reported (Speight, 2006). The composition and properties of crude oils, such as viscosity, vary with the age of the oil field. With an increase in viscosity of crude oils, their density increases. API gravity is $141.5/d - 131.5$, where d is density of crude oil at 60°F. The density of crude oils, expressed in terms of API gravity, relates to specific

TABLE 1.2

Chemistry of Hydrocarbons Found in Crude Oils

Hydrocarbons	Chemistry	Name
Saturated	Straight chain	n-Paraffins
	Branched chain	iso-Paraffins
	Alicyclic	Naphthenes
Aromatics	1-Ring	Benzenes
	2-Ring	Naphthalenes
	3-Ring	Phenanthrenes

Source: From Speight, J.G., *The Chemistry and Technology of Petroleum*, 4th ed., CRC Press, Boca Raton, 2006.

TABLE 1.3

Reservoir Information of the Arabian Saudi Crude Oils

Oil Field/Well	Depth, ft	API Gravity	S, wt%
Safaniya	5500–6500	27–28	3.0
Safaniya	5500	26.9	2.92
Abqaiq	7500	33.2	1.98
Abqaiq	7000	36.1	1.32
Ain-Dar	7520	34.5	1.52
Ain-Dar	6800	33.7	1.70
Wafra/Ratawi	7000	24	3.98
Wafra/Iucene	2300	18.5	3.32
Marjan	6000	32.1	2.42
Zuluf	7852	34.6	1.65

Source: From El-Sabagh, S.M. and Al-Dhafeer, M.M., *Petroleum Science and Technology*, 18, 743, 2000. Reproduced by permission of Taylor & Francis Group, LLC, http://www.taylorandfrancis.com.

gravity. The API and sulfur content of Arabian Saudi crude oils, covering a wide range of maturity, are shown in Table 1.3.

The Arabian Saudi crude oils, having an API gravity in the range of 18.5–36.1 and a sulfur content in the range of 1.32–3.98 wt%, were separated by the urea adduction. *n*-Fatty acids were separated from the fraction containing *n*-alkanes by treatment with an aqueous solution of potassium hydroxide (KOH). According to the literature, gas chromatography (GC) analysis of *n*-paraffins indicated a distribution having three maxima at C_{17}, C_{19}, and C_{31}. GC analysis of *n*-fatty acids indicated a distribution having three different maxima at C_{14}, C_{22}, and C_{24}. The literature reported that the presence of low molecular weight fatty acids was abundant as compared to *n*-paraffins and the results were interpreted in terms of origin and maturation of crude oils (El-Sabagh and Al-Dhafeer, 2000). With an increase in the viscosity of crude oils, their density increases and API decreases. Conventional or "light" crude oils have API gravities >20. Heavy oils have API gravities <20 and are darker in color. The term "heavy oil" is usually used to describe the oils which require thermal stimulation of recovery from the reservoir (Speight, 1999). Crude oils vary in properties and composition with the age and also the location of the oil field and only some crude oils are suitable to produce certain petroleum products, such as lube oil base stocks. The location and viscosity of crude oils suitable to produce lube oil base stocks, for use in lubricating oils, are shown in Table 1.4.

Arab light, Arab heavy, and Kirkuk are often referred to as Middle East crude oils. The literature reported on Bombay High and Assam Mix as Indian crude oils and Daquin, Liaohe, Shengli, and Liaoshu as China crude oils. BSM, Fulmar, and Forties are often referred to as North Sea crude oils. Many different field locations are found around the world and crude oils from different parts of the world, or even the same oil field, are never quite the same. The variation in crude oil

TABLE 1.4

Location and Viscosity of Crude Oils Suitable for Lube Oil Base Stock Production

Crude Oil	Viscosity	Viscosity	Viscosity
Arabian	Light	Medium	Heavy
Basrah			
Citronelle			
Iranian	Light		
Kirkuk			
Kuwait			
Louisiana	Light		
Mid-Continent Sweet			
Raudhatain			
West Texas Bright			

Source: From Pirro, D.M. and Wessol, A.A., *Lubrication Fundamentals*, 2nd ed., Marcel Dekker, Inc., New York, 2001. Reproduced by permission of Routledge/Taylor & Francis Group, LLC.

properties and composition affects their stability, interfacial properties, and their processing. The many different products made by the petroleum industry include fuel gas, liquefied gas, gasoline, diesel fuel, jet fuel, lube oil base stocks, waxes, greases, asphalts, coke, carbon black, and chemicals used as solvents.

Hydrocarbons, in the form of natural gas, coal, and oil, are used as fuels and in the production of electricity. While coal is the most economical fuel source, the burning of coal leads to the formation of pollutants, such as SOx, NOx, and carbon dioxide (CO_2). Over the past few years, there is an increased trend to use natural gas as the fuel of choice and in the production of electricity. As the oil wells mature and are being exhausted and no new discoveries are being made, the natural gas production will decrease. The literature reported that natural gas also exists trapped in coal beds and the technology for recovering the coal bed methane is being developed (Speight, 2006). At the present time, while the gas-fired plants are cheaper to build, the price of natural gas is higher. The clean coal technology is being developed and its main focus is to gasify the coal first and use the resulting gases as fuel rather than burning the coal directly.

1.1.2 SOLVENCY

It is generally accepted that crude oil contains three different fractions. The oil fraction contains hydrocarbons which are paraffins, naphthenes, and aromatics, containing sulfur, oxygen, and nitrogen. Asphaltenes are the most polar and the heaviest fraction of crude oils and can be precipitated by the addition of nonpolar solvents, such as low boiling petroleum naphtha, petroleum ether, *n*-pentane, isopentane, n-hexane, or n-heptane. It is also commonly accepted that asphaltenes present in crude oils form micelles which are stabilized by adsorbed resins. Asphaltenes are defined as a solubility class that precipitates from oils by the addition of an excess of liquid paraffinic hydrocarbons. Defined as nonhydrocarbon components of crude oils, "true" asphaltenes are insoluble in low molecular weight alkanes, such as n-pentane (C5 insolubles) and n-heptane (C7 insolubles), but are soluble in benzene and toluene. In many cases, the literature uses the C5 or C7 insolubles as the asphaltene content. The literature reported that the use of solid adsorbents can separate crude oils into saturates, aromatics, and resins. The portion of crude oil that is soluble in n-pentane or n-heptane is known as maltenes and contains oil and resins. The nonpolar n-heptane solvent is used to desorb saturates followed by polar benzene or toluene required to desorb aromatics. The mixture of benzene and methanol is required to desorb the most polar fraction of deasphalted oils, which are resins (Speight, 2006).

Silica, which is silica oxide, has a high surface area and is known for years to separate polar and aromatic molecules from petroleum oils. Alumina, which is aluminum oxide, also has a high surface area. Mineral clays were also reported effective in separating petroleum oils. Mineral clays, such as Attapulgite clay, are natural compounds of silica and alumina also containing oxides of sodium, potassium, magnesium, calcium, and other alkaline earth metals. The literature reported that the asphaltene fraction was separated from Venezuela crude oil using 30:1 volume ratio of n-heptane solvent followed by the clay treatment used to separate resins (Leon et al., 2001). The resin and asphaltene content, separated from Venezuelan crude oil, are shown in Table 1.5.

The Venezuela crude oil was reported to contain 36.9 wt% of saturates, 37.9 wt% of aromatics, 19.4 wt% of resins, and 5.8 wt% of asphaltenes (Leon et al., 2001). The literature reported that the resins, adsorbing onto asphaltenes, are hydrocarbons which contain sulfur, oxygen, and nitrogen. They peptized the asphaltenes into the crude oil (Ellis and Paul, 2000). The separation, characterization, and the

TABLE 1.5
Resin and Asphaltene Content Separated from Venezuelan Crude Oil

Properties	Venezuela Crude Oil
API gravity	22.4
Saturates, wt%	36.9
Aromatics, wt%	37.9
Resins, wt%	19.4
Asphaltenes, wt%	5.8

Source: Reprinted with permission from Leon, O. et al., *Energy and Fuels*, 15, 1028, 2001. Copyright American Chemical Society.

TABLE 1.6

Composition of Resins and Asphaltenes Separated from Venezuelan Crude Oil

Composition	Resins	Asphaltenes
Carbon	84.6	84.4
Hydrogen	10.37	6.75
Sulfur	0.31	3.5
Nitrogen	1.15	1.31
Oxygen	3.57 (calculated)	4.04 (calculated)

Source: Reprinted with permission from Leon, O. et al., *Energy and Fuels*, 15, 1028, 2001. Copyright American Chemical Society.

effect of petroleum resins on the colloidal stability of crude oils were studied. The addition of adsorbents to *n*-pentane solutions was used to separate resins from oils. Resins adsorbed onto adsorbents were eluted with polar solvents and characterized (Andersen and Speight, 2001).

The colloidal dispersions which are clear and transparent have particles which are less than 100 nm. The particles above 400 nm are opaque and can be visible under a microscope. According to the early literature, asphaltenes may be aggregates or single molecules but, in either case, they are no larger than 35 nm (Speight et al., 1985). More recent literature reported that the asphaltene particles, in the range of 125–400 nm, presented a small positive charge and were not affected by the addition of resins (Neves et al., 2001). The colloidal stability of asphaltenes in crude oils was reported to be affected by the ratio of aromatics to saturates and that of resins to asphaltenes. A decrease in the aromatics or resins can lead to asphaltene aggregation (Demirbas, 2002). The composition of resins and asphaltenes, separated from Venezuelan crude oil, is shown in Table 1.6.

The composition of resins and asphaltenes separated from Venezuelan crude oil was reported and, while the oxygen content of petroleum oils is difficult to measure, it can be calculated. The resins, separated from Venezuelan crude oil, were reported to contain 84.6 wt% of C, 10.37 wt% of H, 0.31 wt% of S, 1.15 wt% of N, and 3.57 wt% of calculated O. The asphaltenes, precipitated from Venezuelan crude oil, were reported to contain a similar C content of 84.4 wt%, a significantly lower H content of 6.75 wt%, a significantly higher S content of 3.5 wt%, a similar N content of 1.31 wt%, and 4.04 wt% of calculated O (Leon et al., 2001). The resin fraction was found to be more aliphatic in nature while asphaltenes were more aromatic. A significantly higher S content in asphaltenes indicates that the majority of S molecules are aromatic in nature. The elemental analysis of C7 insolubles, precipitated from five different Kuwaiti crude oils with *n*-heptane (40:1), is shown in Table 1.7.

The literature reported on the chemical characterization of asphaltenes and resins precipitating from Mexican crude oils during the recovery operation. Their elemental analysis for C, H, N, S, O, trace metal content, FTIR and NMR and

TABLE 1.7
Composition of Asphaltenes Precipitated from Different Kuwaiti Crude Oils

Kuwaiti	Oil # 1	Oil # 2	Oil # 3	Oil # 4	Oil # 5
C, wt%	83.49	78.19	79.01	82.88	82.57
H, wt%	7.34	7.2	7.19	6.41	6.3
S, wt%	7.12	6.67	6.98	4.07	3.98
N, wt%	1.69	2.87	1.27	0.78	0.76
O, wt%	0.41	5.07	5.55	5.87	6.4

Source: From Ibrahim, Y.A. et al., *Petroleum Science and Technology*, 21, 825, 2003. Reproduced by permission of Taylor & Francis Group, LLC, http://www.taylorandfrancis.com.

molecular weight determination was reported (Buenrostro-Gonzalez et al., 2001). Different heteroatom content of Arab crude oils was reported to affect the hydrogen bonding in *n*-pentane insolubles. The phenolic $-OH-N$ hydrogen bonding was reported at about 3200 cm^{-1} while the $-OH-O$ and $-OH-S$ hydrogen bonding was reported at about 3300 cm^{-1} (Siddiqui, 2003). The literature also reported on the characterization of asphaltenes obtained from crude oils by nuclear magnetic resonance (NMR), x-ray diffractometry (XRD), fluorescence spectroscopy, and column chromatographic GPC techniques (Bansal et al., 2004). The chemistry of asphaltenes, precipitated using C7 solvent, from Kuwaiti crude oils is shown in Table 1.8.

The chemistry of asphaltenes, precipitated from Kuwaiti crude oils, analyzed using FTIR, H1-NMR, and C13-NMR, indicated the presence of 5–9 condensed aromatics having 4–6 carbon alkyl side chains. Their molecular weight was

TABLE 1.8
Chemistry of Asphaltenes Precipitated from Kuwaiti Crude Oils

Kuwaiti Crude Oil	Chemistry of C7 Insolubles
Hydrocarbons	Alicyclic rings
	5–9 Condensed aromatics
	4–6 Carbon alkyl side chains
H/C ratio	0.91–1.1
Functional groups	OH, NH, SH, C=O
Molecular weight	4200–6500

Source: From Ibrahim, Y.A. et al., *Petroleum Science and Technology*, 21, 825, 2003.

TABLE 1.9

Effect of Pressure on Metal Content of C7 Insolubles

C7 Insolubles	Pressure			
Metals, ppm	5000 psia	3000 psia	783 psia	14.7 psia
Ti	258	252	256	370
V	219	191	199	237
Cr	27	5	11	38
Mn	18	20	15	53
Fe	641	699	408	1151
Ni	429	385	344	420
Cu	91	116	88	117
Zn	47	118	144	226
Ga	76	75	70	109

Source: From Pasadakis, N., Varotsis, N., and Kallithrakas, N., *Petroleum Science and Technology*, 19, 1219, 2001. Reproduced by permission of Taylor & Francis Group, LLC, http://www.taylorandfrancis.com.

reported to vary in the range of 4200–6500 (Ibrahim et al., 2003). The literature also reported on interrelationships between asphaltenes' characteristics and their precipitation behavior during induced *n*-heptane precipitation (Ibrahim and Idem, 2004).

In its natural state, crude oil contains about 10 wt% of dissolved gases which are held by high reservoir pressure. When this "live" crude oil is extracted and passed into the low pressure environment of gas/oil separator, the natural gas is liberated and separated from "stabilized" crude oil. Light crude oils were reported to be susceptible to asphaltene precipitation as the reservoir pressure decreases during the production. The colloidal instability of asphaltenes and their flocculation, as a result of decrease in the pressure of crude oil, is expected. As the pressure decreases, the amount of dissolved asphaltenes in the oil is reduced (Laux et al., 2001). The influence of pressure on the onset of flocculation of asphaltenes was calculated in the region of 1–300 bar and at temperatures of 50°C–100°C (Browarzik et al., 2002). The effect of reservoir pressure on the metal content of asphaltenes, dissolved in live heavy oil and precipitated using *n*-heptane, is shown in Table 1.9.

Based on many articles published in the literature, asphaltenes consist of condensed polynuclear aromatic hydrocarbons and based on their analyses, asphaltenes consist of aromatics containing heteroatoms, such as S, N, and O, and contain metals. The literature reported that some vanadium and nickel can be loosely held between the asphaltene molecules, known as intercalation (Ellis and Paul, 2000). The literature also reported that asphaltenes from live oil samples, containing soluble gases, and dead samples of the same crude oil are different. Asphaltenes from a live oil sample were reported to have more functional groups and to be less saturated (Aquino-Olivos et al., 2003). The API gravity, composition, metal content, and asphaltene content of crude oils vary as shown in Table 1.10.

TABLE 1.10

Composition and Asphaltene Content of Different Crude Oils

Crude Oils	Olmeca	Isthumus	Maya Heavy
API gravity	38.7	33.1	21.3
Carbon, wt%	85.9	85.4	84.0
Hydrogen, wt%	12.8	12.7	11.8
H/C ratio	1.79	1.78	1.69
Sulfur, wt%	0.99	1.45	3.57
Nitrogen, wt%	0.07	0.14	0.32
Oxygen, wt%	0.23	0.33	0.35
Nickel, ppm	2	10	53
Vanadium, ppm	8	53	298
C5 insolubles, wt%	1.1	3.6	14.1
C7 insolubles, wt%	0.8	3.3	11.3

Source: Reprinted with permission from Ancheyta, J. et al., *Energy and Fuels*, 16, 1121, 2002. Copyright American Chemical Society.

With an increase in viscosity of crude oils, their H/C atomic ratio decreased from 1.79 to 1.69 indicating an increase in their aromaticity. The S content also increased from 0.99 to 3.57 wt%, N content increased from 0.07 to 0.32 wt%, and the O content increased from 0.23 to 0.35 wt%. With an increase in the viscosity of crude oils, their metal content also increased. The nickel content increased from 2 to 53 ppm and the vanadium content increased from 8 to 298 ppm (Ancheyta et al., 2002). The asphaltene content was also reported to increase with an increase in viscosity of different crude oils. The C5 insoluble content increased from 1.05 to 14.1 wt% and C7 insoluble content increased from 0.75 to 11.32 wt%. The use of different paraffinic solvents affects the asphaltene precipitation and the use of *n*-heptane solvent leads to a decrease in C7 insolubles.

The use of heavy crude oils leads to an increase in the asphaltene content measured as C5 and C7 insolubles (Ancheyta et al., 2002). A novel method for measuring the asphaltene content of viscous crude oil by K-ratio dual wavelength spectrophotometry at 750 and 800 nm was reported (Liu et al., 2002). Studies of petroleum asphaltenes have increased over the past years due to an increase in the use of heavy crude oils. The precipitation of asphaltenes from crude oils can cause reservoir plugging and many other problems. An asphaltene deposit was found in the tubing of some wells and the bottom sediment of the tank on site (Andersen et al., 2001). Asphaltene precipitation can cause well bore plugging requiring expensive treatment and cleanup procedures.

1.1.3 EFFECT OF TEMPERATURE

Crude oil is a mixture of hydrocarbons and almost all alkanes, higher than C_{20}, are solids at room temperature. The literature reported on the use of a light

TABLE 1.11

Melting and Boiling Points of Different Hydrocarbons

Hydrocarbons	Carbon Number	Melting Point, °C	Boiling Point, °C
n-Octane	8	−57	126
2,2,4-Octane	8	−107	99
2,2,3,3-Octane	8	104	107
n-Decane	10	−30	174
n-Cetane	16	18	291
n-Eicosane	20	37	343
n-Triacontane	30	64	454
2,6,10,14,18,22-Triacontane	30	−35	435

Source: From Gary, J.H. and Handwerk, G.E., *Petroleum Refining, Technology and Economics*, 4th
ed., Marcel Dekker, Inc., New York, 2001. Reproduced by permission of Routledge/Taylor
& Francis Group, LLC.

transmittance method to evaluate the wax appearance temperature (WAT) and
wax disappearance temperature (WDT) of model paraffin compounds, such as
n-$C_{24}H_{50}$ and n-$C_{36}H_{74}$ in n-decane solutions (Wang et al., 2003a). The literature
also reported on the effect of wax inhibitors on wax deposits. Most wax inhibitors, tested at a dosage of 100 ppm, were reported to suppress the WAT of lower
molecular weight C_{24} paraffin but had no significant effect on the WAT of higher
molecular weight C_{36} paraffin. The effect of wax inhibitors was reported to
increase with an increase in the length of their side chains (Wang et al., 2003b).
Most commercial wax inhibitors were also reported effective in reducing the wax
deposition of only low molecular weight paraffins, such as C_{34} and below. In
some cases of high molecular weight paraffins, the use of wax inhibitors was
reported to increase the wax deposition (Wang et al., 2003b). The effect of
chemistry and molecular weight on the melting and boiling points of different
hydrocarbons is shown in Table 1.11. The melting and boiling points of normal
hydrocarbons increase with their molecular weight. Hydrocarbon isomers, having
the same molecular weight, iso-paraffins, are known to have lower melting and
boiling points. Most crude oils contain heavy hydrocarbons which precipitate as
paraffin deposits. The cloud point is the first sign of the wax formation while the
pour point (PP) is the lowest temperature at which oil will flow under specific
testing conditions. The literature reported on different terminology and techniques
used to measure cloud point of petroleum fluids, such as wax appearance temperature (WAT) and wax precipitation temperature (WPT) (Hammami et al.,
2003). In crude oil recovery, the cloud point measurement is widely used to
evaluate the wax precipitation potential. Many different techniques based on
viscosity, filter plugging, and differential scanning calorimetry (DSC) are used
and the results vary (Coutinho and Daridon, 2005).

TABLE 1.12

Variation in Sulfur and Pour Points of Different Crude Oils

Crude Oil Origin	Viscosity, cSt at 40°C	Sulfur, wt%	Pour Point, °C
India	2.3	0.1	27
Kuwait	8.7	2.0	−23
Venezuela	10.2	1.1	−15
North Sea	4	0.3	−3
Middle East	8	2.5	−15
Indonesia	12	0.2	39
Venezuela	19,000	5.5	9

Source: From Khan, H.U. et al., *Petroleum Science and Technology*, 18, 889, 2000; Al-Kandary, J.A.M. and Al-Jimaz, A.S.H., *Petroleum Science and Technology*, 18, 795, 2000; Speight, J.G., *The Chemistry and Technology of Petroleum*, 3rd ed., Marcel Dekker, Inc., New York, 1999; Prince, R.J., *Chemistry and Technology of Lubricants*, 2nd ed., Mortier, R.M. and Orszulik, S.T., Eds., Blackie Academic & Professional, London, 1997.

The effect of *n*-paraffin weight distribution on the wax precipitation from highly paraffinic crude oils was studied. High molecular weight normal paraffins were reported responsible for wax precipitation. While the cloud point indicates the first sign of wax formation, the pour point of the crude oil is a rough indicator of the relative paraffinicity and aromaticity of the crude oil. There are many test methods, such as ASTM D 97, used to test the pour point of petroleum oils. After the preliminary heating, the sample is cooled at a specified rate and examined at intervals of 3°C for flow characteristics. The lowest temperature at which the movement of the specimen is observed is recorded as the pour point. A lower crude pour point usually indicates a lower paraffinic content. Some crude oils have a high sulfur content and flow easily while others might contain a low sulfur content and have a high content of wax. The variation in sulfur and pour points of different crude oils is shown in Table 1.12.

The literature reported on Indian Mukta crude oil having a low viscosity and a low sulfur content of 0.14 wt% and a high pour point of 27°C (Khan et al., 2000). The properties and NMR analysis of highly waxy Indian Mukta crude oil were reported. The solid paraffins, boiling above 270°C, were separated from Mukta crude oil by the urea adduction and analyzed by gas chromatography, IR, and NMR. With an increase in the boiling range, their average carbon number was reported to increase from 17 to 29.6 and the CH_2/CH_3 ratio increased from 7.9 to 13.8 (Khan et al., 2000). The literature reported on light Kuwaiti crude oil having a higher sulfur content of 2 wt% and a lower pour point of −23°C (Al-Kandary and Al-Jimaz, 2000). Light Tia Juana crude oil, from Venezuela, was reported to have a sulfur content of 1.1 wt% and a pour point of −15°C (Speight, 1999). Middle

TABLE 1.13

Effect of Temperature on Crude Oil Stability

Crude Oils	Mukta	Assam
Density at 15°C, g/cm³	0.8305	0.8498
API gravity at 15.5°C	38.8	34.9
Kin. viscosity at 40°C, cSt	2.25	6.2
Pour point, °C	27	30
Wax content, wt%	9	11
Carbon residue, wt%	0.79	1.53

Source: From Khan, H.U. et al., *Petroleum Science and Technology*, 18, 889, 2000; Kandwal, V.C. et al., *Petroleum Science and Technology*, 18, 755, 2000.

East, Venezuela, North Sea, and Indonesia crude oils, used to produce lube oil base stocks, were reported to have varying sulfur content in a range of 0.2–5.5 wt% and varying pour points from as low as −15°C to as high as 39°C (Prince, 1997).

Crude oils are unstable at low and high temperatures. Carbon residue of an oil, which is what remains after it is evaporated, varies depending on the crude oil. Carbon residue measurements have used Conradson carbon residue (CCR) test procedure which requires a weighted petroleum product to be heated at 500°C, under the nitrogen blanket, for a specific time. The sample undergoes the coking reaction, the volatiles are removed, and the carbonaceous type residue is weighted and reported as a percent of the original sample. The carbon residue value of the various petroleum products is used as an indication of their tendency to form carbonaceous type deposits under degrading conditions. The effect of temperature on the wax content and carbon residue of different crude oils is shown in Table 1.13.

The literature reported on the wax present in Mukta crude oil causing the wax deposition in the pipeline. Mukta crude oil, having a pour point of 27°C, was reported to have a wax content of 9 wt% (Khan et al., 2000). The literature also reported on the properties and wax precipitation from Assam crude oil which was reported to have a higher pour point of 30°C and a higher wax content of 11 wt% (Kandwal et al., 2000). Some crude oils contain a high content of wax and, despite the wide range of hydrocarbons and other organic molecules that are found in crude oils, the main differences between crude oils are not in type of molecules but rather the relative amounts of each type that occur in the oil (Prince, 1997). Mukta crude oil, having a lower viscosity, was reported to have a lower CCR of 0.79 wt%. Assam crude oil, having a higher viscosity, was reported to have a higher CCR of 1.53 wt%. The effect of flow rate and the temperature of oil and cold surface on wax deposition from Assam crude oil are shown in Table 1.14.

With a decrease in the Assam crude oil temperature, from 50°C to 40°C, and in the surface temperature, from 25°C to 20°C, an increase in the wax deposition is observed (Kandwal et al., 2000). The pour point is the lowest temperature at

TABLE 1.14

Effect of Flow Rate and Temperature on Wax Deposition from Assam Crude Oil

Flow Rate	200 L/h	130 L/h	165 L/h
Oil temp.	50°C	50°C	40°C
Cold surface temp.	25°C	25°C	20°C
Wax deposit, g			
After 1 h	2.5	2.9	3.9
After 3 h	4.2	4.3	4.9
After 6 h	4.6	4.8	5.6

Source: From Kandwal, V.C. et al., *Petroleum Science and Technology*, 18, 755, 2000. Reproduced by permission of Taylor & Francis Group, LLC, http://www.taylorandfrancis.com.

which the oil can be made to flow under gravity. During pipeline flow of waxy crude oils, wax gets deposited leading to reduction in pipeline capacity and need for increased pumping pressure. Deposition of waxes in production wells leads to the obstruction of the fluid flow systems and many different predictive models for wax deposition were developed (Ramirez-Jaramillo et al., 2001). The prediction of wax and asphaltene precipitation from live crude oils, containing soluble gases, was reported to be affected by their composition, temperature, and pressure (Wiehe, 2003). In some cases, blending of different crude oils and a wider molecular weight distribution were found effective in decreasing their cloud points indicating a decrease in the wax precipitating tendency (Garcia and Urbina, 2003). The offshore exploration, where the oil fields are under water, leads to lower temperature of oil and an increase in wax precipitation. Problems related to deposition of wax increase as offshore deep water exploration continues to grow.

1.2   INTERFACIAL PROPERTIES

1.2.1   Oil/Air Interface

The literature reported that pure liquids allow entrained air to escape with no delay and no foam is produced (Ross, 1987). Crude oils are mixtures of different hydrocarbons and heteroatom containing molecules, having various functional groups, and foam when mixed with gas or air. Foaming of crude oils can cause major problems for gas/oil separation plants causing a loss of crude oil in the separated gas stream. During the primary separation of crude oils, stable foams complicate the liquid level control inside the separators, require longer residence times, and could result in liquid being carried over into the gas stream (Pacho and Davies, 2003). To prevent foaming which can lead to a loss of crude oils, large separators are required, antifoaming agents are injected, or other physical means

TABLE 1.15

Surface Tensions of Some Inorganic and Organic Liquids

Liquids	Gas Phase	Temperature, °C	ST, mN/m
Mercury	Vacuum	20	484
Mercury	Air	20	435
Water	Air	20	72.8
Benzene	Air	20	28.9
Toluene	Air	20	31.1
Alkylated benzene	Air	20	32.7
n-Hexane	Air	20	18.4
Heptadecane	Air	20	30.1

Source: From *CRC Handbook of Chemistry and Physics*, 86th ed., Lide, D.R., Editor-in-Chief, CRC Press, Boca Raton, 2005. Reproduced by permission of Routledge/Taylor & Francis Group, LLC.

are built (Callaghan et al., 1986). It was reported that the liquids with high surface tension (ST), such as liquid mercury, have strong forces binding the molecules together in the condensed liquid state. Hydrocarbon liquids, such as benzene, have weak dispersive interactions binding the molecules together in the liquid state (Dee and Sauer, 1997). The ST of different inorganic and organic liquids is shown in Table 1.15.

An increase in the molecular weight of the aromatic chemical structure leads to an increase in the ST of benzene from 28.9 to 32.7 mN/m. A saturated hydrocarbon structure, such as n-hexane, has a lower ST of 18.4 mN/m which increases with an increase in molecular weight to 30.1 mN/m in the case of heptadecane. The effect of chemistry on the ST of heteroatom containing molecules is shown in Table 1.16.

The presence of heteroatoms, such S, N, and O, generally leads to an increase in the ST of saturated and aromatic hydrocarbons except for ethers and aliphatic amines. The combination of two heteroatoms, such as sulfates or acids, further increases the ST of molecules. While the aromaticity increases the ST of molecules, the presence of longer aliphatic groups might lead to a decrease in the ST. The ST of liquids is important during the foam inhibition process. When it comes to foaming, it is the purity of the liquid which matters. The literature reported that pure organic liquids, such as heptane, octane, and toluene, do not foam when mixed with air (Poindexter et al., 2002). The absence of foaming of pure organic solvents, measured at 20°C, is shown in Table 1.17.

Crude oils foam when mixed with air. Nonaqueous foaming of crude oils was reported to be a significant problem during decompression and handling on offshore platforms due to a high rate of processing and limited space (Zaki et al., 2002). Foaming of crude oils produced from offshore platforms occurs regularly and the literature reported on efforts to establish the factors affecting

TABLE 1.16
Effect of Chemistry on Surface Tension of Heteroatom Containing Molecules

Chemistry	Formula	Temp., °C	ST, mN/m
Ethyl mercaptan	C_2H_6S	20	21.8
Ethyl alcohol	C_2H_6O	20	22.8
Dimethyl sulfate	$C_2H_6O_4S$	18	40.1
1-Propanethiol	C_3H_8S	25	24.2
Propylamine	C_3H_9N	20	22.4
Diethyl sulfide	$C_4H_{10}S$	25	24.6
Ethyl ether	$C_4H_{10}O$	20	17.0
n-Butyl alcohol	$C_4H_{10}O$	20	24.6
Butyric acid	$C_4H_8O_2$	20	26.8
Thiophene	C_4H_4S	25	30.7
Pyridine	C_5H_5N	20	38
Quinoline	C_9H_7N	20	45
Benzene	C_6H_6	20	28.9
Aniline	C_6H_7N	20	42.9
Benzaldehyde	C_7H_6O	20	40.0
Benzyl alcohol	C_7H_8O	20	39.0
Acetophenone	C_8H_8O	20	39.8
Diethylaniline	$C_{10}H_{15}N$	20	34.2

Source: From *CRC Handbook of Chemistry and Physics*, 86th ed., Lide, D.R., Editor-in-Chief, CRC Press, Boca Raton, 2005. Reproduced by permission of Routledge/Taylor & Francis Group, LLC.

their foaming (Poindexter et al., 2002). The viscosity, asphaltene content, and foamability of different crude oils are shown in Table 1.18.

Lower viscosity crude oils, containing asphaltenes, foam and their foaming tendencies increase with an increase in their viscosity. Arab heavy crude oil, containing a high asphaltene content, was found to have a high foaming tendency

TABLE 1.17
Absence of Foaming of Pure Organic Solvents

Organic Solvents	Purity, %	Foamability, mL at RT
Heptane	100	0
Octane	100	0
Toluene	100	0

Source: Reprinted with permission from Poindexter, M.K. et al., *Energy and Fuels*, 16, 700, 2002. Copyright American Chemical Society.

TABLE 1.18

Viscosity, Asphaltene Content, and Foamability of Different Crude Oils

Crude Oils	Viscosity at 37.8°C, cP	Asphaltenes, wt%	Foamability, mL
Statjord	3	0.09	120
Arab Berri	4	0.79	140
Alaska North Slope	13	3.38	30
Arab Heavy	34	6.68	200
Malu Isan	38	0.18	75
Canadon Seco	70	7.5	45
THUMS 1	152	5.09	0
Hondo	363	14.81	5
San Joaquin Valley	1390	4.56	0

Source: Reprinted with permission from Poindexter, M.K. et al., *Energy and Fuels*, 16, 700, 2002. Copyright American Chemical Society.

of 200 mL at RT. However, with a further increase in viscosity and an increase in asphaltene content, the foaming tendency of crude oils decreases. Very high viscosity crude oils, containing asphaltenes, do not foam indicating the temperature–viscosity effect. Different viscosity Gulf of Mexico crude oils, containing varying asphaltene content, were found to foam. Heavy mineral oil, containing no asphaltenes, was also reported to foam; however, its foaming tendency was lower. The literature reported that the ST of the petroleum products is affected by their molecular weight and the temperature but shows little variation with a change in their composition (Speight, 1999). The foamability and the ST of Gulf of Mexico crude oils and heavy mineral oil are shown in Table 1.19.

With a gradual increase in viscosity of Gulf of Mexico crude oils, from 5 to 68 cP at 37.8°C, their ST also gradually increased from 26.3–26.4 to 28.9 mN/m. The crude oil, having a viscosity of 68 cSt at 37.8°C and containing a high asphaltene content of 8.71 wt%, was reported to have a high foaming tendency of 390 mL and an ST of 28.66 mN/m. The heavy mineral oil, also having a viscosity of 68 cSt at 37.8°C and containing no asphaltenes, was reported to have a lower foaming tendency of 126 mL and a higher ST of 31.32 mN/m. The ST of heavy mineral oil, having the same viscosity, was reported to increase from 28.66 to 31.32 mN/m, indicating some differences in composition. While there is no effect of ST on foaming tendency of crude oils, the value of ST is important during the mechanism of foam prevention. For an antifoaming agent to be surface active, their ST needs to be lower, by several units of mN/m, than that of the foaming liquid (Ross, 1987). Early literature reported on the low ST of polymethylsiloxanes, having a low viscosity of 5–35 cSt (Bondi, 1951). The ST of different chemistry silicone polymers is shown in Table 1.20.

With a change in the chemistry of silicone polymers, from polymethylsiloxane to polymethylphenylsiloxane, an increase in ST, from 19–20 to 28.6 mN/m, was

TABLE 1.19
Foamability and Surface Tension of Crude Oils and Heavy Mineral Oil

Gulf of Mexico Crude Oils	Viscosity at 37.8°C, cP	Asphaltenes, wt%	Foamability, mL	ST, mN/m
# 1	5	0.08	24	26.42
# 2	6	0.21	30	26.30
# 3	7	0.62	39	26.65
# 4	9	0.44	97	27.07
# 5	9	0.36	91	27.20
# 6	10	0.5	105	27.74
# 7	19	4.16	257	27.91
# 8	20	3.61	365	27.49
# 9	24	0.33	127	27.71
# 10	48	9.1	103	28.41
# 11	68	8.71	390	28.66
Heavy mineral oil	68	0	126	31.32

Source: Reprinted with permission from Poindexter, M.K. et al., *Energy and Fuels*, 16, 700, 2002. Copyright American Chemical Society.

reported (Bondi, 1951). There are many different polyorganosiloxane polymers available but the most common is polydimethylsiloxane. The methyl radical in the polydimethylsiloxane polymer might be substituted by many organic groups, such as hydrogen, alkyl, allyl, trifluoropropyl, glycol ether, hydroxyl, epoxy, alkoxy, carboxy, and amino (Rome, 2001). Polymethylsiloxanes, known as silicones, are the most surface active antifoaming agents due to their low ST. According to the literature, silicone must be insoluble in oil and is usually used as a dispersion containing as little as 5 wt% of silicone. If the ST of the process fluid has a lower

TABLE 1.20
Surface Tension of Different Chemistry Silicone Polymers

Silicone Polymers	Viscosity	ST, mN/m at 20°C
Polymethylsiloxane	5 cSt	19.0
	35 cSt	19.9
Polymethylphenylsiloxane	27 cSt	28.6

Source: From Bondi, A., *Physical Chemistry of Lubricating Oils*, Reinhold Publishing Corporation, New York, 1951.

ST than the silicone or it dissolves the silicone, an increase in foaming might be observed (ProQuest, 1996). The use of crude oils, having a lower ST, or the use of an antifoaming agent, having a higher ST, will result in a decrease in their surface activity and no foam prevention will take place.

The performance of silicone antifoaming agents in crude oils was reported to be affected by their solubility and the gas/oil ratio. With an increase in gas content, the lower viscosity silicone antifoaming agents were reported to loose their efficiency to prevent foaming and became a foam promoter. With an increase in gas/oil ratio, higher viscosity silicones are more effective. The literature reported that the effective treat rates of silicone in crude oils vary from 1 to 10 ppm (Rome, 2001). The formation of aqueous foams also occurs in the refining of crude oil. Aqueous foams are formed in drilling muds, deaeration of casing cements, and deaeration of water for well injections. Under these conditions, silicone fluids are not effective and silicone emulsions, containing dispersed hydrophobic silica, are used (Rome, 2001).

1.2.2 OIL/WATER INTERFACE

Interfacial phenomena in porous rocks determine the fraction of oil that moves from the swept region toward a producing well. The initial pressure is sufficient to move oil to the production wells and is called the primary oil recovery. When the initial pressure decreases, it is necessary to increase the pressure by injecting the water which is called the secondary recovery or water flooding. Additional water might be added to aid in secondary oil recovery which might result in emulsion formation. The effect of brine composition on oil recovery by water flooding was studied. The injection of brine containing 2% of KCl and 2% of NaCl increased the oil recovery by 18.8% over the injection of distilled water (Bagci et al., 2001). Surface tension at the interface between two liquids, when each liquid is saturated with the other, is known as interfacial tension (IFT). The IFT between oil and water provides an indication of compounds in the oil that have an affinity for water. The early literature reported that the liquid/water IFT can vary from near 0 for the almost water-soluble polyether to 55 mN/m for the hydrocarbons (Bondi, 1951). The IFT between different inorganic and organic liquids, measured at 20°C, is shown in Table 1.21.

Certain polar compounds adsorb at the oil/water interface leading to reduction in the IFT. Oxidation by-products are known to decrease the oil/water IFT of petroleum oils (Speight, 1999). The oxidation of heavy crude oil was reported to lead to formation of carboxylic acids, aldehydes, and ketones with a higher affinity toward water. The oil/water IFT of oxidized heavy crude oil was reported to decrease from 32 to 7.7 mN/m (Escobar et al., 2001). The literature also reports on asphaltenes adsorbing at the oil/water interface of crude oil. The asphaltenes were reported to decrease the oil/water IFT of crude oils and increase the stability of emulsions (Zaki et al., 2000a). Interfacial properties of crude oils at the oil/ water interface lead to emulsions when mixed with water. The Venezuelan crude oil was reported to form stable emulsions which, under gravity, would not

TABLE 1.21
IFT between Different Inorganic and Organic Liquids

Liquid # 1	Liquid # 2	Temperature, °C	IFT, mN/m
Mercury	Ethyl ether	20	379
Mercury	Water	20	375
Mercury	Benzene	20	357
n-Hexane	Water	20	51.1
n-Octane	Water	20	50.8
Benzene	Water	20	35.0
Carbon tetrachloride	Water	20	45.0
Ethyl ether	Water	20	10.7
n-Octyl alcohol	Water	20	8.5
Heptylic acid	Water	20	7.0

Source: From *CRC Handbook of Chemistry and Physics*, 86th ed., Lide, D.R., Editor-in-Chief, CRC Press, Boca Raton, 2005. Reproduced by permission of Routledge/Taylor & Francis Group, LLC.

separate for weeks. Dilution with *n*-heptane caused the flocculation of asphaltenes and changed the emulsion stability (Kumar et al., 2001). The properties of waxy and asphaltenic type crude oils are shown in Table 1.22.

A decrease in API gravity, from 29.1 to 21.7, corresponds to an increase in specific gravity and a decrease in Reid vapor pressure, from 6.9 to 5.2 psi, is observed. The waxy type crude oil, containing a high wax content of 16.5 wt%, was reported to have a lower sulfur content of 1.97 wt% and a lower asphaltene/resin content of 1.4 wt%. The asphaltenic type crude oil, containing a higher sulfur content of 2.9 wt% and a high asphaltene/resin content of 8.8 wt%, was reported to have a low wax content of 1.6 wt%. Crude oil is a mixture of organic

TABLE 1.22
Properties of Waxy and Asphaltenic Type Crude Oils

Crude Oil Type	Waxy	Asphaltenic
API gravity at 15.5°C	29.1	21.7
Reid vapor pressure, psi	6.9	5.2
Sulfur, wt%	2.0	2.9
Wax content, wt%	16.5	1.6
Asphaltenes/resins, wt%	1.4	8.8

Source: From Al-Sabagh, A.M. et al., *Polymers for Advanced Technologies*, 13, 346, 2002. Copyright John Wiley & Sons Limited. Reproduced with permission.

TABLE 1.23

Effect of Different Type Crude Oils on Emulsion Stability

Crude Oil Type	Waxy	Asphaltenic
Salt content, wt%	2.99	5.16
IFT, mN/m at 25°C	14.6	18.0
Water-in-oil emulsion		
Water content, vol%	20	20
Stability	Less stable	More stable

Source: From Al-Sabagh, A.M. et al., *Polymers for Advanced Technologies*, 13, 346, 2002. Copyright John Wiley & Sons Limited. Reproduced with permission.

compounds ranging in size from simple gaseous molecules, such as methane, to high molecular weight wax and asphaltenes. The proportion of certain types of hydrocarbons varies and some crude oils contain a larger amount of high molecular weight wax or asphaltenes. While instability of crude oils can lead to wax or asphaltene precipitation, the presence of water can lead to stable emulsions. After the addition of 20 vol% of water, the asphaltic crude oil was reported to produce a more stable emulsion (Al-Sabagh et al., 2002). The effect of different types of crude oils on water-in-oil emulsion stability is shown in Table 1.23.

Waxy type crude oil, having a lower viscosity, was reported to have a lower IFT of 14.6 mN/m measured at 25°C. It contained a lower salt content of 2.99 wt% and produced a less stable emulsion. Asphaltic type crude oil, having a higher viscosity, was reported to have a higher IFT of 18 mN/m also measured at 25°C. It contained a higher salt content of 5.16 wt% and produced a more stable emulsion. Ordinary soaps, used as cleaning detergents, are sodium or potassium salts of C_{16} or C_{18} acids. The salts of long chain carboxylic acids are only moderately soluble in water. At a very low concentration, they adsorb at the surface of liquid. At a high concentration and, after the surface becomes saturated, they form micelles or increase the surface leading to an increase in the stability of emulsions. Other alkali metals can also form useful soaps. The formation of stable emulsions requires the use of surface active chemical compounds, known as demulsifiers, to separate water from crude oils. The different types and chemistry of demulsifiers are shown in Table 1.24.

The early work in the emulsion technology was centered on calcium salts of fatty acids which were derived from saponified products of animal fats. These fatty acid salts were heat dried until they formed powders and used as laundry soaps. They were called lye soap (Becker, 1997). The fatty acid salts are easily dispersed in oil systems and tend to form micelles (suspensions) with a highly concentrated internal phase of calcium ions surrounded by radially projected nonpolar fatty acids. Although they have been dried, they contain coordination

TABLE 1.24
Different Types and Chemistry of Demulsifiers

Demulsifier Type	Example	Chemistry
Ionic surfactant	Salts	Ca salts of fatty acids
		Acid sulfates
Nonionic surfactant	Polyether products	Multifunctional
Nucleophiles	Formaldehyde phenols	Multifunctional

Source: From Becker, J.R., *Crude Oil Waxes, Emulsions and Asphaltenes*, PennWell Publishing Company, Tulsa, 1997.

water and act as additional containers for water that normally diffuses between emulsion aggregates (Becker, 1997). Later developments in the demulsification technology were focused on acid sulfates and their salts. This class of chemicals was found to have a stronger anionic functionality than the fatty acids and, with the appropriate alkyl substituents, improved their dispersion in the oil systems (Becker, 1997). With the large-scale production of condensed polyether, new classes of nonionic detergents were introduced. The variation in chemistry and molecular weight of demulsifiers used in crude oils is shown in Table 1.25.

The condensation products of ethylene oxide were found to be water soluble, nearly molecularly dispersible, and a high reactivity of oxirane ring made it very useful to produce various derivatives. The oxirane group can react with acid or base and form various derivatives such as anion salts, carboxylates, phenolates, alkoxides, amines, and amides (Becker, 1997). The literature reported on the

TABLE 1.25
Variation in Chemistry and Molecular Weight
of Demulsifiers

Chemistry of Demulsifiers	Molecular Weight
Phenolic resins derivatives	3000–3700
Polyamines	—
Glycols	—
Alkylaryl sulfonates	—
Aryl sulfonates	—
Ethylene oxide/propylene oxide copolymers	8000
Ethylene oxide/propylene dieoxide	8000

Source: Reprinted with permission from Kim, Y.H. and Wasan, D.T., *Industrial and Engineering Chemistry Research*, 35, 1141, 1996. Copyright American Chemical Society.

factors affecting the demulsification and the performance of different demulsifiers using model water-in-oil and water-in-crude oil emulsions (Kim and Wasan, 1996).

The characteristics of water/oil and oil/water emulsions were studied using the chemiometry technique. The density, pour point, the initial temperature of crystal formation (TIAC), and nitrogen content were reported to affect the diameter of the emulsion drops (Spinelli et al., 2004). Water or brine typically accompanies crude oil during its recovery. The recent literature reported that the stability of water-in-crude oil emulsions increases in the presence of NaCl but decreases with an increase in water concentration and an increase in temperature (Ghannam, 2005).

1.2.3 OIL/METAL INTERFACE

Interfacial properties of crude oils at oil/metal interface lead to rust and corrosion. The presence of water, containing dissolved oxygen, causes rust formation on the metal surface containing iron. The presence of acids leads to corrosion. The literature reported that rust and corrosion are serious problems affecting the pipelines (Vieth et al., 1997). Internal corrosion of carbon steel pipelines from remote wells, where it is more economical to transport oil and gas combined, results in a decrease in pipe lifetime and even shutdown of the line. The corrosion rates were reported related to the type of flow regime (Heeg et al., 1998). It was reported on March 2, 2006 that 270,000 gallons of crude oil spilled from a ruptured pipeline in Alaska due to an accumulation of corrosive sludge in Prudhoe Bay's pipeline. The ST of liquid metals is shown in Table 1.26.

The literature reported on the presence of naphthenic acids in different crude oils. The naphthenic acid content of crude oils, from the same Russia origin, was reported to vary from as low as 0.2 wt% to as high as 1.05–1.1 wt%, found in some light and heavy type crude oils. Balakhany light crude oils were classified as naphthenic while Balakhany heavy was classified as asphaltic type crude oil.

TABLE 1.26
Surface Tension of Liquid Metals

Liquid Metals	Gas Phase	Temperature, °C	ST, mN/m
Iron	Vacuum	1535 (m.p.)	1700
Copper	Vacuum	1083	1300
Copper	Vacuum	1150	1145
Vanadium	Vacuum	1710 (m.p.)	1950
Nickel	Vacuum	1455 (m.p.)	1756
Aluminum	Vacuum	800	845
Molybdenum	Vacuum	2620 (m.p.)	2250

Source: From *CRC Handbook of Chemistry and Physics*, 86th ed., Lide, D.R., Editor-in-Chief, CRC Press, Boca Raton, 2005. Reproduced by permission of Routledge/Taylor & Francis Group, LLC.

TABLE 1.27

Naphthenic Acid Content of Some Crude Oils

Crude Oil Origin	Crude Oil Name	Naphthenic Acids, wt%
Russia	Balakhany light	1.05
Russia	Balakhany heavy	1.1
Russia	Binagady	0.85
Russia	Ramain	0.4
Russia	Surakhani	0.2

Source: From Speight, J.G., *The Chemistry and Technology of Petroleum*, 2nd ed., Marcel Dekker, Inc., 1991. Reproduced by permission of Routledge/Taylor & Francis Group, LLC.

Another Binagady crude oil, classified as asphaltic, was also reported to have a high naphthenic acid content of 0.85 wt% (Speight, 1991). The naphthenic acid content of some Russian crude oils is shown in Table 1.27.

Although the alicyclic naphthenic acids appear to be more prevalent, the alkaline extracts from petroleum were also found to contain aliphatic acids and phenols. The carboxylic acids with less than eight carbon atoms per molecule are almost entirely aliphatic in nature (Speight, 1991). The literature reported that fatty acids are present in crude oils and can be extracted with alkaline solutions (El-Sabagh and Al-Dhafeer, 2000). The total acid number (TAN) value of a petroleum product is the weight of KOH required to neutralize 1 g of oil. The usual components of TAN are naphthenic acids but also organic soaps, soaps of heavy metals, organic nitrates, other compounds used as additives, and oxidation by-products. While petroleum acids found in crude oils include naphthenic acids, fatty acids, and phenols, there are many other compounds present which have acidic properties. The effect of the primary and secondary oil recovery on the composition and TAN of crude oils is shown in Table 1.28.

While the pour point of crude oils was reported to vary, their S and metal contents were reported to be similar. Crude oils recovered during primary and secondary oil recovery were found to have a high TAN in the range of 2.5–3.8 mg KOH/g. The literature also reported on an increase in CCR of crude oils obtained during secondary recovery. The ash residue left after the burning of a crude oil is due to the presence of the metal containing compounds. According to the literature, corrosion, scale, and dissolved metals are the by-products of the presence of emulsions in crude oils (Becker, 1997). The emulsion stability of crude oils was tested by mixing crude oils and brine, stirring, and recording the fraction of water separating with time (Goldszal and Bourrel, 2000). The literature reported on surface active molecules present in crude oils which cause stable emulsions and the difficult task of finding an effective demulsifier.

TABLE 1.28

Effect of Primary and Secondary Oil Recovery on Properties and TAN of Crude Oils

Properties	Crude Oil # 1		Crude Oil # 2	
	Primary	Secondary	Primary	Secondary
Gravity, API	10.4	24.5	15	12.9
Pour point, °C	18	13	−4	10
S, wt%	2.5	2.1	1.2	1.3
N, wt%	1.2	0.9	0.8	0.9
Ni, ppm	51	58	49	66
V, ppm	127	104	22	37
Fe, ppm	8	12	30	29
TAN, mg KOH/g	3.8	3.4	3.2	2.5
Water, wt%	<0.1	<0.1	<0.1	<0.1
CCR, wt%	10.1	17.1	9.4	16
Ash, wt%	<0.01	<0.01	<0.01	<0.01

Source: From Speight, J.G., *The Chemistry and Technology of Petroleum*, 3rd ed., Marcel Dekker, Inc., New York, 1999. Reproduced by permission of Routledge/Taylor & Francis Group, LLC.

1.3 DESALTING

During the desalting step, the crude oil is washed with 3–10 vol% of water and a combination of a demulsifier and the electrostatic desalting system is used to separate the salty water (Gary and Handwerk, 2001). To separate the emulsion, demulsifiers are used. The literature reported that asphaltenes can adsorb at the oil/water interface leading to an increase in emulsion stability and a decrease in the efficiency of demulsifiers to destabilize the oil/water interface (Liu et al., 2003). It was reported that the aromatic type demulsifiers, capable of interacting with asphaltenes, were the most effective in breaking the emulsions. The literature reported on the use of phenolic and polyhydric demulsifiers in the desalting of asphaltic crude oils; however, in some cases, the combination of both demulsifiers was found to be more effective (Liu et al., 2003). Heavy naphthenic crude oils were also reported to form stable emulsions. The demulsification process was reported to be affected by the viscosity of crude oils, the temperature, and the concentration of the demulsifier (Al-Sabagh et al., 2002). The effect of temperature and demulsifier on the emulsion stability of light and heavy crude oils is shown in Table 1.29.

At 50°C the rate of water coalescence in light crude oil, containing min 300 ppm of demulsifier, was reported to vary in a range of 80%–89%. With an increase in temperature to 70°C and a decrease in oil viscosity, the rate of water coalescence increases to 89%–95% and less demulsifier is required. At 50°C, the rate of water coalescence in heavy crude oil, containing min 300 ppm of demulsifier, was reported to be significantly lower in the range of 51%–60%. With an increase in

TABLE 1.29
Effect of Temperature and Demulsifier on Water Coalescence of Crude Oils

Crude Oils Coalescence, %	Light	Light	Heavy	Heavy
Temp., °C	50	70	50	70
Demulsifier				
200 ppm	40	80	23	28
300 ppm	80	89	51	57
400 ppm	85	89	55	61
500 ppm	89	95	60	75

Source: From Al-Sabagh, A.M. et al., *Polymers for Advanced Technologies*, 13, 346, 2002. Copyright John Wiley & Sons Limited. Reproduced with permission.

temperature to 70°C and a decrease in oil viscosity, the rate of water coalescence also increases but only to 57%–75% since it takes longer to separate water from heavy oils. The salt content of crude oils can vary from 50 to 300 lbs per 1000 barrels of crude oil, affecting the emulsion stability (Gary and Handwerk, 2001). An increase in viscosity of crude oils requires an increase in temperature in order to improve rates of water coalescence. The effect of crude oil viscosity on the temperature of the desalting process is shown in Table 1.30.

While many different chemistry demulsifiers were developed and some products, such as derivatives of phenolic resins, have multiple reactive sites, desalting of heavy crude oils is difficult due to their high viscosity. The literature also reported on the effect of solid particles on emulsion stability of crude oils (Sullivan and Kilpatrick, 2002). The presence of asphaltenes and clay particles, at a specific ratio, was reported to increase the stability of emulsions and, in some cases, wetting agents are added to improve the water wetting properties of solids

TABLE 1.30
Effect of Crude Oil Viscosity on Desalting Process

API of Crude Oil	Water Wash, vol%	Temperature, °C
>40	3–4	115–125
30–40	4–7	125–140
<30	7–10	140–150

Source: From Gary, J.H. and Handwerk, G.E., *Petroleum Refining, Technology and Economics*, 4th ed., Marcel Dekker, Inc., 2001. Reproduced by permission of Routledge/Taylor & Francis Group, LLC.

TABLE 1.31

Metals, Salt, and Asphaltene Contents of Some Crude Oils before Desalting

Crude Oils	Luning	Rocalya	Shengli
Density at 20°C, g/cm^3	0.9108	0.9263	0.9156
Freezing point, °C	21	<-30	15
Wax content, wt%	12.34	10.78	9.66
Sulfur, wt%	0.21	0.47	0.91
Fe, ppm	10	38	53
Ca, ppm	31	5	438
Mg, ppm	1	1	13
Na, ppm	19	<1	1
Ni, ppm	16	28	63
V, ppm	4	4	2
Salt content, mg/L	6.97	14.12	14.96
Asphaltenes, wt%	2.52	0.94	1.82

Source: Reprinted with permission from Liu, G., Xu, X., and Gao, J., *Energy and Fuels*, 18, 918, 2004. Copyright American Chemical Society.

(Gary and Handwerk, 2001). The literature reported that asphaltenes converted water-wet particles of clay into oil-wet particles, thus making them act as emulsifying agents. Polyoxyalkylenated amines were reported effective in breaking water-in-crude oil emulsions stabilized by asphaltenes and clay (Zaki et al., 2000b). The viscosity, metal, salt, and asphaltene contents of crude oils were reported to affect the efficiency of the desalting process. The metal, salt, and asphaltene contents of some crude oils, before desalting, are shown in Table 1.31.

Almost all crude oils contain detectable amounts of metals and their solubility and concentrations can vary. Metals in crude oils can be organometallic or inorganic which affects their solubility in water. In crude oils, vanadium and nickel containing compounds occur as oil-soluble porphyrins and high sulfur crude oils were reported to contain higher content of vanadyl porphyrins than nickel porphyrins (Speight, 1999). Inorganic irons might originate from pipelines and reserve tanks while organic irons might originate from feedstock and reaction of corrosive components such as naphthenic acids. The analysis of Luning, Rocalya, and Shengli crude oils indicated the presence of 3–6 wt% of inorganic iron, 49–75 wt% of organoacid iron, and 7–39 wt% of iron complex, respectively (Liu et al., 2004). Many other inorganic metals are present in crude oils as a result of additives used during the oil recovery. The removal rates of different metals from crude oils, during the desalting process, are shown in Table 1.32.

The removal rates for nickel and vanadium were relatively low when compared to removal rates of Ca, Na, and Mg thus confirming their organochemistry and oil-soluble properties. Only a small amount of iron can be removed from

TABLE 1.32
Removal Rates of Metals during the Desalting Process of Crude Oils

Desalted Crude Oils Removal Rates	Luning %	Rocalya %	Shengli %
Fe	20	3	16
Ca	76	79	88
Mg	44	68	65
Na	95	99	97
Ni	21	25	24
V	43	38	40

Source: Reprinted with permission from Liu, G., Xu, X., and Gao, J., *Energy and Fuels*, 18, 918, 2004. Copyright American Chemical Society.

crude oils during the desalting step and using only a demulsifier. The literature reported that iron in fine particles, usually iron sulfide, is difficult to remove in a desalter (Ellis and Paul, 2000). It was reported that iron can exist as ferric chloride, oxide, sulfide, oil-soluble naphthenate, and form complexes with porphyrins (Liu et al., 2004). The deferric agents, organophosphates and polyamine carboxylates, were reported effective in decreasing the iron content of desalted crude oils. The efficiency of deferric agents was reported to vary depending on their chemistry and concentration (Liu et al., 2004). The oil-soluble and water-soluble metals found in crude oils are shown in Table 1.33.

There are two groups of metals in desalted crude oils. The metals, such as vanadium (V), nickel (Ni), copper (Cu), and iron (Fe), form oil-soluble

TABLE 1.33
Typical Oil-Soluble and Water-Soluble Metals Found in Crude Oils

Crude Oil Metals	Typical Metal	Typical Content, ppm
Oil-soluble complexes	V	5–1500
	Ni	3–120
	Fe	0–120
	Cu	0–12
Water-soluble soaps	Ca	1–3
	Mg	1–3

Source: From Speight, J.G., *The Chemistry and Technology of Petroleum*, 3rd ed., Marcel Dekker, Inc., New York, 1999.

TABLE 1.34

Surface Tension of Molten Salts

Molten Salts	Gas Phase	Temperature, °C	ST, mN/m
AgCl	Air	452	126
NaF	Air	1010	260
NaCl	Air	1000	98
NaBr	Air	1000	88

Source: From Moore, W.J., *Physical Chemistry*, 4th ed., Prentice-Hall, Inc., Englewood Cliffs, New Jersey, 1972.

compounds and their content was reported to vary. An increased iron content may also be due to contamination by iron-containing equipment (Speight, 1999). The inorganic water-soluble salts, mainly chlorides and sulfates of sodium (Na), potassium (K), magnesium (Mg), and calcium (Ca), are removed or their level is significantly decreased during the desalting operation. Metals, such as zinc (Zn), titanium (Ti), calcium (Ca), and magnesium (Mg), appear in the form of organometallic soaps and they adsorb at the oil/water interface acting as emulsion stabilizers (Speight, 1999). Many other metals, such as aluminum (Al), cobalt (Co), and silicon (Si), were also found in crude oils. The presence of silicone might be related to the use of silicone antifoaming agents. The ST of molten salts is shown in Table 1.34.

Crude oil contains about 0.2 wt% of water in which soluble salts, such as sodium chloride and other metal salts, are dissolved. If the salt content of the crude oil is high, expressed as sodium chloride, it is necessary to desalt the crude oil before processing. The total ash from desalted crude oils was reported to be in the range of 0.1–100 mg/L (Speight, 1999). Metals and salts not separated during the desalting step will end up in desalted crude oil affecting their quality and processing equipment.

1.4   PETROLEUM SLUDGE

Organic solid deposition was reported to be the most difficult problem related to crude oil operation and it can occur during transportation or storage. It is affected by the fluid composition and surrounding conditions. During the oil recovery, asphaltenes and the presence of gases dissolved in oil cause the sediment formation. Asphaltenes form sediment when gases are dissolved in the oil and this affects the light and heavy oil recovery (Speight, 1999). The literature reported that the asphaltic type deposits are a common problem when light crude oils are produced near the flocculation onset pressure. Wax deposition is observed when paraffinic oils are produced and when low temperatures are involved which is the case of the offshore production (Garcia and Carbognani, 2001). The literature reported that the tank sludges of Bombay High crude oil contained high melting point crystalline waxes. High temperature gas chromatography (HTGC) analysis

TABLE 1.35

Factors Affecting Crude Oil Stability

Crude Oil Properties	Crude Oil Instability
API gravity	Low API, high asphaltene content
Viscosity	High viscosity, high asphaltene content
Asphaltene/resin ratio	>1 (oil is stable)
	<1 (oil is unstable)
Heteroatom content	Provides polarity
	Preferential reaction with oxygen
Oxidation	Changes functional group composition

Source: From Speight, J.G., *The Chemistry and Technology of Petroleum*, 3rd ed., Marcel Dekker, Inc., New York, 1999. Reproduced by permission of Routledge/Taylor & Francis Group, LLC.

indicated the presence of alkanes, ranging from C_{21} to C_{75}, along with many different isomers. The most predominant alkanes were identified to be C_{40} and C_{67} (Kumar et al., 2003). The asphaltene agglomeration and wax precipitation leading to deposits were reported to affect both upstream and downstream sectors of the oil industry (Lira-Galeana, 2004). The literature reported on many factors affecting crude oil stability as listed in Table 1.35.

Crude oils vary widely in viscosity and composition. Stable crude oils were reported to contain asphaltenes having a lower density and a lower aromaticity when compared to unstable crude oils (Rogel et al., 2003). Thermal methods can change the oil/resin ratios and further affect the crude oil stability (Speight, 1999). The thermal methods can also affect the oil stability by changes in oil medium. Crude oil instability can lead to different deposit formation called precipitate, sediment, or sludge. The Kuwaiti crude oil lakes, subjected to fires and extreme weather conditions, were reported to contain a sludge and have different properties (Al-Kandary and Al-Jimaz, 2000). An increase in temperature, in the presence of air, leads to formation of sludge affecting other properties of crude oils. The properties of "weathered" Kuwaiti light crude oils, containing sludges, are shown in Table 1.36.

The viscosity of weathered crude oils, containing sludge, increased from 8.7 to 15.8 cSt at 40°C, and an increase in pour point, from less than −23°C to −1°C or −2°C, was observed. Some increase in S content, from 2 to 3–3.5 wt%, with a drastic increase in salt content was reported. The volumetric measure of refinery feedstocks is usually in barrels. The salt content is expressed as sodium chloride equivalent in pounds per thousand barrels (PTB) of crude oil. Typical values range from 1 to 20 PTB and 1 PTB is about 3 ppm (Gary and Handwerk, 2001). The salt content of weathered crude oil increased from 7.2 PTB to 7200–10072 PTB, and a significant increase in water content, from 0.1 to 10–18 wt%, was also reported which might lead to stable emulsions. The asphaltene content of weathered crude oils increased,

TABLE 1.36
Properties of Weathered Crude Oils Containing Sludge

Kuwaiti Crude Oils	Kuwaiti Light	Weathered Oil	
Properties	Crude Oil	With Sludge A	With Sludge B
Gravity, API	37.01	15.75	15.77
Kin. viscosity at 40°C, cSt	8.7	180	275.9
Pour point, °C	<−23	−1	−2
Sulfur, wt%	2.01	2.98	3.45
Salt content, PTB	7.2	10,072	7,200
Water, vol%	0.96	10	18
Asphaltenes, wt%	2.1	5.4	4
Carbon residue, wt%	5.6	9.2	9.1

Source: From Al-Kandary, J.A.M. and Al-Jimaz, A.S.H., *Petroleum Science and Technology*, 18, 795, 2000. Reproduced by permission of Taylor & Francis Group, LLC, http://www.taylorandfrancis.com.

from 2.1 to 4–5.4 wt%, and a significant increase in carbon residue, from 5.6 to 9.1–9.2 wt%, was reported. Petroleum sludges were also reported to form as a result of waste water treatment, maintenance of the equipment, and cleaning of tanks. The composition of different petroleum sludges is shown in Table 1.37.

The composition of petroleum sludges varies depending on their source and the literature reported that the petroleum sludge stored for years in tanks is different from fresh sludge. The use of many different demulsifiers, such as polyalkyl phenol formaldehyde monoethanol amine ethoxylate, polyalkyl phenol

TABLE 1.37
Composition of Different Petroleum Sludges

Composition	Sludge Pond	Sludge Flotation Unit	Sludge Buffer Pond
Oil phase, wt%	15.6	5.2	11.2
Aqueous phase, wt%	75	92	68
pH	8.4	8.6	5.8
Ca, mg/L	328	496	360
Mg, mg/L	110	53	130
Cl, mg/L	511	497	1243
SO_4, mg/L	173	189	2150
Solids, wt%	9.4	2.8	20.8

Source: From Mazlova, E.A., Meshcheryakov, S.V., and Klimova, L.Z., *Chemistry and Technology of Fuels and Oils*, 36, 431, 2000. Reproduced with permission of Springer Science and Business Media.

TABLE 1.38
Paraffin and Asphaltene Contents of Different
Petroleum Sludges

Composition	Sludge Settling Pond	Sludge Collector
Paraffins, wt%	23.8	25.7
Resins, wt%	12.5	14.6
Water, wt%	13.8	14
Cl, mg/L	1464	909
Particulates, wt%	1.81	2.54
Asphaltenes, wt%	23.1	17.3

Source: From Elasheva, O.M. et al., *Chemistry and Technology of Fuels and Oils*, 39, 151, 2003. Reproduced with permission of Springer Science and Business Media.

formaldehyde diethanol amine ethoxylate, and polyalkyl phenol formaldehyde triethanol amine ethoxylate, was reported effective in demulsifying water-in-crude oil emulsions. The use of the same demulsifiers, on their own, was reported ineffective in separating petroleum sludge. The literature reported on the need to use demulsifiers and flocculants to treat different petroleum sludge and sediments from cleaning equipment and slop oils. The combination of demulsifier, flocculant, and solvents, such as xylene, was also reported effective in separating the petroleum sludge (Mazlova et al., 2000). The paraffin and resin contents of different petroleum sludges are shown in Table 1.38.

The literature reported on low sediment but high content of paraffins, resins, and chlorides in some petroleum sludges (Elasheva et al., 2003). One of the issues in the refining industry is utilization of liquid and solid crude oil–containing wastes stored in sludge pits and tanks. Storing them is expensive and can lead to soil or groundwater contamination. According to the literature, the solid deposits that precipitate in crude oil production facilities are complex materials which include large alkanes, insoluble asphaltenes, and also different inorganic species. The corrective action involves the analysis of deposits, use of asphaltene precipitation inhibitors, and redissolution of the solids (Carbognani, 2001). The solubility of asphaltenes in heptane–toluene mixtures was studied at different temperatures. Similar to nonpolar waxes, a significant increase in asphaltenes' solubility with an increase in the temperature, from 0°C to 20°C, was observed (Neves et al., 2001). The literature reported that stable emulsions and difficulties in separating petroleum sediments and sludges are caused by the presence of surfactants.

1.5   USE OF SURFACTANTS

A wide range of different chemistry additives are used in drilling fluids and during the crude oil recovery and transportation to improve their stability and interfacial

TABLE 1.39

n-Pentane Insoluble Content of Different Crude Oils and Blend

Crude Oil Properties	Iranian Light	Arabian Heavy	Russian Blend
Density at 25°C, g/cm³	0.9127	0.9404	0.9066
Average molecular mass	394	427	350
C5 insolubles, wt%	3.44	9.23	2.12

Source: From Laux, H., Rahimian, I., and Browarzik, D., *Petroleum Science and Technology*, 19, 1155, 2001. Reproduced by permission of Taylor & Francis Group, LLC, http://www. taylorandfrancis.com.

properties. The drilling rig is used to handle the drill pipe and bit and to set casing to complete the well. Drilling fluid, called mud, is used to maintain hydrostatic pressure for well control, carry drill cuttings to the surface, and cool and lubricate the drill bit. Drilling fluids may be fresh water–based, salt water–based, oil-based, or synthetic based (Railroad Commission of Texas, 2001). Water used to make up the drilling fluid (makeup water) may require treatment to remove dissolved calcium and magnesium. Soda ash may be added to precipitate calcium carbonate and caustic soda may be added to form magnesium hydroxide. Solid additives are usually introduced into the mud system in a mixing hopper. Other additives for control of mud viscosity and gel strength are mixed in tanks connected to the mud stream (Railroad Commission of Texas, 2001). Drilling fluid additives include acids, caustics, bactericides, antifoaming agents, emulsifiers, thinners, dispersants, and others. Precipitation of asphaltenes from crude oils leads to plugging and production loss. The literature reported on the effect of average molecular mass on n-pentane insoluble content of different crude oils, stabilized at 50 mbar and 180°C in a rotary evaporator (Laux et al., 2001). The n-pentane insoluble content of different crude oils and Russian blend is shown in Table 1.39.

With an increase in average molar mass of crude oils, from 350 to 427, an increase in n-pentane insolubles, from 2.12% to 9.23%, is observed (Laux et al., 2001). Several methods are used for predicting the precipitation of asphaltenes on blending of dead crude oils not containing soluble gases (Wiehe, 2003). The literature reported on blending of Arabian Medium and Gippsland crude oils to investigate various testing procedures. While different tests can establish an unstable region, the onset of the asphaltene flocculation was reported to vary (Schermer et al., 2004). A novel technique for breaking asphaltenes using the laser energy was reported (Zekri et al., 2003). The literature also reported that blending of some crude oils might lead to a decrease in their colloidal stability leading to asphaltene precipitation. Some crude oil blends were reported to contain two or even four different crude oils ranging in volume from as low as 10 to 70 vol% and having varying composition and properties. Major refining problems related to crude oils include line blockage caused by deposition of asphaltenes, wax, scale, and hydride. Additional crude oil production issues

---

TABLE 1.40

Use of Additives in Drilling Fluids and Crude Oils

Drilling Fluids	Crude Oils
Acids	Asphaltene stabilizers
Caustics	Pour point depressants
Bacteriocides	Antifoaming agents
Defoamers	Demulsifiers
Emulsifiers	Wetting agents
Thinners	Corrosion inhibitors
Dispersants	Other

Source: From Railroad Commission of Texas, Waste Minimization in the Oil Field, 2001.

---

involve the interfacial properties such as foaming, stable emulsions, rusting, and corrosion (Poindexter et al., 2002). The use of additives in drilling fluids and crude oils is shown in Table 1.40.

Wax precipitation may take place at temperatures far above the freezing point of water and consists of closed packed lattices of aligned paraffinic and naphthenic molecules. Wax crystallization and deposition of crude oils can lead to severe pipeline restrictions and eventually block the line or well. Remediation might involve mechanical pigging through solvent washes or hot oiling and use of additives (Groffe et al., 2001). Pour point depressants, also known as wax inhibitors, are usually used to prevent wax deposition. The literature reported on the use of ethylene-co-vinyl acetate copolymers as effective modifiers of oil wax crystallization in crude oils. The ethylene-vinyl acetate (EVA) copolymers include in their structure polymethylene segments identical to those of the paraffins forming the wax phase (Pedersen et al., 1989). The ethylene-co-vinyl acetate polymers, containing 32 wt% of vinyl acetate, were found the most effective in decreasing the wax precipitation from Brazilian crude oils (Machado and Lucas, 2001). Different crude oils were reported to have pour points varying in a range, from as low as −1°C to as high as 30°C. The effect of EVA polymer on pour points of different crude oils is shown in Table 1.41.

The use of 25 ppm of EVA polymer was reported effective in decreasing the pour point of crude oil # 1, from −1°C to −26°C, by 25°C. The use of the same treat rate of EVA polymer in crude oil # 2, having a higher pour point of 8°C, decreased the pour point by only 10°C. A significantly higher treat rate of 100 ppm of EVA polymer was required to decrease the pour point of crude oil # 3, having a high pour point of 30°C, by 21°C. The use of 200 ppm of EVA polymer in crude oil # 4 resulted in an increase in the pour point from 21°C to 25°C. The use of different crude oils affects the effective treat rates and performance of pour point depressant and an overtreatment might actually increase their pour point. The use of polymers, such as EVA as pour point depressant, leads to differences in their efficiency

TABLE 1.41

Effect of EVA Polymer on Pour Points of Different Crude Oils

Crude Oils EVA Polymer, ppm	Oil # 1 PP, °C	Oil # 2 PP, °C	Oil # 3 PP, °C	Oil # 4 PP, °C
0	−1	8	30	21
25	−26	−18		
100			9	
200				25
PP difference (°C)	(−25)	(−26)	(−21)	(+4)

Source: From Pedersen, K.S., Fredenslund, Aa., and Thomassen, P., *Properties of Oils and Natural Gases*, Gulf Publishing Company, Houston, 1989. Reproduced with permission of GPC Books.

to depress the pour points of crude oils depending on their composition and their wax content. The literature reported that pour point depressants were developed to interfere with the growth of n-paraffinic type wax and do not entirely prevent wax crystal growth. They only lower the temperature at which a rigid wax structure is formed (Pirro and Wessol, 2001).

Precipitation of wax can be due to a drop in the temperature or a result of evaporation of volatile light ends acting as natural solvents. The literature also reported that the presence of asphaltenes affects the paraffin–oil gel deposits on the pipe wall during transportation of crude oils. The gel deposit was reported to contain only 3–4 wt% of solid wax with the remainder being occluded oil (Venkatesan et al., 2003). Wax inhibitors and antisticking agents, which are blends of ionic surfactants, can be used to prevent the wax deposition in more problematic crude oils. While wax inhibitors act as wax modifiers, the antisticking agents can adsorb onto bare surfaces of tubings and make them oleophobic and thus retard the wax deposition (Groffe et al., 2001). Many other additives, used in drilling fluids and crude oils, are surfactants. The incompatibility within the same molecule is the unique property of the surfactant molecule. The types and chemistry of different surfactant molecules are shown in Table 1.42.

The surfactant molecule will change its orientation to place its hydrophobic group in hydrophobic medium, form the adsorbed layer at the interface, and agglomerate into spherical, cylindrical, or lamellar micelles above the critical micelle concentration (CMC). According to the literature, sulfates are classified as anionic, quaternary ammonium salts and imidazolinium salts are classified as cationic, imidazolines are amphotheric while ethoxylates, PEG esters, and EO/PO copolymers are classified as nonionic surfactants (Karsa, 1986). The surface active properties of surfactants at the oil/air, oil/water, and oil/metal interfaces can lead to foaming or defoaming, emulsification or demulsification, wetting or nonwetting properties. The main solution behavior affected by the presence or absence of surfactant molecules can lead to variation in solubilizing, dispersing, or detergency properties (Karsa, 1986). The addition of amphiphiles is

TABLE 1.42

Types and Chemistry of Different Surfactant Molecules

Chemistry of Molecule	Surfactant Types	Examples
Hydrophobe–hydrophile	Anionic	Alkyl benzene sulfonates
Hydrophobe–hydrophile	Cationic	Quaternary ammonium salts
Hydrophobe–hydrophile	Amphoteric	Imidazolines
Hydrophobe–hydrophile	Nonionic	Alkyl phenol ethoxylates

Source: From Karsa, D.R., *Industrial Applications of Surfactants*, The Royal Society of Chemistry, Burlington House, London, 1986.

frequently used to prevent asphaltene precipitation in reservoir rocks and well bore tubing. It was reported that the adsorption of amphiphiles on the asphaltenes makes them more soluble under the production conditions (Rogel et al., 2002). The effect of surfactants on the interfacial properties and the solution behavior of liquids is shown in Table 1.43.

The total recovery by the primary and secondary oil recovery is usually less than 40% of the original oil and the remaining oil is recovered during the tertiary oil recovery also known as enhanced oil recovery (EOR). The literature reports on several proven techniques for the EOR such as surfactant-polymer flooding, foam flooding, CO_2 flooding, caustic solution flooding, microbial method, steam injection, and thermal combustion (Ling et al., 1986). According to the literature, the most promising EOR technique is the micellar-polymer flooding process which uses surfactants to decrease IFT and allow the oil to flow through the porous media. The IFT was reported to increase with pressure but decrease with temperature, salinity, and surfactant concentration (Al-Sahhaf et al., 2002). The literature reported on the IFT obtained through alkaline flooding and when synthetic surfactant, dodecyl benzene sulfonic acid sodium salt, is used in the

TABLE 1.43

Effect of Surfactants on Interfacial and Solution Properties of Liquids

Interfacial Properties	Solution Properties
Foaming or defoaming	Solubility
Emulsification or demulsification	Dispersing
Wetting or rewetting	Detergency

Source: From Karsa, D.R., *Industrial Applications of Surfactants*, The Royal Society of Chemistry, Burlington House, London, 1986.

TABLE 1.44

Typical Use of Surfactants in Crude Oil Refining

Surfactant Applications	Surfactant Additives
Drilling mud	Demulsifiers
EOR process	Wetting agents
Desalter	Corrosion inhibitors
Oil spill	Dispersants

Source: From Karsa, D.R., *Industrial Applications of Surfactants*, The Royal Society of Chemistry, Burlington House, London, 1986.

EOR process (Elkamel et al., 2002). Reducing the mobility of CO_2 by generating in situ foam is an effective method for improving the oil recovery in CO_2 flooding processes. It involves a coinjection of CO_2 and surfactant solution into the porous medium (Asghari and Khalil, 2005). The typical use of surfactant molecules in crude oil refining is shown in Table 1.44.

After the primary recovery and water flooding stages, a large amount of residual oil remains in the reservoir which leads to the use of EOR methods to make sure that all the oil is recovered. Adsorption of surfactants and polymers at liquid and solid interfaces is widely used to modify the interfacial properties of many industrial processes including the EOR process. The literature reported on the chemistry of surfactants used in drilling mud, the EOR process, and as additives (Karsa, 1986). The behavior of surfactants and polymers at interfaces was reported to vary and be affected by many different interactions, such as electrostatic attraction, covalent bonding, hydrogen bonding, hydrophobic bonding, solvation, and dissolvation of different molecules. Many different techniques, including fluorescence and ESR, can be used to study the chemistry of surfaces (Somasundaran and Huang, 2000). Depending on the crude oil and the processing requirements, different chemistry surfactants are used. The stability of crude oil emulsions was reported to vary depending on the chemistry and the surfactant concentrations. The nonionic surfactants were found to produce emulsions having a lower viscosity while the anionic surfactants were reported to stabilize emulsions at lower concentrations (Ahmed et al., 1999). Some anionic and nonanionic surfactants were reported to form oil-in-water emulsions of asphaltic heavy crude oils without any deposition of asphaltenes (Zaki et al., 2001). The literature reported on asphaltenic crude oil, having practically no salt content and a relatively low TAN, which formed a stable emulsion (Liu et al., 2003). The composition of asphaltenic crude oil, containing no salt and forming a stable emulsion, is shown in Table 1.45.

The early literature reported that certain surface active agents capable of reducing the IFT to <2 mN/m can lead to emulsion formation during very slight agitation. Under certain circumstances, such an emulsification can occur

TABLE 1.45
Composition and Emulsion Stability of Asphaltenic Crude Oil

Asphaltenic Crude Oil	Properties
Density, g/cm^3 at 20°C	0.893
Freezing point, °C	−1
Paraffin content, wt%	5.6
Sulfur, wt%	9
Salt content, mg/L	8.85
TAN, mg KOH/g	0.12
Water-in-oil emulsion	Stable
Asphaltenes, wt%	2.3

Source: Reprinted with permission from Liu, G., Xu, X., and Gao, J., *Energy and Fuels*, 17, 543, 2003. Copyright American Chemical Society.

simultaneously (Bondi, 1951). The literature reported on the effect of pressure, temperature, and alkali concentration on the IFT of Arabian Heavy crude oil. The alkali reacts with acidic components of crude oil which leads to the formation of surface active soaps, lowering the IFT and the mobilization of residual oil (Elkamel et al., 2002). The literature also reported on the effect of amines and the presence of NaCl on the interfacial properties of acidic crude oil emulsions. The addition of low molecular weight aliphatic amines decreased the IFT of crude oils, from 20 to 0.5 mN/m, and after the addition of NaCl an increase in the IFT and a spontaneous emulsification were observed (Gutierrez et al., 2003).

Some molecules found in crude oils can behave as surfactants if they contain a hydrophobic (water-repellant) hydrocarbon and a hydrophilic (water-compatible) head group. The addition of some additives, while not surfactants, might create surfactant molecules by interacting with some molecules present in crude oils. Crude oils cause rust and corrosion in pipelines and inhibitors are added to minimize the reaction between acid and the metal surfaces. The presence of surfactant type corrosion inhibitors, such as oleic imidazoline, was reported to improve the performance of some wax inhibitors in decreasing the wax deposits (Wang et al., 2003b). The presence or formation of surfactants, having emulsifying properties, can increase the stability of crude oil emulsions and decrease the efficiency of the desalter to remove salts and metals.

REFERENCES

Ahmed, N.S. et al., *Petroleum Science and Technology*, 17(5&6), 553, 1999.
Al-Kandary, J.A.M. and Al-Jimaz, A.S.H., *Petroleum Science and Technology*, 18(7&8), 795, 2000.

Al-Sabagh, A.M. et al., *Polymers for Advanced Technologies*, 13, 346, 2002.

Al-Sahhaf, T., Suttar Ahmed, A., and Elkamel, A., *Petroleum Science and Technology*, 20(7&8), 773, 2002.

Ancheyta, J. et al., *Energy and Fuels*, 16, 1121, 2002.

Andersen, S.I. and Speight, J.G., *Petroleum Science and Technology*, 19(1&2), 1, 2001.

Andersen, S.I. et al., *Petroleum Science and Technology*, 19(1&2), 55, 2001.

Aquino-Olivos, M.A., Andersen, S.I., and Lira-Galeana, C., *Petroleum Science and Technology*, 21(5&6), 1017, 2003.

Asghari, K. and Khalil, F., *Petroleum Science and Technology*, 23(2), 189, 2005.

Bagci, S., Kok, M.V., and Turksoy, U., *Petroleum Science and Technology*, 19(3&4), 359, 2001.

Bansal, V., Patel, M.B., and Sarpal, A.S., *Petroleum Science and Technology*, 22(11&12), 1401, 2004.

Becker, J.R., *Crude Oil Waxes, Emulsions and Asphaltenes*, PennWell Publishing Company, Tulsa, 1997.

Bondi, A., *Physical Chemistry of Lubricating Oils*, Reinhold Publishing Corporation, New York, 1951.

Browarzik, D. et al., *Petroleum Science and Technology*, 20(3&4), 233, 2002.

Buenrostro-Gonzalez, E. et al., *Petroleum Science and Technology*, 19(3&4), 299, 2001.

Callaghan, I.C. et al., *The Symposium on Industrial Applications of Surfactants*, Special Publication No. 59, Royal Society of Chemistry, April 15–17, 1986.

Carbognani, L., *Energy and Fuels*, 15, 1013, 2001.

Coutinho, J.A.P. and Daridon, J.-L., *Petroleum Science and Technology*, 23(9&10), 1113, 2005.

CRC Handbook of Chemistry and Physics, 86th ed., Lide, D.R., Editor-in-Chief, CRC Press, Boca Raton, 2005.

Dee, G.T. and Sauer, B.B., *TRIP*, 5(7), 23, 1997.

Demirbas, A., *Petroleum Science and Technology*, 20(5&6), 485, 2002.

Elasheva, O.M. et al., *Chemistry and Technology of Fuels and Oils*, 39(3), 151, 2003.

Elkamel, A., Al-Sahhaf, T., and Suttar Ahmed, A., *Petroleum Science and Technology*, 20(7&8), 789, 2002.

Ellis, P.J. and Paul, C.A., *Delayed Coking Fundamentals*, AIChE 2000 Spring National Meeting, Atlanta, May 5–9, 2000.

El-Sabagh, S.M. and Al-Dhafeer, M.M., *Petroleum Science and Technology*, 18(7&8), 743, 2000.

Escobar, G. et al., *Petroleum Science and Technology*, 19(1&2), 107, 2001.

Garcia, M.C. and Carbognani, L., *Energy and Fuels*, 15, 1021, 2001.

Garcia, M.C. and Urbina, A., *Petroleum Science and Technology*, 21(5&6), 863, 2003.

Gary, J.H. and Handwerk, G.E., *Petroleum Refining, Technology and Economics*, 4th ed., Marcel Dekker, Inc., New York, 2001.

Ghannam, M.T., *Petroleum Science and Technology*, 23(5&6), 649, 2005.

Goldstein, R.F., *The Petroleum Chemicals Industry*, 2nd ed., E. & F.N. SPON, London, 1958.

Goldszal, A. and Bourrel, M., *Industrial and Engineering Chemistry Research*, 39, 2746, 2000.

Groffe, D. et al., *Petroleum Science and Technology*, 19(1&2), 205, 2001.

Gutierrez, X. et al., *Petroleum Science and Technology*, 21(7&8), 1219, 2003.

Hammami, A., Ratulowski, J., and Coutinho, J.A.P., *Petroleum Science and Technology*, 21(3&4), 345, 2003.

Heeg, B., Moros, T., and Klenerman, D., *Corrosion Science*, 40(8), 1303, 1998.

Ibrahim, H.H. and Idem, R.O., *Energy and Fuels*, 18, 1038, 2004.
Ibrahim, Y.A. et al., *Petroleum Science and Technology*, 21(5&6), 825, 2003.
Kandwal, V.C. et al., *Petroleum Science and Technology*, 18(7&8), 755, 2000.
Karsa, D.R., Industrial applications of surfactants—an overview, in *Industrial Applications of Surfactants*, Karsa, D.R., Ed., The Royal Society of Chemistry, Burlington House, London, 1986.
Khan, H.U. et al., *Petroleum Science and Technology*, 18(7&8), 889, 2000.
Kim, Y.H. and Wasan, D.T., *Industrial and Engineering Chemistry Research*, 35, 1141, 1996.
Kumar, K., Nikolov, A.D., and Wasan, D.T., *Industrial and Engineering Chemistry Research*, 40, 3009, 2001.
Kumar, S., Gupta, A.K., and Agrawal, K.M., *Petroleum Science and Technology*, 21(7&8), 1253, 2003.
Laux, H., Rahimian, I., and Browarzik, D., *Petroleum Science and Technology*, 19(9&10), 1155, 2001.
Leon, O. et al., *Energy and Fuels*, 15, 1028, 2001.
Ling, T.F., Lee, H.K., and Shah, D.O., Surfactants in enhanced oil recovery, in *Industrial Applications of Surfactants*, Karsa, D.R., Ed., The Royal Society of Chemistry, Burlington House, London, 1986.
Lira-Galeana, C., *International Conference on Heavy Organic Depositions Announcement*, 2004.
Liu, G. et al., *Petroleum Science and Technology*, 20(1&2), 43, 2002.
Liu, G., Xu, X., and Gao, J., *Energy and Fuels*, 17, 543, 2003.
Liu, G., Xu, X., and Gao, J., *Energy and Fuels*, 18, 918, 2004.
Machado, A.L.C. and Lucas, E.F., *Petroleum Science and Technology*, 19(1&2), 197, 2001.
Mazlova, E.A., Meshcheryakov, S.V., and Klimova, L.Z., *Chemistry and Technology of Fuels and Oils*, 36(6), 431, 2000.
Miller, C.A. and Qutubuddin, S., Enhanced oil recovery with microemulsions, in *Interfacial Phenomena in Apolar Media*, Eicke, H.F. and Parfitt, G.D., Eds., Marcel Dekker, Inc., New York, 1987.
Moore, W.J., *Physical Chemistry*, 4th ed., Prentice-Hall, Inc., Englewood Cliffs, New Jersey, 1972.
Morrow, N.R., *Interfacial Phenomena in Petroleum Recovery*, Marcel Dekker, Inc., New York, 1991.
Neves, G.B.M. et al., *Petroleum Science and Technology*, 19(1&2), 35, 2001.
Pacho, D. and Davies, G., *Industrial and Engineering Chemistry Research*, 42, 636, 2003.
Pasadakis, N., Varotsis, N., and Kallithrakas, N., *Petroleum Science and Technology*, 19(9&10), 1219, 2001.
Pedersen, K.S., Fredenslund, Aa., and Thomassen, P., *Properties of Oils and Natural Gases*, Gulf Publishing Company, Houston, 1989.
Pirro, D.M. and Wessol, A.A., *Lubrication Fundamentals*, 2nd ed., Marcel Dekker, Inc., New York, 2001.
Poindexter, M.K. et al., *Energy and Fuels*, 16, 700, 2002.
Prince, R.J., Base oils from petroleum, in *Chemistry and Technology of Lubricants*, 2nd ed., Mortier, R.M. and Orszulik, S.T., Eds., Blackie Academic & Professional, London, 1997.
ProQuest Document ID 10552585, *Chemical Engineering*, 103(12), 99, 1996.
Railroad Commission of Texas, Waste Minimization in the Oil Field, 2001, http://www.rrc.state.tx.us (accessed June 2002).

Ramirez-Jaramillo, E., Lira-Galeana, C., and Brito, M.O., *Petroleum Science and Technology*, 19(5&6), 587, 2001.

Roberts, W.G., *The Quest for Oil*, S.G. Phillips, New York, 1977.

Rogel, E., Contreras, E., and Leon, O., *Petroleum Science and Technology*, 20(7&8), 725, 2002.

Rogel, E. et al., *Energy and Fuels*, 17(6), 1583, 2003.

Rome, C.F., Hydrocarbon Asia, 42 May/June 2001, http://www.hcasia.safan.com (accessed June 2002).

Ross, S., Foaminess and capilarity in apolar solutions, in *Interfacial Phenomena in Apolar Media*, Eicke, H.F. and Parfitt, G.D., Eds., Marcel Dekker, Inc., New York, 1987.

Schermer, W.E.M., Melein, P.M.J., and van den Berg, F.G.A., *Petroleum Science and Technology*, 22(7&8), 1045, 2004.

Siddiqui, M.N., *Petroleum Science and Technology*, 21(9&10), 1601, 2003.

Somasundaran, P. and Huang, L., *Advances in Colloid and Interface Science*, 88(1&2), 179, 2000.

Speight, J.G., *The Chemistry and Technology of Petroleum*, 2nd ed., Marcel Dekker, Inc., New York, 1991.

Speight, J.G., *The Chemistry and Technology of Petroleum*, 3rd ed., Marcel Dekker, Inc., New York, 1999.

Speight, J.G., *The Chemistry and Technology of Petroleum*, 4th ed., CRC Press, Boca Raton, 2006.

Speight, J.G. et al., *Oil & Gas Science and Technology*, 40, 51, 1985.

Spinelli, L.S. et al., *Petroleum Science and Technology*, 22(9&10), 1199, 2004.

Sullivan, A.P. and Kilpatrick, P.K., *Industrial and Engineering Chemistry Research*, 41, 3389, 2002.

Venkatesan, R. et al., *Energy and Fuels*, 17, 1630, 2003.

Vieth, P.H. et al., *Oil and Gas Journal*, 95(6), 52, 1997.

Wang, K.-S. et al., *Petroleum Science and Technology*, 21(3&4), 359, 2003a.

Wang, K.-S. et al., *Petroleum Science and Technology*, 21(3&4), 369, 2003b.

Wiehe, I., *Petroleum Science and Technology*, 21(3&4), 1, 2003.

Zaki, N., Schorling, P.C., and Rahimian, I., *Petroleum Science and Technology*, 18(7&8), 945, 2000a.

Zaki, N.N., Maysour, N.E.-S., and Abdel-Azim, A.-A.A., *Petroleum Science and Technology*, 18(9&10), 1009, 2000b.

Zaki, N., Butz, T., and Kessel, D., *Petroleum Science and Technology*, 19(3&4), 425, 2001.

Zaki, N.N., Poindexter, M.K., and Kilpatrick, P.K., *Energy and Fuels*, 16, 711, 2002.

Zekri, A.Y., Shedid, S.A., and Alkashef, H., *Petroleum Science and Technology*, 21(9&10), 1409, 2003.

2 Conventional Refining

2.1 PETROLEUM FRACTIONS

2.1.1 ATMOSPHERIC AND VACUUM DISTILLATION

Pipelines transport natural gas from the wellhead to gas processing plants and crude oils to storage tanks. The first step in refining of crude oils is desalting which is required to remove water soluble inorganic salts and metals. The chemical demulsification process requires the addition of demulsifiers to destabilize the emulsions. Crude oil is a mixture of compounds having different boiling points and distillation is the simplest way to separate them into different fractions. The distillation of crude oil occurs in large fractionating towers.

The desalted crude oil is heated in a tube furnace to 385°C, just below the temperature at which cracking of the oil can occur, then flashed into a distillation column. Crude oil, heated and partially vaporized in a furnace, enters near the bottom of the distillation tower. As the vapors rise through the tower, the various fractions condense and are drawn off. The presence of alkali chlorides in desalted crude oils and an increase in temperature will lead to corrosive acids, such as HCl, and cause corrosion during the processing of crude oils. Magnesium chloride causes most of the corrosion because it breaks down at lower temperatures during the distillation of crude oils forming hydrochloric acid (Ellis and Paul, 2000).

The literature also reported that individual crude oils have a different stability and blending of some crude oils can lead to asphaltene destabilization causing fouling of equipment during the refinery operation. Blending of incompatible crude oils was reported to lead to problems, such as desalter upset and fouling in preheat train heat exchangers during processing (Stark and Asomaning, 2003).

Foaming problems are known to occur in a wide range of industrial processes including filtration, painting, printing, moulding, and distillation. In the process of fractionation or stripping the conditions are such that they will lead to a stable foam which lasts for a very short time, no more than several seconds, and are the cause of severe foam-flooding in distillation or fractionation towers (Ross, 1987). In many distillation and adsorption processes, the foam formation can cause the loss of throughput, reduction in separation efficiency, and decrease in product quality and product contamination (Thiele et al., 2003). The literature reports on preflash drum foaming which leads to a decrease in distillate production

TABLE 2.1
Atmospheric Distillation of Arab Light Crude Oil

Distillation Temperature, °C	Crude Oil Fraction	Yield on Arab Light Crude Oil, vol%
<25	Gases	1.1
25–200	Naphtha	25.3
200–350	Middle distillate	28.8
>350	Atmospheric residue	45.1

Source: From Hamid, S.H., *Petroleum Science and Technology*, 18, 871, 2000. Reproduced by permission of Taylor & Francis Group, LLC, http://www.taylorandfrancis.com.

(Golden, 1997). Silicones are used in the petroleum industry as foam control agents during different processing steps including well drilling, gas separation, gas scrubbing, and distillation (Rome, 2001). According to the literature, dealing with foaming is difficult and it requires the use of antifoaming agents and installation of foam-destroying devices (Chernozemov, 2001). The material balance obtained from atmospheric distillation of Arab light crude oil is shown in Table 2.1.

The gas fraction boils below 25°C. The fraction having a boiling range of 25°C–200°C is known as naphtha. The fraction known as the middle distillate (MD) fraction boils in a higher boiling range of 200°C–350°C. The Arab light crude oil was reported to contain 1.1 vol% of gases, 25.3 vol% of naphtha, 28.8 vol% of MD, and 45.1 vol% of atmospheric residue (Hamid, 2000). Atmospheric residua were reported to be black and viscous materials (Speight, 1999). Crude oil separation is actually accomplished in two steps. The first step involves the fractionation of the crude oil at atmospheric pressure. Distillation of crude oil under atmospheric pressure produces an atmospheric residue which is a liquid having a bp > 350°C. The next step is the fractionation of the high boiling fraction, atmospheric residue, under vacuum. The atmospheric residue (reduced crude) is heated to 395°C and flashed into a vacuum distillation column that is operated at low pressures, usually 25–100 mm Hg (Ellis and Paul, 2000). The reduced crude is fractionated in a vacuum distillation unit into light and heavy vacuum gas oil (VGO). Distillation of crude oil under reduced pressure produces the vacuum residue which is a solid and has a bp > 565°C (Speight, 1999). The true boiling points (TBPs) of atmospheric and vacuum distillation crude oil fractions are shown in Table 2.2.

The two-step distillation process of crude oil leads to atmospheric distillation products, light and heavy VGO, and vacuum residue. To meet the viscosity requirements of lubricant products, VGO needs to have a narrow boiling range. The atmospheric residue (reduced crude) is fractionated in a vacuum distillation unit into fractions of the desired viscosity and flash point. It is desirable to lift the maximum amount of oil boiling below 565°C into heavy vacuum gas oil (HVGO) reducing the production of vacuum reduced crude (Ellis and Paul, 2000). VGO is separated into light, medium, and heavy lube oil distillates. The effect of the boiling range on the pour point of MD is shown in Table 2.3.

TABLE 2.2
Boiling Points of Atmospheric and Vacuum Distillation Crude Oil Fractions

Distillations	Fractions	TBP Cut Point, °F
Atmospheric	Light naphtha	50–200
	Heavy naphtha	200–375
	Kerosene	375–450
	Diesel heating oil	450–550
	Atmospheric gas oil	550–650
	Atmospheric residue	>650
Vacuum	Light vacuum gas oil	650–750
	Heavy vacuum gas oil	750–1000
	Light lube distillate	625–725
	Medium lube distillate	725–825
	Heavy lube distillate	825–1050
	Vacuum residue	>1000–1050

Source: From Sequeira, A. Jr., *Lubricant Base Oil and Wax Processing,* Marcel Dekker, Inc., New York, 1994. Reproduced by permission of Routledge/Taylor & Francis Group, LLC.

The cloud point is the first sign of the wax formation while the pour point is the lowest temperature at which oil will flow under specific testing conditions. The pour point of MD was reported to increase with an increase in their boiling range. While the pour point increases with an increase in the boiling point of the petroleum fractions, their paraffinic content was reported to decrease. In the gasoline range, the paraffinic content was reported to be as high as 80 wt% but

TABLE 2.3
Effect of Boiling Range on the Pour Point of Middle Distillate Fraction

Middle Distillate	Boiling Range, °C	Pour Point, °C
Broad cut	250–375	15
Subfraction	250–275	−9
Subfraction	275–300	3
Subfraction	300–325	15
Subfraction	325–350	27
Subfraction	350–375	36

Source: From Purohit, R.C., Srivastava, S.P., and Verma, P.S., *Petroleum Science and Technology,* 21, 1369, 2003. Reproduced by permission of Taylor & Francis Group, LLC, http://www.taylorandfrancis.com.

TABLE 2.4
Different Techniques Used to Analyze Light Gas Oil

Characteristics	Analytical Methods
Boiling point (bp)	Distillation
	GC-SimDis, GC–MS
H/C ratio	Elemental analysis
Compound class	HPLC
	PIONA (paraffin, iso-paraffin, olefin, naphthene, aromatic)
	SARA (saturates, aromatics, resins, asphaltenes)
Molecular weight	VPO cryoscopy, GPC, FIMS
Atomic connectivity	H-NMR, C-NMR

Source: From Klein, M.T. et al., *Molecular Modeling in Heavy Hydrocarbon Conversions*, CRC Press, Boca Raton, 2006. Reproduced by permission of Routledge/Taylor & Francis Group, LLC.

only 30 wt% in lube distillate (Speight, 2006). Crude oil can be separated into a variety of different generic fractions by distillation and the main distillation products are straight-run gasoline or naphtha, kerosene, atmospheric gas oil, and atmospheric reduced crude. The different techniques available to analyze the properties and the composition of light gas oil are shown in Table 2.4.

Many different analytical techniques and different predictive tools have been developed to analyze the chemistry of petroleum products and their use varies depending on their boiling range. The literature reported on the prediction of molecular weight of different petroleum fractions, containing C_5–C_{120} carbons, by using the 50 wt% TBP and density (Goossens, 1996). The analysis of light gas oil, using all available techniques, is shown in Table 2.5.

Distillation separates crude oil into useful components which may be end products, such as stove oil, or feedstocks for other refinery processes, such as lube oil base stocks. Different analytical techniques are used to study the composition of petroleum fractions, used as fuels, VGO, and petroleum residua. The analysis of the atmospheric residue, from Arab light crude oil, was reported to contain a high content of sulfur, nitrogen, and metals. The asphaltene content was found to vary depending on the precipitating solvent and the use of *n*-heptane decreased the insoluble precipitation. The mass spectrometry (MS) and Fourier transform infrared (FTIR) spectroscopy were used to characterize asphaltenes present in the atmospheric residue (Hamid, 2000). The composition and heteroatom content of MD and vacuum gas oil (VGO) are shown in Table 2.6.

An increase in the boiling range of MD, from 144°C–379°C to 182°C–385°C, leads to an increase in the density and viscosity. With an increase in the boiling range, the sulfur content of MD increased, from 0.1 to 0.25 wt%, and some increase in the nitrogen content, from 0.05 to 0.08 wt%, was reported. An increase in the boiling range of MD, from 144°C–379°C to 182°C–385°C, also leads to a decrease in the saturate content and an increase in the aromatic content, from 42.4

TABLE 2.5
Analysis of Light Gas Oil Fraction

Light Gas Oil	Properties and Composition
Simulated Distillation	
10% cut off, °C	211.4
30% cut off, °C	248.0
50% cut off, °C	271.0
70% cut off, °C	295.6
90% cut off, °C	331.7
Final boiling point, °C	403.5
PIONA Fractions	
Paraffin, wt%	4.4
Iso-paraffin, wt%	9.9
Naphthenic, wt%	32.0
Aromatic, wt%	53.7
H/C ratio	1.67
Molecular weight	181.4

Source: From Klein, M.T. et al., *Molecular Modeling in Heavy Hydrocarbon Conversions*, CRC Press, Boca Raton, 2006. Reproduced by permission of Routledge/Taylor & Francis Group, LLC.

TABLE 2.6
Heteroatom Content of Middle Distillates and Polar Content of VGO

Petroleum Fractions	MD #1	MDMD #2	VGO
Boiling range, °C	144–379	182–385	>400
Density at 15°C, g/mL	0.8548	0.8659	0.8720
Flash point, °C	60	86	—
C, wt%	86.6	86.9	85.1
H, wt%	13.1	12.7	12.5
S, wt%	0.1	0.25	2.3
N, wt%	0.05	0.08	0.1
Saturates, vol%	57.6	40.2	49.9
Aromatics, vol%	42.4	59.8	41.1
Olefins, vol%	0	0	—
Polars, wt%	—	—	9.0

Source: From Severin, D. and David, T., *Petroleum Science and Technology*, 19, 469, 2001. Reproduced by permission of Taylor & Francis Group, LLC, http://www.taylorandfrancis.com.

TABLE 2.7
Properties and Composition of Atmospheric and Vacuum Residue

Tia Juana Crude Oil	Atmospheric Residue, bp > 345°C	Vacuum Residue, bp > 565°C
API gravity	17.3	7.1
Viscosity at 100°C, cSt	35	7959
Pour point, °C	7	49
Sulfur, wt%	1.8	2.6
Nitrogen, wt%	0.3	0.6
Nickel, ppm	25	64
Vanadium, ppm	185	450
Iron, ppm	28	48
Carbon residue, wt%	9.3	21.6

Source: From Speight, J.G., *The Chemistry and Technology of Petroleum*, 3rd ed., Marcel Dekker, Inc., New York, 1999. Reproduced by permission of Routledge/Taylor & Francis Group, LLC.

to 59.8 wt%, was observed. During distillation some cracking might take place; no presence of olefins in MD was reported (Severin and David, 2001).

With a further increase in the boiling point of VGO, over 400°C, a further increase in the heteroatom content to 2.3 wt% of sulfur and 0.1 wt% of nitrogen was reported. VGO was also reported to have a high polar content of 9 wt% indicating the presence of other heteroatom molecules. The oxygen content of petroleum oils is difficult to measure but based on the high polar content of VGO, the oxygen content of VGO could be significant. The properties and composition of atmospheric and vacuum residue, from Tia Juana crude oil, are shown in Table 2.7.

With a decrease in the API gravity, from 17.3 to 7.1, and an increase in the viscosity, a significant increase in the pour point of vacuum residue from 7°C (45°F) to 49°C (120°F) was reported. The atmospheric residue was reported to contain 1.8 wt% of sulfur and 0.3 wt% of nitrogen. The sulfur content of vacuum residue increased to 2.6 wt% and the nitrogen content increased to 0.6 wt%. The distillation process was reported to concentrate the metallic constituents of crude oils in the residue (Speight, 1999). The atmospheric residue was reported to contain 25 ppm of Ni, 185 ppm of V, and 28 ppm of Fe. The vacuum residue was reported to contain 64 ppm of Ni, 450 ppm of V, and 48 ppm of Fe. The carbon residue of vacuum residue was reported to significantly increase from 9.3 to 21.6 wt%.

Other metals, not removed during the desalting step, will also concentrate in the residue. The literature reported on the use of low vacuum scanning electron microscopy (LV-SEM) and energy dispersive spectroscopy (EDS) to analyze the morphology and composition of vacuum residue obtained from Mexican crude oil. The vacuum residue contained irregular size and shape particles rich in S and Si. The particles were also found to contain Fe, K, Mg, and Cl (Bragado et al., 2001).

2.1.2 CRUDE OIL TYPES

The majority of crude oils contain a certain proportion of hydrocarbon molecules which may precipitate as a waxy solid phase. Some crude oils contain as much as 60 wt% of wax. The seepage of crude oil led to evaporation of lower boiling point fractions leaving a dark colored waxy residue known as "ozokerite" or "earth wax." The term mineral wax applies to wax mined from the earth such as ozokerite and also its refined form, ceresine. Ozokerite wax is a valuable form of paraffin wax and, in its natural dark form, is known to be used in shoe polishes (Tooley, 1971). Natural deposits of paraffin wax are found around some oil fields in Europe, Asia, and Texas. Mineral wax occurs naturally as a yellow to brown deposit composed mostly of paraffins. Their melting points can vary in a range of 60°C–95°C (Speight, 1999).

In addition to having different physical properties and quantities of hydrocarbons within a given boiling range, crude oils have different heteroatom, wax, and asphaltene contents. Asphaltenes are dark solids with no definite melting point and their melting point can vary in a range of 115°C–330°C. Asphaltite occurs naturally as brown to black solids containing a high content of n-pentane insolubles (Speight, 1999). Regional variations in the paraffin, sulfur, and resin contents of crude oils, based on statistical analysis, were reported (Polishchuk and Yashchenko, 2001). The crude oil classification, based on the composition of 250°C–300°C fraction, is shown in Table 2.8.

Crude oils contain different hydrocarbons, heteroatom components, resins, contents of wax, and asphaltenes. Wax precipitated from crude oils consists of n-paraffins, low branching density iso-paraffins, and naphthenic molecules having long straight side chains. Crude oils and different petroleum fractions contain wax type molecules reducing their flowability or pumpability. The cloud point of

TABLE 2.8
Crude Oil Classification According to Chemical Composition

Crude Oil Classification	250°C–300°C Paraffins, %	250°C–300°C Naphthenes, %	250°C–300°C Aromatics, %	250°C–300°C Wax, %	250°C–300°C Asphalt, %
Paraffinic	46–61	22–32	12–25	1.5–10	0–6
Paraffinic–naphthenic	42–45	38–39	16–20	1–6	0–6
Naphthenic	15–26	61–76	8–13	Trace	0–6
Paraffinic–naphthenic–aromatic	27–35	36–47	26–33	0.05–1	0–10
Aromatic	0–8	57–78	20–25	0–0.5	0–20

Source: From Speight, J.G., *The Chemistry and Technology of Petroleum*, 3rd ed., Marcel Dekker, Inc., New York, 1999. Reproduced by permission of Routledge/Taylor & Francis Group, LLC.

TABLE 2.9

Composition, Cloud Point, and Pour Point of Paraffinic Gas Oil

Paraffinic Gas Oil	Test Method	Properties
API gravity	D 287-92	30
Kin. viscosity at 4°C, cSt	IP 71/80	3.8
Molecular weight	D 2503-92	241
n-Paraffins, wt%	GLC	18.4
Iso-paraffins, wt%	GLC	1.01
Sulfur, wt%	IP 266/87	0.82
Cloud point, °C	IP 219/82	19
Pour point, °C	IP 15/67 (86)	9
Flash point, °C	IP 34/85 (87)	120

Source: From Khidr, T.T. and Mohamed, M.S., *Petroleum Science and Technology*, 19, 547, 2001. Reproduced by permission of Taylor & Francis Group, LLC, http://www.taylorandfrancis.com.

petroleum products is used to measure the temperature when the wax crystals first appear upon cooling, usually visible as a milky cloud. With a decrease in the temperature, below the cloud point, the clusters of wax crystals will grow until no flow is observed which is reported as the pour point. Using the urea adduction method, the paraffinic content of gas oil, produced from paraffinic crude oil, was separated and analyzed. The paraffinic gas oil was reported to contain 18.4 wt% of n-paraffins and only 1.01 wt% of iso-paraffins (Khidr and Mohamed, 2001). The composition, cloud point, and a pour point of gas oil, produced from paraffinic crude oil, are shown in Table 2.9.

The crude oils require desalting to minimize corrosion and fouling caused by salt deposition and acids formed by decomposition of the chloride salts (Gary and Handwerk, 2001). Many petroleum fractions contain wax and are also corrosive when in contact with ferrous pipeline metals during their transport and storage. The use of additives which prevent wax deposition and corrosion is required. The paraffinic gas oil was reported to have a high cloud point of 19°C and a high pour point of 9°C. A high cloud point of paraffinic gas oil would be related to crystallization of n-paraffinic hydrocarbons. Many different pour point depressants were developed to interfere with the crystallization of n-paraffinic hydrocarbons. Different chemistry pour point depressants, such as polymethacrylates, polyacrylates, ditetraparaffins phenol phthalates, highly branched poly-alfa-olefins, alkylesters of unsaturated carboxylic acids can be used. The literature reported on the performance of alkenylsuccinic anhydrides and their derivatives, used as corrosion inhibitors and flow improvers in paraffinic gas oil (Khidr and Mohamed, 2001). The effect of different alkenyl bis-succinimides and their derivatives on the cloud point and the pour point of paraffinic gas oil is shown in Table 2.10.

TABLE 2.10
Performance of Corrosion Inhibitors as Flow Improvers
in Paraffinic Gas Oil

Paraffinic Gas Oil	Cloud Point, °C	Pour Point, °C
Neat gas oil	19	9
Tetradecene Bis-Succinimides		
250 ppm	19	6
500 ppm	19	6
1000 ppm	18	3
2000 ppm	18	0
3000 ppm	18	−3
Hexadecene Bis-Succinimides		
250 ppm	19	6
500 ppm	19	3
1000 ppm	18	0
2000 ppm	17	−3
3000 ppm	17	−6
58% Tetradecene/40% Hexadecene Bis-Succinimides		
250 ppm	19	9
500 ppm	19	6
1000 ppm	19	3
2000 ppm	19	3
3000 ppm	19	0
40% Tetradecene/58% Hexadecene Bis-Succinimides		
250 ppm	19	9
500 ppm	19	9
1000 ppm	19	6
2000 ppm	19	3
3000 ppm	19	3

Source: From Khidr, T.T. and Mohamed, M.S., *Petroleum Science and Technology*, 19, 547, 2001. Reproduced by permission of Taylor & Francis Group, LLC, http://www.taylorandfrancis.com.

The typical pour point depressants were reported usually effective in depressing the pour point by 11°C–17°C and, in some cases, the pour point depression by 28°C was observed (Pirro and Wessol, 2001). At the treat rate of 3000 ppm, the most effective hexadecene bis-succinimides depressed the pour point of paraffinic gas oil, from 9°C to −6°C, by 15°C. While different alkenyl bis-succinimides and their mixtures were found effective in depressing the pour point, no significant effect in depressing the cloud point was observed. After only 1 week of storage time, an increase in the depressed pour point of paraffinic gas oil, containing a less effective additive, was observed. The wax content of petroleum oils and the chemistry of pour point depressants can affect their efficiency to depress the pour point and their storage stability. While paraffinic crude oils contain a higher content

TABLE 2.11

Effect of Crude Oils on Naphthenic Acid Content of Petroleum Fractions

Crude Oils	Crude Oil Type	Fraction	Naphthenic Acids, %
Pennsylvania	Paraffinic	Kerosene	0.006
		Gas oil	0.01
California	Naphthenic	Kerosene	0.06
		Gas oil	0.36
Texas Heavy	Naphthenic	Kerosene	0.08
		Gas oil	0.35

Source: From Speight, J.G., *The Chemistry and Technology of Petroleum*, 2nd ed., Marcel Dekker, Inc., New York, 1991. Reproduced by permission of Routledge/Taylor & Francis Group, LLC.

of *n*-paraffinic hydrocarbons, precipitating as wax, naphthenic type crude oils contain a higher content of naphthenic acids. The effect of different crude oils on the naphthenic acid content of different petroleum fractions is shown in Table 2.11.

The literature reported that while the distribution of naphthenic acids throughout the crude oil varies, higher boiling gas oil fractions were reported to contain a higher content of naphthenic acid (Speight, 1991). The use of paraffinic crude oil, from Pennsylvania, was reported to increase the naphthenic acid content of kerosene fraction from 0.006 wt% to 0.01 wt% in the gas oil fraction. The use of naphthenic crude oil, from California, was reported to increase the naphthenic acid content of the kerosene fraction to 0.06 and to 0.36 wt% in the gas oil fraction. The use of heavy naphthenic crude oil, from Texas, was reported to further increase the naphthenic acid content of the kerosene fraction to 0.08 and to 0.35 wt% in the gas oil fraction. The hydrocarbon composition of the VGO was also reported to be affected by the crude oil type. The effect of paraffinic and aromatic base crude oils on the composition of VGO, having the same boiling range, is shown in Table 2.12.

TABLE 2.12

Effect of Paraffinic and Aromatic Base Crude Oils on the Composition of VGO

Properties	Paraffinic Base VGO	Aromatic Base VGO
Boiling range, °C	370–550	370–550
Total saturates, wt%	76.9	46.5
n-Paraffins, wt%	53.4	29.5
Aromatics, wt%	23.1	53.5

Source: From Srivastava, S.P. et al., *Petroleum Science and Technology*, 20, 269, 2002a. Reproduced by permission of Taylor & Francis Group, LLC, http://www.taylorandfrancis.com.

TABLE 2.13

Typical Characteristics and Use of Different Type Crude Oils

Crude Oil Type	Typical Content	Typical Use
Paraffin base	Wax and no asphalt	Paraffinic base stocks
		Wax products
Naphthene base	No wax and no asphalt	Naphthenic base stocks
Mixed base	Wax and asphalt	Low yield base stocks
Asphalt base	High S and N	Mineral base stocks
	Asphaltic residue	Asphalt

Source: From Sequeira, A. Jr., *Lubricant Base Oil and Wax Processing*, Marcel
          Dekker, Inc., New York, 1994. Reproduced by permission of
          Routledge/Taylor & Francis Group, LLC.

The urea adduction method was reported effective in separating the *n*-paraffins from iso-paraffins and cyclo-paraffins present in different VGOs (Srivastava et al., 2002a). The VGO, from paraffinic base crude oil, was reported to have a high saturate content of 76.9 wt%, including 53.4 wt% of *n*-paraffins. The VGO, from aromatic base crude oil, was reported to have a significantly lower saturate content of 46.5 wt%, including a significantly lower *n*-paraffinic content of 29.5 wt%. The use of aromatic base crude also increased the aromatic content of VGO from 23.1 to 53.5 wt%. While a high saturate content of VGO is more desirable, a high *n*-paraffinic content will lead to high pour points. The literature reported that the crystallization of *n*-paraffins and agglomeration of high viscosity molecules, at low temperatures, led to the formation of gel. The DSC technique was used to study the wax precipitation from VGO (Srivastava et al., 2002b) and from middle distillates (Srivastava et al., 2002c). The use of different type crude oils affects the properties and composition of different petroleum products. The typical characteristics and the use of different type crude oils are shown in Table 2.13.

Crude oils are usually classified as paraffin base, naphthene base, mixed base, and asphalt base and only selected crude oils are suitable to produce lube oil base stocks (Sequeira, 1994). Paraffin base crude oils contain wax and little or no asphalt and are suitable to produce solvent refined lube oil base stocks, used in various lubricating oils and wax products. Naphthene base crude oils, which contain little or no wax and contain no asphalt, are suitable to produce naphthenic lube oil base stocks. Naphthene base crude oils, containing little or no wax and asphalt, are suitable for specialty oil manufacture. Mixed base crude oils contain wax and asphalt and are used to produce low yield base stocks. Asphalt base crude oils contain high S and N contents and asphaltic residue. The asphalt type crude oils must be thoroughly refined to remove aromatics and sulfur compounds before a kerosene fraction can be obtained (Speight, 1999). Asphaltic type crude oils are used to produce base stocks and asphalts. Asphaltenes and resins, consisting of aromatic molecules classified as asphaltics, are usually found in very heavy VGO and in residue (Sequeira, 1994). If crude oils are not suitable to produce lube oil base stocks, they are used to produce asphalts.

2.2 LUBE OIL BASE STOCKS

2.2.1 VISCOSITY INDEX

The composition of crude oils affects the composition of VGO and the properties of lube oil base stocks. Viscosity index (VI) describes the relationship of viscosity to temperature and high VI base stocks tend to display less change in viscosity with temperature. The literature reported that VI of hydrocarbons can decrease from as high as 175 for normal paraffins to 155 for iso-paraffins, 70–142 for naphthenes, and a low VI of 50 for aromatics (Sequeira, 1994). Wax precipitated from crude oils was reported to consist of n-paraffins, iso-paraffins, and naphthenic molecules. Iso-paraffins were shown to have a low branching density and naphthenic molecules were shown to consist of a 1-ring naphthene hydrocarbon having a long straight side chain (Pedersen et al., 1989). The pour point of iso-paraffins, having a high VI of 125, was reported to vary depending on the structure of the isomers (Mang, 2001). Aromatics present in heavier fractions of petroleum, such as lubricating oils, usually contain naphthene rings and they also contain sulfur and nitrogen containing molecules (Speight, 1999). Some polyaromatic molecules were reported to have a negative VI of −60 and high pour points (Mang, 2001). The preferred molecules for the lube oil manufacture are isoparaffins having a high VI and low pour point. VI and pour points of some hydrocarbon molecules are shown in Table 2.14.

The lube oil base stocks, from Pennsylvanian crude oils, predominantly paraffinic, were found to show a small reduction in viscosity with an increase in the temperature. The lube oil base stock, from a Pennsylvanian crude oil, was assigned the VI of 100 (Phillips, 1999). The VI of lube oil distillates vary depending on the composition of crude oils. The lube oil base stocks, from Gulf crude oils, predominantly naphthenic, were found to show a significant reduction

TABLE 2.14
Viscosity Index and Pour Points of Some Hydrocarbons

Hydrocarbons	VI	Pour Point, °C
n-Paraffins	175	—
Iso-paraffins (C_{26})	125	−40 to 20
1-Ring naphthenes with alkyl side chains	130	−10
Polynaphthenes	20	20
Polyaromatics	−60	50

Source: From Mang, T., *Lubricants and Lubrications*, Mang, T. and Dresel, W., Eds., Wiley-VCH, Weinheim, 2001.

TABLE 2.15
VI and Composition of Distillates from Different Crude Oils

Lube Oil Distillates	Viscosity at 40°C, cSt	VI	Pour Point, °C	Aromatics, wt%	Sulfur, wt%
Paraffinic	103	86	24	34	1.1
Naphthenic	108	17	−37	34	0.1
Tia Juana	23	10	−48	21	1.6
Forcados	18	42	18	28	0.3
Arabian	14	70	19	18.5	2.6
Forties	16	92	25	20	0.3

Source: From Sequeira, A. Jr., *Lubricant Base Oil and Wax Processing*, Marcel Dekker, Inc., New York, 1994; Prince, R.J., *Chemistry and Technology of Lubricants*, 2nd ed., Mortier, R.M. and Orszulik, S.T., Eds., Blackie Academic & Professional, London, 1997.

in viscosity with an increase in the temperature. The lube oil base stock, from a Gulf crude oil, was assigned the VI of 0 (Phillips, 1999). The literature reported on properties and composition of lube oil distillates produced from different crude oils. Paraffinic distillate was reported to have a VI of 86, a pour point of 24°C, an aromatic content of 34 wt%, an S content of 1.1 wt%, and a flash point of 193°C. The same viscosity naphthenic distillate was reported to have a low VI of 17, a low pour point of −37°C, the same aromatic content of 34 wt%, a lower S content of 0.1 wt%, and a lower flash point of 179°C (Sequeira, 1994). The literature reported on VI, pour point, aromatic content, and S content of distillates from Venezuela, Nigeria, Middle East, and North Sea crude oils which are used to produce lube oils (Prince, 1997). The VI and composition of distillates produced from different crude oils are shown in Table 2.15.

Both distillates from Tia Juana and Forcados crude oils were reported to have a relatively low VI, in a range of 10–42, similar to distillates from naphthenic type crude oils. However, a significant difference in pour points, ranging from −48°C to 18°C, is reported indicating that only Tia Juana crude oil is more typical of naphthenic distillate having a low VI and a low pour point. Both distillates from Arabian and Forties crude oils have a relatively high VI, in the range of 70–92, similar to distillates from paraffinic type crude oils. Both distillates have a high pour point, in the range of 19°C–25°C, indicating that they are rich in alkanes and are examples of paraffinic crude oils. The literature reported on the characterization of distillates, from West Siberia crude oil, for production of high viscosity base stocks. Distillates were adsorbed on silica gel and different hydrocarbons were selectively removed using different solvents. Isooctane was used to dissolve and desorb paraffins and naphthenes. Toluene was used to dissolve and desorb aromatics. The resin fractions of about 3 wt% were removed by dissolving in acetone (Starkova et al., 2001). The yields, VI, and solid points of different hydrocarbons, separated from medium viscosity distillate produced from West Siberian crude oil, are shown in Table 2.16.

TABLE 2.16
VI and Solid Points of Hydrocarbons from Medium Viscosity Distillate

West Siberia Crude Oil	Yield, wt%	VI	Solid Point, °C
Medium viscosity distillate	100	85	14
Methano-naphthenes	47.1	114	30
Aromatics light	31.7	85	23
Aromatics middle	4.6	72	18
Aromatics heavy	13.6	41	16

Source: Modified from Starkova, N.N. et al., *Chemistry and Technology of Fuels and Oils*, 37, 191, 2001.

Medium viscosity distillate, having a viscosity of 21.72 mm^2/s at 50°C, was reported to have a VI of 85, a solid point of 14°C, and contain 49.9 wt% of aromatics (Starkova et al., 2001). Medium viscosity distillate was reported to contain 47% of saturate hydrocarbons, methano-naphthenes having a high VI of 114, and a high solid point of 30°C. The VI and solid points of aromatics were reported to vary. Medium viscosity lube oil distillate was found to contain 32% of light aromatics, having a lower VI of 85 and a lower solid point of 23°C, and 14% of heavy aromatics, having a low VI of 41 and a low solid point of 16°C. After the hydrocarbon separation, heavy aromatics were found to have a lower VI and a decrease in their solid point was reported (Starkova et al., 2001). The literature also reported on the VI and solid points of different hydrocarbons separated from heavy viscosity distillate obtained from West Siberian crude oil (Starkova et al., 2001). The yields, VI, and solid points of different hydrocarbons, separated from high viscosity distillate produced from West Siberian crude oil, are shown in Table 2.17.

High viscosity distillate, having a viscosity of 53.59 mm^2/s at 50°C, was reported to have a lower VI of 64, a solid point of 12°C, and a higher aromatic content of 59.4 wt% (Starkova et al., 2001). High viscosity distillate was reported to contain 38 wt% of saturate hydrocarbons, methano-naphthenes also having a high VI of 114 but a higher solid point of 36°C. The VI and solid points of aromatics were reported to vary. High viscosity lube oil distillate was found to contain 27 wt% of light

TABLE 2.17
VI and Solid Points of Hydrocarbons from High Viscosity Distillate

West Siberia Crude Oil	Yield, wt%	VI	Solid Point, °C
High viscosity distillate	100	64	12
Methano-naphthenes	38	114	36
Aromatics light	27	80	24
Aromatics middle	13	67	17
Aromatics heavy	19	36	14

Source: Modified from Starkova, N.N. et al., *Chemistry and Technology of Fuels and Oils*, 37, 191, 2001.

aromatics, having a lower VI of 80 and a lower solid point of 23°C, and 19 wt% of heavy aromatics, having a low VI of 36 and a low solid point of 14°C (Starkova et al., 2001). Heavy viscosity lube distillate was reported to have a lower VI and a higher aromatic content. After the hydrocarbon separation, heavy aromatics were found to have a lower VI and a decrease in their solid point was reported. To remove low VI components of crude oils, conventional refining uses solvent extraction.

2.2.2 SOLVENT REFINING OF MINERAL BASE STOCKS

The use of different crude oils affects the quality of VGO, used as feed to lube oil production, and will affect the VI of base stocks. The literature reported on VGO, solvent dewaxed to a pour point of −12°C, produced from different crude oils and having a different VI (Henderson, 2006). Different solvent based processes have been developed for the dewaxing of lubricating oils but they are all based on three basic steps. The first step is the dilution and chilling of the feedstock with solvent (crystallization); the second step is the filtration of the wax; and the last step is the solvent recovery from the wax cake and the filtrate. The cold settling–pressure filtration processes and centrifuge dewaxing processes have been replaced by solvent dewaxing. These older processes might be used but to a limited degree (Sequeira, 1994). After the solvent dewaxing, Arab light was reported to have a VI of 60 while Arab heavy was reported to have a lower VI of 55. Iranian light was reported to have a VI of 55 and, similarly, Iranian heavy was reported to have a lower VI of 40. Alaska North Slope crude oil was reported to have a very low VI of 15 and the dewaxed VI of VGO, from California light crude oil, was reported to be only 8 (Henderson, 2006). The dewaxed VI and wax content of VGO, produced from different crude oils, are shown in Table 2.18.

TABLE 2.18
Dewaxed VI of VGO Produced from Different Crude Oils

Crude Oils	VGO Solvent Dewaxed VI	VGO Wax, LV%
Pennsylvania	100	0
Ordovician	85	13
Brent	65	19
Arab Light	60	9
Arab Heavy	55	9
Iranian Light	55	16
Lagomedio	45	10
Urals	40	11
Iranian Heavy	40	16
Alaska North Slope	15	8

Source: From Mang, T., *Lubricants and Lubrications*, Mang, T. and Dresel, W., Eds., Wiley-VCH, Weinheim, 2001. Reproduced by permission of Wiley-VCH Verlag GmbH & Co KG.

The lube oil base stocks need to have a high VI and good low temperature fluidity. The majority of lubricating oils use mineral base stocks produced from selected crude oils using conventional refining based on solvent extraction and solvent dewaxing. The literature reported on processing conditions used during the solvent extraction and the use of different solvents (Sequeira, 1994). Extraction solvents in commercial use include sulfur dioxide, phenol, furfural, and N-methylpyrrolidone (NMP). The use of sulfur dioxide is rare and the use of phenol solvent is declining. Furfural is the most widely used solvent; however, the use of NMP is increasing (Prince, 1997). More recent literature reported on the modification to the furfural extraction unit to improve the quality and yield of the refined oils (Yan et al., 2002). Acids present in lube oil distillates cause severe corrosion of pipelines and equipment used to carry out the liquid extraction process.

Dewaxing solvents in commercial use include propane, ketones, and mixed solvents. When using propane as a dewaxing solvent, the feedstock is diluted with propane and heated to a temperature where all the wax is dissolved. The mixture of propane and feedstock is then cooled and additional cold propane is added to replace propane vaporized during the chilling. The wax slurry is filtered under pressure and cold propane is used to remove oil from the filter cake (Sequeira, 1994). When using propane as a dewaxing solvent, dewaxing aids are required to obtain good filtration rates. The chemistry of solvent dewaxing aids can vary from N-alkylated naphthalene polymers, N-alkyl polymethacrylates to N-alkyl polyaromatics. Asphaltenes and microcrystalline waxes were also reported to be used as dewaxing aids (Sequeira, 1994). The literature reported that while solvent dewaxing is used to remove wax and lower the pour point of base stocks, a small decrease in VI is also observed (Gary and Handwerk, 2001). The mineral base stocks are usually solvent dewaxed to meet the selected pour point and the effect of different solvent dewaxing conditions on VI is shown in Table 2.19.

The literature reported on the effect of solvent dewaxing on VI of oil, obtained from a crude oil blend containing 78% Arabian Berri and 22% of Arabian Medium. The refined waxy distillate was reported to have a VI of 110 and a

TABLE 2.19
Effect of Different Solvent Dewaxing Conditions on VI

Crude Oil Blend	Pour Point, °C	Viscosity at 38°C, cSt	VI
Waxy raffinate feed	27	93	110
Solvent dewaxed	16	106	106
Solvent dewaxed	4	105	105
Solvent dewaxed	−7	104	102
Solvent dewaxed	−18	106	100
Solvent dewaxed	−21	107	101

Source: Reprinted with permission from Taylor, R.J. and McCormack, A.J., *Industrial and Engineering Chemistry Research*, 31, 1731, 1992. Copyright American Chemical Society.

TABLE 2.20

Effect of Crude Oil Type on VI and Pour Point of Base Stocks

Base Stock	Paraffinic 100N	Naphthenic 100N
API gravity	32.7	24.4
Viscosity, SUS at 100°F	100	100
Viscosity, cSt at 38°C	20.53	20.53
VI	100	15
Pour point, °F (°C)	0 (−18)	−50 (−46)
Color (ASTM)	0.5	1.5
Flash point, °F (°C)	390 (199)	340 (171)

Source: From Pirro, D.M. and Wessol, A.A., *Lubrication Fundamentals*, 2nd ed., Marcel Dekker, Inc., New York, 2001. Reproduced by permission of Routledge/Taylor & Francis Group, LLC.

pour point of 27°C and it contained 21 wt% of aromatics, less than 0.5 wt% of S, and 50 ppm of N (Taylor and McCormack, 1992). With a decrease in wax content of waxy raffinate, a decrease in dewaxed pour point leads to an increase in viscosity at 40°C and a decrease in VI is observed. The selection of crude oil type can have a significant effect on the VI and the pour point of base stocks. The use of paraffinic base crude oils requires dewaxing. The wax content of refined oils removed during the dewaxing step is called slack wax. The base stocks produced from naphthenic crude oils have low pour points, excellent solvency properties, but a low VI which limits their application. Mineral base stocks are traditionally classified according to Saybolt Universal Seconds (SUS) and a 100N (neutral) base stock has a viscosity of 100 SUS at 100°F. At the present time, the viscosity is usually reported in cSt measured at 38°C or 40°C. The effect of crude oil type on VI and properties of base stocks, having a similar viscosity, is shown in Table 2.20.

The typical conventional refining of mineral base stocks involves vacuum distillation, solvent extraction, solvent dewaxing, and finishing (Pirro and Wessol, 2001). The VGO are derived directly from crude oils and are the feeds for lube oil refining. A low VI of VGO leads to high extraction losses in solvent extraction. A high wax content of VGO increases the cost of the dewaxing. The wax content of paraffinic base stocks is related to the crude oil source and affects the dewaxed lube oil yield. A high sulfur content of VGO requires hydrotreating of finished base stocks to eliminate sulfur (Mang, 2001). A significant difference in the sulfur content of VGO will affect the severity of processing and the composition of the base stocks. The paraffinic base stocks which require the removal of aromatics, usually by solvent extraction, and dewaxing are referred to as solvent neutrals (SN), where solvent means that the base stock is solvent refined and neutral means that the oil is of neutral pH (Prince, 1997). The VI, properties, and composition of different viscosity mineral base stocks are shown in Table 2.21.

The refined paraffinic oils contain waxes which crystallize out at low temperatures and require dewaxing to lower their pour points. The majority of

TABLE 2.21

VI and Properties of Different Viscosity Paraffinic Base Stocks

Base Stocks	SNO-100	SNO-150	SNO-320	SNO-850
API gravity	32.4	30.9	29.3	26.8
Viscosity, cSt at 38°C	107	155	332	844
VI	95	96	97	89
Dewaxed pour point, °C	−12	−15	−12	−9
Aromatics, wt%	16.14	23.39	25.44	27.60
S, wt%	0.14	0.27	0.31	0.38
ASTM color	0.5	0.5	L1.5	2.5
COC flash point, °F	380	404	440	505
Carbon residue, wt%	0.02	0.02	0.03	0.38

Source: From Sequeira, A., *Pre-Prints Division of Petroleum Chemistry*, ACS, 37, 1286, 1992.
Reproduced by permission of Chevron Corporation.

refiners use ketone solvents or their mixtures. The ketone dewaxing processes use a mixture of methyl ethyl ketone (MEK) with methyl isobutyl ketone (MIBK) or MEK with toluene. The use of mixed solvents helps to control the oil solubility and wax crystallization properties better than the use of single solvents (Prince, 1997). The literature reported on a membrane process, called Max–Dewax, which leads to an increase in lube and wax production (Bhore et al., 1999). Naphthenic base stocks are made in small quantities from selected crude oils, having low or no wax content, and might not require dewaxing. The slack waxes, from the solvent dewaxing, are fed to other refinery units or deoiled to produce wax. Clay treatment or hydrofinishing is used to remove the molecules that affect the color and the stability of lube oil base stocks. During processing, color is a useful guide indicating whether processes are operating properly (Pirro and Wessol, 2001). Spent clay disposal problems and other operating restrictions have caused the clay treating processes to be replaced by hydrofinishing.

2.2.3 RESIDUAL OILS

The vacuum residue contains a heavy residual oil which is mixed with asphalt and resins. To recover the heavy oils, most refiners use a propane de-asphalting process. The precipitation of asphaltenes with *n*-pentane reduces the vanadium content of the oil by up to 95% with a significant reduction in the amounts of iron and nickel. The literature reported on the propane de-asphalting and the majority of the vanadium, nickel, iron, and copper in residual stocks may be precipitated along with asphaltenes by the hydrocarbon solvents (Speight, 1999). It was also reported that the asphaltene fraction, precipitated from crude oils by propane, might contain acid material known as asphaltic or asphaltogenic acid. These acids appear in the resin fraction and are considered to be cyclic or noncyclic organic acids of high molecular weight (Speight, 1999). Propane is a gaseous paraffinic

TABLE 2.22

Composition of VGO and Residual Oil from Daqing Crude Oil

Composition, wt%	VGO # 1	VGO # 2	Residual Oil
Total saturates	69.8	54.1	28
Naphthenics	41.4	35.8	18.2
1-Ring aromatics	8.6	14	19.2
2-Ring aromatics	3.9	7.5	10.9
3-Ring aromatics	2.3	4.2	4.9
4-Ring aromatics	1.2	1.7	2.6
5-Ring aromatics	0.8	1.5	1.1
Unidentified	3	5.4	1
Gum	9.3	9.9	27.9
Asphaltenes	0.1	0.3	0.5

Source: From Kai-Fu, H. et al., *Petroleum Science and Technology*, 18, 815, 2000. Reproduced by permission of Taylor & Francis Group, LLC, http://www.taylorandfrancis.com.

hydrocarbon present in natural gas and crude oil, also termed, along with butane, as liquefied petroleum gas (LPG). Asphaltenes are insoluble in liquefied petroleum gas and propane is used commercially in the processing of residues. Liquid propane is used for its high selectivity and solubility in oil compared to the heavier material known as asphalt. Typically, the amount of propane required is 2–4 times of the residue. If not enough is used, the extraction will result in low yield and low viscosity oil. If too much is used, the extraction leads to high yields of heavy and poor quality oil (Shell, 2001). De-asphalting reduces the aromatic, acidic, and basic contents of de-asphalted oils (DAOs). The composition of VGO and residual oil, produced from Daqing crude oil, is shown in Table 2.22.

The use of the same crude oil leads to VGO and residual oils having different viscosity and composition. The use of the same crude oil decreases the saturate and naphthenic content of residual oil while increasing its aromatic content, gum forming content, and asphaltene content. Processing of paraffinic and naphthenic crude oils leads to vacuum residue, containing high viscosity lube oil molecules called residual oils, which can be used to produce heavy base stocks. The early literature reported on a high VI of 95 of lube oil distillate and de-asphalted oil, from Ordovician crude oil, suitable to produce lubricating oils (Beuther et al., 1964). The properties and VI of lube oil distillate and de-asphalted oil from Ordovician crude oil are shown in Table 2.23.

The early literature reported that de-asphalted oils, produced from Ordovician crude oil, were found to have an S content of 0.2 wt% and an ASTM color of 8+. The de-asphalted oil, from West Texas crude oil, was reported to have a VI of 84, a pour point of 38°C, and a higher S content of 1.4 wt% (Beuther et al., 1964). The recent literature reported on VGO, solvent dewaxed to a pour point of −12°C, produced from West Texas Intermediate crude oil having a VI of only 55

TABLE 2.23

VI of Lube Oil Distillate and De-Asphalted Oils from Ordovician Crude Oil

Ordovician Crude Oil	Lube Oil Distillate	De-Asphalted Oil #1	De-Asphalted Oil #2
Yield, % of crude oil	9.8	4.9	3.6
API gravity	25.2	23.2	24
Viscosity, cSt at 100°C	55	149	161
VI	95	95	96
Pour point, °C	38	32	27
Carbon residue, wt%	0.2	1.6	1.9

Source: Reprinted with permission from Beuther, H., Donaldson, R.E., and Henke, A.M., *I&EC Product Research and Development*, 3, 174, 1964. Copyright American Chemical Society.

(Henderson, 2006). The early literature reported on de-asphalted oil, from Kuwaiti crude oil, having a VI of 81, pour point of 32°C, and containing 2.94 wt% of S (Beuther et al., 1964). The recent literature reported on VGO, solvent dewaxed to a pour point of -12°C, produced from Kuwaiti crude oil, having a VI of only 50 (Henderson, 2006). The lube oil distillates are solvent extracted to remove undesirable aromatics and to increase their VI. The literature reported on the properties of de-asphalted oil, from Middle East crude oil, having an API gravity of 19.8, a viscosity of 231 cSt at 100°C, a VI of 74, and a high pour point of over 54°C (Sequeira, 1994). In some cases, de-asphalted oils are also solvent extracted to further increase their VI. Bright stocks (BS) and cylinder oils are residual oils produced from paraffinic and naphthenic vacuum residue (Sequeira, 1994). The properties and composition of high viscosity bright stock and cylinder oil are shown in Table 2.24.

TABLE 2.24

Properties and Composition of Bright Stock and Cylinder Oil

Properties	Bright Stock BS-150	Cylinder Oil
API gravity	26.5	20.4
Viscosity, cSt at 38°C	2586	9440
VI	95	70
Dewaxed pour point, °C	−12	−7
Aromatics, wt%	32.5	36.61
S, wt%	0.52	0.7
ASTM color	L4.5	8+
COC flash point, °F	545	585
Carbon residue, wt%	0.65	2.9

Source: From Sequeira, A., *Pre-Prints Division of Petroleum Chemistry*, ACS, 37, 1286, 1992. Reproduced by permission of Chevron Corporation.

Bright stocks are usually manufactured using propane de-asphalting, solvent extraction, and solvent dewaxing (Sequeira, 1994). After solvent extraction and dewaxing, bright stock BS-150 was reported to have a high VI of 95 and a dewaxed pour point of $-12°C$. It was reported to contain an aromatic content of 32.5 wt%, an S content of 0.52 wt%, and a carbon residue of 0.65 wt%. The slack wax from dewaxing of the bright stock was reported to contain naphthenic and aromatic molecules having alkyl side chains which are long enough to give these molecules a high VI and pour point characteristics of n-paraffins (Sequeira, 1994). High viscosity bright stocks are blended with lower viscosity base stocks to meet the viscosity requirements of many different lubricant products. Cylinder oils are produced using propane de-asphalting with solvent dewaxing sometimes used to reduce their pour point (Sequeira, 1994). After solvent dewaxing, the cylinder oil was reported to have a low VI of 70 and a dewaxed pour point of $-7°C$. Higher viscosity cylinder oil was also reported to have a higher aromatic content of 36.6 wt%, a higher S content of 0.7 wt%, and a higher carbon residue of 2.9 wt%. Cylinder oils are used to blend some high viscosity lubricant products.

2.3   WAX PRODUCTS

2.3.1   COMPOSITION OF WAX

There are many different compounds known as wax which have different composition. The literature reported on petroleum and natural waxes, such as insect, animal, and vegetable wax (Sequeira, 1994). The beeswax and carnauba wax (palm tree) were reported to be composed of high molecular weight esters with some acids, alcohols, and hydrocarbons (Exxon, 2002). Synthetic waxes are manufactured by chemical synthesis from a variety of chemical compounds, such as alcohols, polyethylene glycols, esters, hydrocarbons, and chlorinated hydrocarbons. The carbowaxes, based on polyethylene glycol with a molecular weight from 200 to 700, are liquids used as waxes and can be esterified to form wax emulsifiers. The "pseudo-ester" wax is produced from fatty acids and amines instead of fatty acids and alcohols (Tooley, 1971). Synthetic hydrocarbon waxes may be produced by the Fischer–Tropsch process and by the polyethylene manufacturing process. Low molecular weight grades of polyethylene are used as wax modifiers. The term mineral wax may be applied to any wax of mineral origin, such as montan wax, lignite wax, and peat wax. The term mineral wax also applies to petroleum wax derived from crude oils. Types of different waxes are shown in Table 2.25.

The crude oil from the well may be at an elevated temperature and on cooling the higher melting waxes tend to solidify. Tank bottom waxes, separating and accumulating on the bottoms of tanks used for storing waxy crude oils, are high melting and microcrystalline in type. The typical lube oil distillates contain 15–20 wt% of wax and the wax type molecules, undesirable as lube oil molecules due to their high pour points, are removed during the dewaxing step and used to produce the petroleum wax. The viscosity and wax content of the feed affect the solvent dilution ratio, refrigeration requirements, filtration rate, and the size of

TABLE 2.25
Types of Different Waxes

Wax Type	Example
Insect wax	Beeswax
Animal wax	Spermaceti
Vegetable wax	Carnauba (palm tree)
Synthetic wax	Carbowaxes
Mineral wax	Peat wax
Mineral wax	Petroleum wax

Source: From Sequeira, A. Jr., *Lubricant Base Oil and Wax Processing*, Marcel Dekker, Inc., New York, 1994.

the solvent recovery facility. The residence time at the required temperature is important in the case of heavy feedstocks (Sequeira, 1994). The feedstock entering the solvent dewaxing unit must be heated to ensure that the wax is in complete solution. The boiling range and type of the wax affect the filtration rate. Filtration rate is the lowest for the most viscous feedstock and also requires the highest solvent dilution (Sequeira, 1994). The solvent dewaxing processes are suitable for dewaxing the entire range of different viscosity oils and solvent dewaxing aids are used to improve the filtration rates. Lubricating oils are not the only materials produced from the high boiling fraction of crude oils and the composition of wax varies depending on the viscosity of waxy raffinate. The wax composition obtained from different viscosity waxy raffinates is shown in Table 2.26.

TABLE 2.26
Wax Composition from Different Viscosity Waxy Raffinates

Feedstock Wax Composition	SNO-150, wt%	SNO-320, wt%	SNO-850, wt%	Bright Stock, wt%
Saturate fraction				
n-Paraffins	59	31	20	0
Iso-paraffins	24	35	29	32
1-Ring naphthenes	10	15	17	16
2-Ring naphthenes	3	6	9	11
3-Ring naphthenes	0	5	7	10
Other	0	0	9	1
Aromatic fraction				
Alkylbenzenes	2	4	9	10
Other	2	4	0	20

Source: From Taylor, R.J., McCormack, A.J., and Nero, V.P., *Pre-Prints Division of Petroleum Chemistry*, ACS, 37, 1337, 1992. Reproduced by permission of Chevron Corporation.

TABLE 2.27

Hydrocarbon Content of Different Petroleum Wax Types

Feedstock	Hydrocarbons	Wax Type
Light distillate	Mostly n-paraffins	Paraffin
Medium distillate	Paraffins and mononaphthenes	Semimicrocrystalline
Heavy distillate	Paraffins, naphthenes, aromatics	Microcrystalline
Bright stock	Iso-paraffins, naphthenes, aromatics	Microcrystalline

With an increase in the viscosity of distillate, from 150 SNO to 850 SNO, the n-paraffinic content of slack wax was reported to significantly decrease from 59 to 20 wt% and essentially no n-paraffins were found in slack wax produced from dewaxing the bright stock. With an increase in viscosity of distillate, their iso-paraffinic content was reported to increase from 24 to 32 wt% and their total naphthenic content increased from 13 to 36 wt%. Their aromatic content was also reported to increase from 4 to 30 wt%. The bright stock slack wax was reported to contain no n-paraffins, 32 wt% of iso-paraffins, 37 wt% of naphthenics, and a high content of 30 wt% of aromatics. Slack waxes, from solvent dewaxing of different viscosity waxy raffinates, have different composition and there are two basic types of wax: paraffin and microcrystalline. The hydrocarbon content of different petroleum wax types is shown in Table 2.27.

The paraffin waxes, derived from light distillates, form large crystals. The wax obtained from light paraffin distillate has a distinctive crystalline structure and it is sold as pale raw wax (Speight, 1999). In higher boiling oils, waxy hydrocarbons occur as microcrystalline wax and they filter more slowly. Microcrystalline waxes differ from paraffin waxes in having poorly defined crystalline structure, darker color, higher viscosity, and higher melting points. The wax from intermediate paraffin distillate contains waxes which are a mixture of paraffin and microcrystalline wax and are known as semimicrocrystalline wax. According to the literature, semimicrocrystalline wax, also called "intermediate wax," has certain characteristics of both, paraffin and microcrystalline waxes (Shell, 2001). Microcrystalline waxes, also known as microwaxes, are produced from heavier lube distillates and residual oils. The microcrystalline wax obtained from heavy paraffin distillate has a tendency to form small crystals and it is sold as dark raw wax. Bright stock slack wax or bright stock crude wax, produced during the solvent dewaxing of bright stocks, is microcrystalline type. The waxes derived from residual oils, such as bright stocks, are known as petrolatum.

2.3.2 DEOILING OF WAX

The solvent dewaxing produces different composition slack waxes, also known as crude waxes and raw waxes, depending on whether light, intermediate, or heavy distillate is processed. Slack waxes from solvent dewaxing of waxy raffinates can

TABLE 2.28

Oil Content of Different Wax Products

Wax Processing	Wax Products	Oil Content, wt%
Solvent dewaxing	Slack wax	5–50
Wax sweating	Scale wax	1–5
Wax fractionation	Hard wax	<0.5
Wax fractionation	Soft wax	Wax/oil mixture

contain up to 50 wt% of oil and are deoiled to produce wax. The literature reported that the wax from solvent dewaxing of low viscosity 100N base stock contained 71 wt% n-paraffins, 17 wt% of iso-paraffins, 8 wt% of naphthenes, and 4 wt% of aromatics. It was also reported to contain 8 wt% of oil in the wax (Taylor et al., 1992). Light distillates have a high content of n-paraffins which crystallize easily in the form of large crystals. The waxes derived from solvent dewaxing of bright stock are called malcrystalline waxes because they filter poorly (Sequeira, 1994). The more recent literature reported on commercial slack wax which contains 79–89 wt% of paraffins, has a melting point in the range of 48°C–55°C, and has a flash point of 150°C. The slack wax was reported to contain 10–20 wt% of oil. Another slack wax was reported to have 89 wt% of paraffins, a melting point of 54°C–55°C, and a flash point of 150°C. It was reported to have a max 15 wt% of oil. The slack waxes were also reported to contain 0.5–0.7 wt% of sulfur and have a trace of water (Hinco Group, 2002). To produce wax products, the slack wax requires the use of a wax deoiling process to reduce the oil content below 1 wt%. The oil content of different wax products is shown in Table 2.28.

The commercial wax deoiling processes are based on the wax sweating, the spray deoiling, the recrystallization, and the warm-up deoiling process (Sequeira, 1994). The wax sweating process is an old process which can be used for the deoiling of the paraffin wax. In the wax sweating process, the molten wax is solidified by chilling and then slowly heated. The oil and lower melting waxes are separated (sweated) from higher melting wax. Wax sweating requires that the wax consists of large crystals that have spaces between them, through which the oil and lower melting waxes can percolate (Speight, 1999). Scale wax is a paraffin wax which has been partially deoiled in the initial stages of the sweating process. The term was originally used for waxes with oil contents in the range of 2–6 wt% (Speight, 1999).

The wax recrystallization process was developed to replace the wax sweating process and can be used to deoil all types of waxes. The wax recrystallization process separates wax into fractions by making use of the different solubility of the wax fractions in a solvent. Slack wax is heated, mixed with a solvent, such as ketone, and cooled. After a mixture of ketone and wax is heated, the wax usually dissolves completely. As it is cooled, wax crystallizes out leaving oil in solution.

If the solution is cooled slowly, the wax having the same melting point will crystallize. To obtain low oil content, two or three filtrations are required (Speight, 1999). Wax specification, such as oil content, melting point, and penetration, is controlled by the amount of solvent added, the rate of cooling, and the temperature of the crystallization process.

The wax recrystallization process is also called wax fractionation. The process flow for the wax recrystallization process can vary depending on the operational conditions and can have an effect on the type and filterability of the wax crystals formed (Sequeira, 1994). An overview of the petroleum wax manufacturing process based on recrystallization was published (IGI, 2002). The crystallized wax is filtered from the solvent in enclosed, inert gas blanketed, rotary drum filters. Two streams come from each drum filter, one containing the wax and some solvent and the other containing extracted oil and solvent. These streams go to a recovery plant where the solvent is removed by continuous distillation in steam heated exchangers and stripping towers. The recovered solvent is recycled to the crystal-lization process and to the drum filters as a wash. The solvent free wax and oil, called "foots oil," is used as feed to the catalytic cracker unit (IGI, 2002).

The warm-up deoiling process is more cost effective than the recrystallization process because energy requirements are lower. It is operated in conjunction with the solvent dewaxing unit. The slack wax from solvent dewaxing is diluted and mixed with warm solvent and filtered at the temperature which provides a hard wax of the desired melting point. The hard wax is not melted and recrystallized before filtration. The recrystallization or warm-up deoiling process is used to produce a hard wax and a soft wax known as foots oil (Sequeira, 1994).

The spray deoiling process can be used to deoil paraffin wax containing up to 15 wt% of oil. Molten slack wax is atomized under pressure into the top of a tower. The finely dispersed droplets of wax fall through a rising stream of air which is cooled below ambient conditions. The solidified wax settles to the bottom of the tower as a dry powder. Any oil which adheres to the wax is removed by the counter flow of $-4°C$ to $16°C$ solvent, dichloromethane, in mixer settlers and separated in a settling tank. The wax is centrifuged and washed with fresh solvent (Sequeira, 1994). The typical composition and melting point of paraffin and microcrystalline waxes are shown in Table 2.29.

The paraffin wax, containing 80–95 wt% of n-paraffins, 2–15 wt% of iso-paraffins, and 2–8 wt% of naphthenes, was reported to have the carbon number range of C_{18}–C_{36} and have an average molecular weight of 350–430. It was reported to have the melting point of 50°C–65°C (Pedersen et al., 1989). Paraffin waxes are soluble in most organic solvents except alcohols. With an increase in the viscosity of feedstock, increase in the polynaphthenes and aromatics leads to different chemistry wax type molecules affecting their physical properties. The microcrystalline wax, containing only 0–15 wt% of n-paraffins, 15–30 wt% of iso-paraffins, and 65–75 wt% of naphthenes, was reported to have a higher carbon number range of C_{30}–C_{60} and a higher average molecular weight of 500–800. It was reported to have a higher melting point in the range of 60°C–90°C (Pedersen et al., 1989).

TABLE 2.29
Typical Composition and Melting Point of Paraffin and Microcrystalline Waxes

Wax Type	Paraffin	Microcrystalline
n-Paraffins, %	80–95	0–15
Iso-paraffins, %	2–15	15–30
Naphthenes, %	2–8	65–75
Carbon number	18–36	30–60
Molecular weight	350–430	500–800
Melting point, °C	50–65	60–90

Source: From Pedersen, K.S., Fredenslund, Aa., and Thomassen, P., *Properties of Oils and Natural Gases*, Gulf Publishing Company, Houston, 1989. Reprinted with permission of GPC Books.

Petrolatum and its manufacture were patented in 1872 under the name "Vaseline." At that time, it was primarily used for treating leather, as hair pomade and for treating chapped hands (Penreco, 2002). In 1875, the American Pharmaceutical Association found petrolatum useful in treating burns and scalds. Due to its pharmacological properties, petrolatum was found to be an effective skin moisturizer. The skin moisturizing properties of petrolatum, coupled with its oxidative stability, are used by cosmetic industry in skin-care products. Petrolatum is a grease-like material and it is often described as an unctuous mass. The wax sweating process cannot be used to deoil microcrystalline wax or petrolatums. Residual wax is a microcrystalline wax produced by a combination of solvent dilution and chilling (Sequeira, 1994). The waxes obtained from residual oils, called petrolatums, are pale to yellow in color. The typical properties of petrolatum are shown in Table 2.30.

Petrolatums are mixtures of pale to yellow in color semisolid hydrocarbons and their melting point can vary in a range of 38°C–60°C (Sequeira, 1994). Petrolatum

TABLE 2.30
Typical Properties of Petrolatum

Properties	Petrolatum
Specific gravity, at 60°C	0.818–0.880
Melting range, °C	38–60
Consistency, mm/10	100–300

Source: From Sequeira, A. Jr., *Lubricant Base Oil and Wax Processing*, Marcel Dekker, Inc., New York, 1994. Reproduced by permission of Routledge/Taylor & Francis Group, LLC.

TABLE 2.31

Properties of Hard and Soft Wax Produced Using Combined Process

Combined Process Melting Point, °C	Waxy Raffinate 8	Waxy Raffinate 35	Waxy Raffinate 38
Dewaxed oil, °C	−51	−36	−36
Hard wax			
Oil content, wt%	0.6	0.4	0.4
Melting point, °C	47–49	53–56	59–61
Soft wax			
Oil content, wt%		2.4	5
Melting point, °C		32	44–46

Source: Modified From Kachlishvili, I.N. and Filippova, T.F., *Chemistry and Technology of Fuels and Oils*, 39, 233, 2003.

is available in different grades and their properties are described by testing different properties including their color, melting point, flash point, viscosity, congealing point, oil content, penetration, density, refractive index, and consistency. Petrolatum, which is solid at ordinary temperatures, is essentially a low-melting, ductile, microcrystalline wax (Speight, 1999). Typical grades of microcrystalline waxes widely vary in their physical characteristics. Some microcrystalline waxes are ductile and others are brittle or crumble easily. The lower melting grades, 57°C–63°C, are flexible and adhesive. The waxes made from petroleum differ in melting points, hardness, color, and oil content. The slack waxes, from the solvent dewaxing of base stocks, are deoiled to produce wax. Another method is to deoil waxy raffinates to produce dewaxed base stocks and hard and soft wax. An extensive review of base stock wax processing, including the use of different dewaxing methods, dewaxing aids and an integrated solvent dewaxing–deoiling process, was published (Sequeira, 1994). The properties of hard and soft wax, produced using a combined dewaxing–deoiling process, are shown in Table 2.31.

Industrial waxes can contain 1.8–3 wt% of oil. Slack waxes, containing less than 10 wt% of oil, are used in the manufacture of candles, chlorinated wax and in some paper-making applications (Exxon, 2002). Production of hard waxes was reported to be expensive and difficult due to stringent requirements for very low oil content. According to the literature, the use of combined dewaxing–deoiling process, while more economical, does not meet the quality requirements of some wax products. To meet a very low oil content of a maximum of 0.45 wt%, the literature reported that additional deoiling of the slack wax with a high purity solvent and a high dilution ratio is required (Kachlishvili and Filippova, 2003). The low oil content waxes, meeting the low oil content requirements, are difficult to produce.

2.3.3 USE OF PETROLEUM WAX

Wax which has been deoiled but has not received any finishing treatment is called unrefined wax or raw wax. Paraffin waxes are pale yellow to white or colorless. Waxes containing high paraffin content are used to produce semirefined or fully refined paraffin wax for use in candles. The removal of oil to 1 wt%, followed by finishing treatment, produces the "semirefined" wax. To produce a "fully refined" wax, it is necessary to remove color and odor. Fully refined paraffin waxes contain a maximum of 0.5 wt% of oil and are white, odorless, and tasteless. Fully refined paraffin waxes are dry, hard, and capable of imparting good gloss. The refined paraffin waxes are widely used in the packaging industry as wax coatings for a variety of paper, film, and foil substrates including corrugated board, cups, containers, and folding cartons.

The refined paraffin waxes are also used to produce textiles, adhesives, PVC lubricants, and ozone inhibitors for the tire and rubber industry. Chlorinated waxes are used as polyvinyl chloride plasticizers, extreme-pressure additives, and components of cutting fluids. The petroleum wax is also used to form emulsions. The oxidation of paraffin wax was reported to decrease the wax/water interfacial tension (Feng et al., 2002). Paraffin wax emulsions are used for specialized purposes including wood products, textile, latex, paint, and coating. High stability emulsions of petroleum waxes are used in many applications related to the production of rubber, leather, shoes, wood, paper, ink, and textiles. The optimum emulsifying conditions required the stirring of 8 wt% of emulsifier, containing a mixture of anionic and nonanionic surfactants, at the temperature of 85°C–95°C (Pang et al., 2004).

Microcrystalline waxes are also good moisture and grease barriers and are widely used in making special types of paper for packaging. Microcrystalline waxes are available in laminating, coating, and hardening grades. Microcrystalline waxes are used in formulated coatings to impart flexibility and improve low temperature properties. Microcrystalline wax based coatings protect against rust and corrosion. Nonpackaging applications of microcrystalline wax include candles, polishes, printing inks, investment casting, and adhesives. They have high adhesive and cohesive strength over a wide range of temperatures. Special formulations of microcrystalline waxes are used for foil/tissue laminations, industrial wraps, bottle sealing, and glassine/laminates.

Petroleum waxes are also blended together to give certain desired properties such as melting point and penetration and are available in a wide range of melting points. The phase transition temperature of hard microcrystalline waxes having a high melting point of 60°C–97°C have been determined by differential scanning calorimetry (DSC) in both heating and cooling mode. The DSC data were compared to ASTM drop melting points and the results indicated that DSC can be used to identify whether the wax sample is a blend of different waxes or not (Kumar et al., 2004). Petroleum waxes are also blended with other products such as resins and polymers to incorporate special properties such as flexibility, toughness, or gloss.

TABLE 2.32
Some Uses of Technical Grade Petroleum Wax

Candles	Polishes
Coatings	Printing inks
Packaging	Casting waxes
Adhesives	Rust preventatives
Textiles	Latex emulsions

The specifications for finished waxes depend on their end use which determines the degree of refining required. The use of petrolatum in the pharmaceutical industry grew mainly due to its ointment-like properties. It is used by the pharmaceutical industry in ophthalmic ointments, topical ointments, dental adhesives, and petroleum gauzes (Morrison, 2002). Technical grade petrolatum is used in carbon papers, buffing and polishing compounds, general purpose lubricants, shoe polishes, printing inks, and solder paste. They are used as rust preventatives (Exxon, 2002). Technical grade petrolatum is also used in plastics and elastomers as rubber processing aids. Some uses of technical grade petroleum wax are shown in Table 2.32.

The finishing process for waxes involves treatment with sulfuric acid, separation of acid sludge, and subsequent neutralization of the wax with activated earth clay. The wax is passed through a bed of clay to remove color and through a vacuum stripping tower for odor removal. The clay treatment of the wax yields 75%–90% of the finished wax. The use of hydrofinishing yields finished waxes in almost 100% (Sequeira, 1994). The recent literature reported on the separation and characterization of macrocrystalline waxes from crude petrolatums using *n*-hexane at different temperatures, dilutions, and washing solvent ratios. The finishing step was carried out using percolation in the molten state through an activated bauxite column (Mohamed and Zaky, 2004).

The highest purity waxes are free of unsaturated components and meet the standards of the US Pharmacopoeia (USP) for food, medicinal, and cosmetic applications. Finished paraffin and microcrystalline waxes are also available in many food grades and need to meet the US Food and Drug Administration (FDA) requirements for applications that involve direct or indirect contact with food. The USDA grade wax needs to meet the requirements of the US Department of Agriculture (USDA) for use in federally inspected meat and poultry plants where incidental or direct contact with food may occur. A severe hydrogen finishing process, known as hydrorefining, is used to remove sulfur, nitrogen, and trace impurities in the manufacture of food and pharmaceutical grade wax. The level of purity is tested following the UV absorbance limits. The effect of finishing on color and properties of petrolatum are shown in Table 2.33.

The technical grade petrolatum was reported to have color varying from amber, red to dark brown while USP grade petrolatum was reported to have a white color (Penreco, 2002). The properties of different amber and snow white

TABLE 2.33

Effect of Finishing on Color and Properties of Petrolatum

Petrolatum	Technical Penreco	USP Grade Penreco	Technical Chevron	USP Grade Chevron
Color	Dark brown	White	Amber	Snow white
Viscosity, SUS at 210°F	70	57–64	78	72
Melting point, °F	115–125	118–130	134	133
Consistency at 77°F	130–170	155–210	195	190

Source: From Penreco, http://www.penreco.com (accessed May 2002); Chevron, http://library.cbest. chevron.com/lubes (accessed May 2002).

color petrolatum reported some small differences in their properties (Chevron, 2002). In highly refined form, petrolatum can be snow white. In the past, the wax in the residual oil was removed by cold settling (Speight, 1999). After the addition of a large volume of naphtha to a residual oil, the mixture was allowed to stand as long as necessary in a tank exposed to low temperatures. The waxy components would congeal and settle to the bottom of the tank. After the removal of naphtha, the steam-refined stock was filtered through charcoal to improve its color and it was called bright stock. The wax product that settled to the bottom of the cold settling tank was crude petrolatum. Further treatments with clay and sulfuric acid produced white grades of petrolatum (Speight, 1999). The color and properties of different petroleum jellies are shown in Table 2.34.

The base ingredient of most antibiotic ointments is petroleum jelly which can be used as an inexpensive wound dressing (Wolfe, 1997). Petroleum jelly can be produced by mixing petroleum wax, such as paraffin wax and microcrystalline wax, and a mineral oil (Panama, 2002). White petroleum jelly is also produced by blending paraffin wax, microcrystalline wax, and liquid paraffin. White petroleum jelly is refined and filtered during its production process to conform to strict purity standards and is free from carcinogenic substances (Kim Chemicals, 2002). In many cases, snow white petroleum jelly is a highly purified petrolatum (Alka Chemicals, 2002). Some uses of FDA/USDA/USP grade petroleum wax are shown in Table 2.35.

TABLE 2.34

Properties of Different Petroleum Jellies

Petroleum Jelly Source	Panama	Panama	Alka Chemicals
Color	Yellow	White	Snow white
Melting point, °C	42–55	44–56	46–52
Consistency at 25°C	150–180	175–200	160–180

Source: From Panama, http://www.panamapetro.com (accessed May 2002); Alka Chemicals, http://www.alkachemical.com (accessed May 2002).

TABLE 2.35

Some Uses of FDA/USDA/USP Grade Petroleum Wax

FDA/USDA/USP Grade Uses	USP Grade Uses
Food packaging	Dental adhesives
Food containers	Pharmaceutical ointments
Fruit and vegetable coating	Petroleum gauzes
Hair products	Petroleum jelly
Cosmetics	Skin-care products

Paraffin and microcrystalline waxes have excellent barrier properties to protect freshness of food and are used in food packaging. White petrolatum USP grade is used by the cosmetic industry as a component of creams, lotions, hair products, lip balms, makeups, and as absorption bases. White petrolatum USP grade is also used by the food industry as animal feed supplements, coatings for fruits and vegetables, and food packaging materials. It is used as trough greases by the baking industry. Pharmaceutical grade petroleum jellies find applications in the manufacture of pharmaceutical ointments and creams such as ophthalmic ointments, dermal ointments, pain balms, protective creams, cold creams, hair vaseline, and makeup preparations. Perfumed and extra soft white petroleum jelly is used in formulations to treat baby's rash.

2.4 ASPHALTS

Refined crude oil is used to produce liquid fuels for use in transportation and heating of buildings, solvents, and lubricants. Additionally, process oils and wax are produced as by-products of base oil manufacture. Process oils are highly aromatic extracts or lightly refined base stocks and used as plasticizers in automotive tires, printing inks, and mold release oils. Vacuum residue is used to produce asphalt. The residue is used to produce asphalts which are a high value product for highway surfaces, roofing materials, and waterproofing uses (Speight, 1999). Typically crude oils are used to produce light distillates, lube oil base stocks, aromatic extracts, wax, and vacuum residue. In cases when crude oils are not suitable to produce lube oil base stocks, the atmospheric residue is used to produce asphalt. Some crude oils are not suitable to produce light distillates and lube oil base stocks and are used to produce asphalts. The definition of "heavy oil" is based on the API gravity which is less than 20 or viscosity. The carbon residue value of petroleum products is used as an indication of the tendency of the material to form carbonaceous type deposits under degradation conditions. In most cases, the lower the carbon residue the more valuable the crude oil. The typical oil content and carbon residue of crude oils, heavy oils, and residue are shown in Table 2.36.

Conventional crude oils contain 67–97 wt% of oil, 3–22 wt% of resins, 0.1–12 wt% of asphaltenes, and have a carbon residue in a range of 0.2–10 wt%.

TABLE 2.36
Typical Oil and Carbon Residue Content of Crude Oils and Residue

Content, wt%	Crude Oils	Heavy Oils	Residue
Oil	67–97	24–64	<49
Resins	3–22	14–39	29–39
Asphaltenes	0.1–12	11–45	11–29
Carbon residue	0.2–10	10–22	18–32

Source: From Speight, J.G., *The Chemistry and Technology of Petroleum*, 3rd ed., Marcel Dekker, Inc., New York, 1999. Reproduced by permission of Routledge/Taylor & Francis Group, LLC.

Heavy oils contain a lower oil content of 24–64 wt%, a higher resin content of 14–39 wt%, a higher asphaltene content of 11–45 wt%, and have a significantly higher carbon residue in a range of 10–22 wt%. After atmospheric crude oil distillation, the residue contains less than 49 wt% of oil, a high resin content in a range of 29–39 wt%, a high asphaltene content in a range of 11–29 wt%, and a high carbon residue in a range of 18–32 wt%. After the vacuum distillation, the use of different crude oils was reported to affect the composition of vacuum residue and lead to differences in the composition of de-asphalted oils and asphalts. The literature reported on the compositional analysis of vacuum distillates, de-asphalted oils, and asphalts from crude oils processed in India. The NMR analysis identified the presence of paraffinic, naphthenic, and aromatic carbons on predominantly polycyclic structures in de-asphalted oils and asphalts. The MS analysis indicated a high content of acids and bases (Kohli et al., 2001). The composition of de-asphalted oil and asphalt produced from Bombay High crude oil is shown in Table 2.37.

TABLE 2.37
Composition of DAO and Asphalt Produced from Bombay High Crude Oil

Bombay High Crude Oil	De-Asphalted Oil (DAO)	Asphalt
Saturates, wt%	38.7	6.1
Aromatics, wt%	48.9	50.9
Bases, wt%	2.9	6.4
Acids, wt%	4.3	11.6
C_5 insolubles, wt%	0.8	2.4

Source: From Kohli, B., Bhatia, B.M.L., and Madhwal, D.C., *Petroleum Science and Technology*, 19, 885, 2001. Reproduced by permission of Taylor & Francis Group, LLC, http://www.taylorandfrancis.com

TABLE 2.38
Composition of DAO and Asphalt Produced from Assam Mix Crude Oil

Assam Mix Crude Oil	De-Asphalted Oil (DAO)	Asphalt
Saturates, wt%	30.2	0
Aromatics, wt%	56.2	62.2
Bases, wt%	4.1	9.4
Acids, wt%	5.7	13.8
C_5 insolubles, wt%	1.1	11.5

Source: From Kohli, B., Bhatia, B.M.L., and Madhwal, D.C., *Petroleum Science and Technology*, 19, 885, 2001. Reproduced by permission of Taylor & Francis Group, LLC, http://www.taylorandfrancis.com.

The de-asphalted oil from Bombay High crude oil was reported to contain 38.7 wt% of saturates and 48.9 wt% of aromatics. The de-asphalted oil was also reported to contain 2.9 wt% of bases, 4.3 wt% of acids, and 0.8 wt% of C_5 insolubles. The asphalt from Bombay High crude oil was found to contain a lower saturate content of only 6.1 wt% but a similar aromatic content of 50.9 wt%. The asphalt was also found to contain a higher base content of 6.4 wt%, a higher acid content of 11.6 wt%, and a higher C_5 insoluble content of 2.4 wt% (Kohli et al., 2001). The composition of de-asphalted oil and asphalt produced from Assam Mix crude oil is shown in Table 2.38.

The de-asphalted oil from Assam Mix crude oil was reported to contain 30.2 wt% of saturates and 56.2 wt% of aromatics. The de-asphalted oil was also reported to contain 4.1 wt% of bases, 5.7 wt% of acids, and 1.1 wt% of C_5 insolubles. The asphalt from Assam Mix crude oil was found to contain no saturate content and a higher aromatic content of 62.2 wt%. The asphalt was also found to contain a higher base content of 9.4 wt%, a higher acid content of 13.8 wt%, and a higher C_5 insoluble content of 11.5 wt% (Kohli et al., 2001). The use of different crude oils affects the composition of asphalts and their physical properties, such as the softening point, penetration, and ductility. The literature reported on the composition and the properties of straight-run petroleum asphalt from Shengli and Huansanlian oil fields in China (Qi, 2000). The literature also reported on the effect of oxidation on aging of petroleum asphalts. The oxygen adsorption was reported to increase the presence of —OH, C=O, and S=O groups (Qi and Wang, 2004a) and to increase the molecular weight of asphalt (Qi and Wang, 2004b). The composition and the softening point of Shengli petroleum asphalt (SPA) and Huansanlian petroleum asphalt (HPA) are shown in Table 2.39.

While the majority of China crude oils were reported to be paraffin type, some crude oils from Liaohe, Shengli, and Xinjiang oil fields were reported to be highly viscous and very different. Liaohe crude oils were reported to have a high density, high pour points, and a high carbon residue. The actual wax content was reported

TABLE 2.39

Composition and Properties of Different Petroleum Asphalts

Straight-Run Petroleum Asphalts	Shengli Crude Oil SPA	Huansanlian Crude Oil HPA
Saturates, wt%	23.2	25.9
Aromatics, wt%	37.6	31.4
Resins, wt%	36.3	40.9
Asphaltenes, wt%	3.0	1.8
Softening point, °C	46	48.2
Penetration (0.1 mm)	75	93
Ductility at 25°C, cm	>140	>100

Source: From Qi, Y., *Petroleum Science and Technology*, 18, 929, 2000. Reproduced by permission of Taylor & Francis Group, LLC, http://www.taylorandfrancis.com.

to be low, below 2 wt%, and the crude oil was classified as low sulfur naphthene base crude oil. Due to a high viscosity and a high pour point of 48°C, the transportation and the storage of this crude oil was reported to be very difficult (Sun et al., 2001). The properties and composition of highly viscous Liaohe crude oil are shown in Table 2.40.

The Liaohe crude oil was reported to have a relatively low S content of 0.42 wt% and a relatively high N content of 0.41 wt%, 276 ppm of Ni, and 79 ppm of Fe with practically no V. The crude oil was also reported to have a high acidity of 5.57 mg KOH/g, the salt content of 4.9 mg NaCl/L, and some water of 2.85 wt% (Sun et al., 2001). The total content of asphaltene and resins was very high, about 30 wt%, of the total crude oil. The literature reported that a high metal content and a high carbon residue of 20.9 wt% make this crude oil similar to the atmospheric residue of general crude oils (Sun et al., 2001). The literature also reported that the highly viscous crude oil from the Liaohe oil field contained no gasoline fraction and a low diesel fraction of only 7.19 vol% (Jin and Liao, 2003). The effect of the boiling range on pour point and TAN of light distillates from highly viscous Liaohe crude oil is shown in Table 2.41.

While typical China crude oils were reported to produce about 30 vol% of lower than 350°C distillate, the Liaohe crude oil was reported to produce only 7.19 vol% of lower than 350°C distillate. The acid content of the distillate was reported high and increased with an increase in boiling point leading to a severe problem of equipment corrosion. Liaohe crude oil could not be used to produce gasoline, kerosene, and diesel oil using a normal distillation technology (Sun et al., 2001). The yields and composition of heavy residue distillates from highly viscous Liaohe crude oil are shown in Table 2.42.

The yields of residue distillate above 350°C were reported as high as 92.48 vol% but decreased to 67.11 vol% above 500°C. They were reported to contain

TABLE 2.40
Properties and Composition of Highly Viscous Liaohe Crude Oil

Liaohe Crude Oil	Properties
API gravity	9.94
Density at 20°C, g/cm^3	0.9977
Pour point, °C	48
Wax content, wt%	<2
Resins, wt%	28.2
Sulfur, wt%	0.42
Nitrogen, wt%	0.41
Ni, ppm	276
V, ppm	1
Fe, ppm	79
Cu, ppm	1
TAN, mg KOH/g	5.6
Salt content, mg NaCl/L	4.9
Water, wt%	2.9
Asphaltenes, wt%	2
Carbon residue, wt%	20.6
Ash content, wt%	0.13

Source: From Sun, D., Lu, W., and Liao, K., *Petroleum Science and Technology*, 19, 837, 2001. Reproduced by permission of Taylor & Francis Group, LLC, http://www.taylorandfrancis.com.

high resin and asphaltene contents and have a high carbon residue (Sun et al., 2001). Highly viscous Liaoshu crude oil was reported not suitable to produce lube oil base stocks and its residue fraction was not suitable for catalytic cracking

TABLE 2.41
Pour Point and TAN of Light Distillates from Viscous Liaohe Crude Oil

Distillates	190°C–250°C	250°C–275°C	275°C–300°C	300°C–320°C	320°C–350°C
Yield, vol%	0.94	1.05	1.24	1.14	2.82
Density, g/cm^3	0.8497	0.8745	0.8882	0.9095	0.9231
Pour point, °C	—	—	<−40	−37	−32
TAN, mg KOH/g	0.69	0.95	3.23	4.82	7.96

Source: From Sun, D., Lu, W., and Liao, K., *Petroleum Science and Technology*, 19, 837, 2001. Reproduced by permission of Taylor & Francis Group, LLC, http://www.taylorandfrancis.com.

TABLE 2.42

Yields and Composition of Heavy Residue Distillates from Viscous Liaohe Crude Oil

Distillates	>350°C	>360°C	>370°C
Yield, vol%	92.48	91.23	90.05
Density at 20°C, g/cm^3	1.0015	1.0023	1.0035
Saturates, wt%	33.34	27.13	23.98
Aromatics, wt%	28.53	30.29	31.04
Resins, wt%	28.61	32.72	34.29
Asphaltenes, wt%	9.52	9.86	10.69
Carbon residue, wt%	27.7	27.9	28

Source: From Sun, D., Lu, W., and Liao, K., *Petroleum Science and Technology*, 19, 837, 2001. Reproduced by permission of Taylor & Francis Group, LLC, http://www.taylorandfrancis.com.

(Weizhen et al., 2001). The compositional analysis indicated low wax content but a high content of resins and asphaltenes indicating a low sulfur naphthene based crude oil suitable for production of high quality paving asphalt (Jin and Liao, 2003). The composition and the softening point of the vacuum residue from Liaoshu crude oil are shown in Table 2.43.

The Liaoshu crude oil was reported to be an excellent material to produce paving asphalt. The vacuum distillate, above 350°C, was blended with modifier at 130°C–150°C to produce high quality asphalts having good low temperature and heat resistant antiaging properties (Weizhen et al., 2001). The literature also reported on the composition and the performance characteristics of paving

TABLE 2.43

Composition and Softening Point of Vacuum Residue from Liaoshu Crude Oil

Vacuum Residue	>400°C	>480°C
Saturates, wt%	24.6	18.1
Aromatics, wt%	30.4	29.9
Wax, wt%	1.8	1.9
Resins, wt%	42.4	49.3
Asphaltenes, wt%	2.7	2.7
Softening point, °C	53.5	63.5

Source: From Weizhen, S. et al., *Petroleum Science and Technology*, 19, 1187, 2001. Reproduced by permission of Taylor & Francis Group, LLC, http://www.taylorandfrancis.com.

TABLE 2.44

Effect of Oxidation on Softening Point of Vacuum Residue

Arlan Crude Oil Softening Point	Vacuum Residue, 38°C	Vacuum Residue, 3.5°C
Oxidation	60 min at 220°C	320 at 222°C
Softening point (R & B)	45	44.5
Oxidation	48 min at 245°C	200 at 245°C
Softening point (R & B)	45	43

Source: Modified From Evdokimova, N.G., Lobanov, V.V., and Khivintsev, A.V., *Chemistry and Technology of Fuels and Oils*, 37, 128, 2001.

asphalts produced from Liaoshu crude oil. Their analysis indicated the presence of 20–25 wt% of saturates, 35–40 wt% of aromatics, 25–30 wt% of resins, and 8–10 wt% of asphaltenes (Cong et al., 2004).

The vacuum residue can be used to produce modified asphalts. The vacuum residue blended with aromatic rich oil and styrene–butadiene–styrene (SBS) polymer, used as a modifier, can produce good quality paving asphalts. According to the literature paving and construction asphalts are primarily produced by oxidation of vacuum residue with atmospheric oxygen in column units. Asphalts have a softening point of 25°C–55°C and if the softening point of asphalt is too low, it can be hardened by oxidation (Speight, 1999). The effect of oxidation time and temperature on the softening point of vacuum residue from Arlan crude oil is shown in Table 2.44.

The asphalts produced by the oxidation of vacuum residue at a higher temperature of 245°C were reported to have improved properties in terms of a lower temperature of brittleness and better penetration at 25°C and at 0°C. The properties of asphalts are improved by compounding with additives used as inhibitors, the optimization of the oxydation regime, and regeneration of used asphalts (Evdokimova et al., 2001). The literature also reported that the asphaltene content in asphalt is not that important because the most important property is actually the disperse phase particle size. Certain crude oils and the use of vacuum residue require a carefully selected oxidation temperature to produce good quality asphalt. The use of low temperature limits the reaction for transformation of oil components of vacuum residue in asphalt. With the use of high temperature, the molecules undergo cracking and condensation reactions. The temperature changes during winter and summer, and sunlight were reported to lead to degradation in the asphalt properties.

REFERENCES

Alka Chemicals, http://www.alkachemical.com (accessed May 2002).

Beuther, H., Donaldson, R.E., and Henke, A.M., *I&EC Product Research and Development*, 3(3), 174, 1964.

Bhore, N.A. et al., *Oil and Gas Journal*, 97(46), 67, 1999.

Bragado, G.A.C., Guzman, E.T., and Yacaman, M.J., *Petroleum Science and Technology*, 19(1&2), 45, 2001.

Chernozemov, N.S., *Chemistry and Technology of Fuels and Oils*, 37(4), 252, 2001.

Chevron, http://www.chevron.com (accessed May 2002).

Cong, Y. et al., *Petroleum Science and Technology*, 22(3&4), 455, 2004.

Ellis, P.J. and Paul, C.A., *Delayed Coking Fundamentals*, AIChE 2000 Spring National Meeting, Atlanta, May 5–9, 2000.

Evdokimova, N.G., Lobanov, V.V., and Khivintsev, A.V., *Chemistry and Technology of Fuels and Oils*, 37(2), 128, 2001.

Exxon, http://www.exxon.com (accessed May 2002).

Feng, Y., Shen, B., and Gao, J., *Petroleum Science and Technology*, 20(9&10), 973, 2002.

Gary, J.H. and Handwerk, G.E., *Petroleum Refining, Technology and Economics*, 4th ed., Marcel Dekker, Inc., New York, 2001.

Golden, S.W., *Hydrocarbon Processing* (International Edition), 76, 141, May 1997.

Goossens, A.G., *Industrial and Engineering Chemistry Research*, 35, 985, 1996.

Hamid, S.H., *Petroleum Science and Technology*, 18(7&8), 871, 2000.

Henderson, H.E., Chemically modified mineral oils, in *Synthetics, Mineral Oils, and Bio-Based Lubricants: Chemistry and Technology*, Rudnick, L.R., Ed., CRC Press, Boca Raton, 2006.

Hinco Group, http://www.hinco.com (accessed May 2002).

IGI, http://www.igiwax.com (accessed May 2002).

Jin, Z. and Liao, K., *Petroleum Science and Technology*, 21(7), 1077, 2003.

Kachlishvili, I.N. and Filippova, T.F., *Chemistry and Technology of Fuels and Oils*, 39(5), 233, 2003.

Kai-Fu, H. et al., *Petroleum Science and Technology*, 18(7&8), 815, 2000.

Khidr, T.T. and Mohamed, M.S., *Petroleum Science and Technology*, 19(5&6), 547, 2001.

Kim Chemicals, http://www.kimchemicals.com (accessed May 2002).

Klein, M.T. et al., *Molecular Modeling in Heavy Hydrocarbon Conversions*, CRC Press, Boca Raton, 2006.

Kohli, R., Bhatia, B.M.L., and Madhwal, D.C., *Petroleum Science and Technology*, 19(7&8), 885, 2001.

Kumar, S. et al., *Petroleum Science and Technology*, 22(3&4), 337, 2004.

Mang, T., Base oils, in *Lubricants and Lubrications*, Mang, T. and Dresel, W., Eds., Wiley-VCH, Weinheim, 2001.

Mohamed, N.H. and Zaky, M.T., *Petroleum Science and Technology*, 22(11&12), 1553, 2004.

Morrison, D.S., Petrolatum, http://www.penreco.com (accessed June 2002).

Panama, http://www.panamapetro.com (accessed May 2002).

Pang, H. et al., *Petroleum Science and Technology*, 22(3&4), 439, 2004.

Pedersen, K.S., Fredenslund, Aa., and Thomassen, P., *Properties of Oils and Natural Gases*, Gulf Publishing Company, Houston, 1989.

Penreco, http://www.penreco.com (accessed May 2002).

Phillips, R.A., Highly refined mineral oils, In *Synthetic Lubricants and High-Performance Functional Fluids*, 2nd ed., Rudnick, L.R. and Shubkin, R.L., Eds., Marcel Dekker, Inc., New York, 1999.

Pirro, D.M. and Wessol, A.A., *Lubrication Fundamentals*, 2nd ed., Marcel Dekker, Inc., New York, 2001.

Polishchuk, Yu.M. and Yashchenko, I.G., *Petroleum Chemistry*, 41(4), 247, 2001.

Prince, R.J., Base oils from petroleum, in *Chemistry and Technology of Lubricants*, 2nd ed., Mortier, R.M. and Orszulik, S.T., Eds., Blackie Academic & Professional, London, 1997.

Purohit, R.C., Srivastava, S.P., and Verma, P.S., *Petroleum Science and Technology*, 21(9&10), 1369, 2003.

Qi, Y., *Petroleum Science and Technology*, 18(7&8), 929, 2000.

Qi, Y. and Wang, F., *Petroleum Science and Technology*, 22(3&4), 263, 2004a.

Qi, Y. and Wang, F., *Petroleum Science and Technology*, 22(3&4), 275, 2004b.

Rome, C.F., *Hydrocarbon Asia*, 42, May/June 2001.

Ross, S., Foaminess and capillarity in apolar solutions, in *Interfacial Phenomena in Apolar Media*, Eicke, H.F. and Parfitt, G.D., Eds., Marcel Dekker, Inc., New York, 1987.

Sequeira, A., *Pre-Prints Division of Petroleum Chemistry*, ACS, 37(4), 1286, 1992.

Sequeira, A. Jr., *Lubricant Base Oil and Wax Processing*, Marcel Dekker, Inc., New York, 1994.

Severin, D. and David, T., *Petroleum Science and Technology*, 19(5&6), 469, 2001.

Shell, http://www.shell-lubricants.com (accessed May 2001).

Speight, J.G., *The Chemistry and Technology of Petroleum*, 2nd ed., Marcel Dekker, Inc., New York, 1991.

Speight, J.G., *The Chemistry and Technology of Petroleum*, 3rd ed., Marcel Dekker, Inc., New York, 1999.

Speight, J.G., *The Chemistry and Technology of Petroleum*, 4th ed., CRC Press, Boca Raton, 2006.

Srivastava, S.P. et al., *Petroleum Science and Technology*, 20(3&4), 269, 2002a.

Srivastava, S.P. et al., *Petroleum Science and Technology*, 20(3&4), 291, 2002b.

Srivastava, S.P. et al., *Petroleum Science and Technology*, 20(7&8), 831, 2002c.

Stark, J.L. and Asomaning, S., *Petroleum Science and Technology*, 21(3&4), 569, 2003.

Starkova, N.N. et al., *Chemistry and Technology of Fuels and Oils*, 37(3), 191, 2001.

Sun, D., Lu, W., and Liao, K., *Petroleum Science and Technology*, 19(7&8), 837, 2001.

Taylor, R.J. and McCormack, A.J., *Industrial and Engineering Chemistry Research*, 31, 1731, 1992.

Taylor, R.J., McCormack, A.J., and Nero, V.P., *Pre-Prints Division of Petroleum Chemistry*, ACS, 37(4), 1337, 1992.

Thiele, R. et al., *Industrial and Engineering Chemistry Research*, 42, 1426, 2003.

Tooley, P., *Fats, Oils and Waxes,* William Clowes & Sons, Ltd., London, 1971.

Weizhen, S. et al., *Petroleum Science and Technology*, 19(9&10), 1187, 2001.

Wolfe, Y.L., *Prevention*, 49(1), 55, 1997.

Yan, F., Yang, P., and Liao, K., *Petroleum Science and Technology*, 20(5&6), 571, 2002.

3 Nonconventional Refining

3.1 CATALYTIC PROCESSING

3.1.1 Use of Heavy Oils

The current trend in the refining industry is to utilize the heavier and higher molecular weight feedstocks which are more difficult to process. Sulfur occurs naturally in crude oils and must be removed during the refining process to meet the product specifications. Crude oils containing high sulfur content, frequently termed as "sour" crudes, require more extensive processing than "sweet" crudes which contain little or no dissolved hydrogen sulfide, mercaptans, and other sulfur compounds (Gary and Handwerk, 2001). Crude oil is first distilled in an atmospheric distillation unit to remove gases, gasoline, naphthas, kerosene, and light gas oil. The remaining liquid from the bottom of the crude oil distillation tower is flashed and evaporated in a vacuum. Porphyrins forming complexes with metals, such as vanadium and nickel, appear throughout the entire boiling range of the crude oil but tend to concentrate in the heavier fractions and in nonvolatile residues. Some metallic constituents were found in higher boiling distillates (Speight, 1999). The extraction solvent removes the condensed ring aromatics and polar molecules from the lube oil distillates. The solvent-rich fraction, after evaporation of solvent, is an aromatic-rich extract used as fuel, cracker feed, and process oils (Mang, 2001). The solvent dewaxing process removes the waxy type molecules from the lube oil distillates. Slack wax can be used as feed to slack wax hydroisomerization or, after a deoiling process, can be used to produce petroleum wax. The typical yields of various petroleum fractions in conventional refining of a crude oil, suitable to produce lube oil base stocks, are shown in Table 3.1.

A crude oil whose API gravity has been reduced by distillation of the lighter lower boiling constituents is called reduced crude (Gary and Handwerk, 2001). Sweetening of gasoline distillate by oxidizing thiols to disulfides is used to produce a clean transportation fuel. The Merox process is a combination process for mercaptan extraction and sweetening of gasoline or lower boiling products. The literature reported that the gasoline is washed with alkali and contacted by the catalyst and caustic in the extractor. Air is then injected and the treated product is stored in the regeneration step. The caustic from extractor and air is mixed in the

TABLE 3.1
Typical Yields of Petroleum Fractions in Conventional Refining

Distillation	Temp. Range, °C	Fractions	Typical Yield, %	Products
Atmospheric	<30	Gas	4	Light distillates
	30–175	Naphtha	20	
	175–200	Kerosene	4	
	200–350	Gas oil	23	
Vacuum	350–530	Vacuum gas oil	23	Base stocks
		Distillates		Aromatic extracts
				Wax
	>530	Vacuum residue	26	Deasphalted oils
				Asphalt

Source: From Mang, T., *Lubricants and Lubrications*, Mang, T. and Dresel, W., Eds., Wiley-VCH, Weinheim, 2001.

oxidizer and disulfides are separated. The regenerated caustic solution is recirculated to the top of the extractor (Speight, 2006). The literature reported on the effect of caustic concentration on the liquefied petroleum gas (LPG) sweetening process. It was reported to affect the dispersion of the sulfonated cobalt phthalocyanine catalyst and transformation of mercaptans into mercaptides and further into disulfides (Liu et al., 2005). Typical organic sulfur compounds found in gasoline also include thiophene, methylthiophene, and dimethylthiophene (Speight, 2006). Crude oils have different fractions of naphtha, kerosene, gas oil, and residue as shown in Table 3.2.

TABLE 3.2
Effect of Different Crude Oils on Yields of Petroleum Fractions

Crude Oils	Gas and Gasoline	Kerosene	Gas Oil	Residue, >1000°F
Oklahoma	47.6	10.8	21.6	20.0
Pennsylvania	47.4	17.0	14.3	1.3
Iraq	45.3	15.7	15.2	23.8
Iran	45.1	11.5	22.6	20.8
Kuwait	39.2	8.3	20.6	31.9
California	36.6	4.4	36	23.0
Saudi Arabia	34.5	7.7	29.3	27.5
Bahrain	26.1	13.4	34.1	26.4

Source: From Speight, J.G., *The Chemistry and Technology of Petroleum*, 3rd ed., Marcel Dekker, Inc., New York, 1999. Reproduced by permission of Routledge/Taylor & Francis Group, LLC.

TABLE 3.3

Yields of Petroleum Fractions from Arab Light and Arab Heavy Crude Oils

Boiling Range Temp., °C	Petroleum Fractions	Arab Light Yield, %	Arab Heavy Yield, %
<25	Gases	0.7	0.6
25–200	Naphtha	21.7	19.8
200–350	Middle distillate	28	22.2
>350	Atmospheric residue	49.6	57.4

Source: From Ali, M.A., *Petroleum Science and Technology*, 19, 1063, 2001. Reproduced by permission of Taylor & Francis Group, LLC, http://www.taylorandfrancis.com.

Depending on the crude oil, the yields of gasoline and other petroleum fractions can vary. The kerosene fraction was reported to vary from 17 vol% found in Pennsylvania crude oil to only 4 vol% in California crude oil. Some crude oils have a low yield of kerosene. The desirable components of kerosene are saturated hydrocarbons and only some crude oils contain a high quality kerosene fraction having a boiling range of 205°C–260°C (Speight, 1999). To be used as a burning oil, kerosene must be free of aromatic and unsaturated hydrocarbons as well as free of the more obnoxious sulfur compounds. Excessive amount of aromatics can be removed by extraction, followed by a lye wash or a doctor treatment. The comprehensive review of petroleum products, their processing options, and the use of different treatments was published (Speight, 2006). The use of heavy crude oils decreases the yields of lower boiling petroleum fractions. The yields of different petroleum fractions, from Arab light and Arab heavy crude oils, are shown in Table 3.3.

The use of heavy crude oils, even from the same oil field, affects the volume of different petroleum fractions. With an increase in viscosity of Arab heavy crude oil, the yield of naphtha fraction was reported to decrease from 21.7 to 19.8 vol% and the yield of middle distillate decreased from 28 to 22.2 vol%. The literature also reported on a significant increase in the atmospheric residue from about 50 to above 57 vol% (Ali, 2001). The use of heavy Arab crude oil decreases the volume of light distillates and increases the volume of atmospheric residue. The sulfur content of Kuwaiti crude oils, fractionated into naphtha, kerosene, medium gas oil, and heavy gas oil, was analyzed. The use of capillary gas chromatography, equipped with a sulfur chemiluminescence detector (GC-SCD) indicated the presence of mercaptans, sulfides, thiophenes, benzothiophenes, dibenzothiophenes, benzonaphthothiophenes, and their alkyl derivatives (Behbehani et al., 2005). The properties and composition of atmospheric residue, from Arab light and Arab heavy crude oils, are shown in Table 3.4.

TABLE 3.4
Composition of Atmospheric Residue from Arab Light and Arab Heavy Crude Oils

Properties	Arab Light, Atmospheric Residue	Arab Heavy, Atmospheric Residue
Specific gravity at 15.6°C, g/mL	0.9572	0.995
Viscosity at 90°C, cSt	18.5	108
Pour point, °C	15	21
Saturates, wt%	54.7	29.8
Aromatics, wt%	27.1	42.4
Polars, wt%	12.3	17.8
Sulfur, wt%	3	4.2
Nitrogen, wt%	0.17	0.5
Nickel, ppm	10	32
Vanadium, ppm	34	105
Iron, ppm	1	17
Water and sediment, wt%	0.015	0.022
C6 insolubles, wt%	6	10
Carbon residue (RCR), wt%	7.54	14.2

Source: From Ali, M.A., *Petroleum Science and Technology*, 19, 1063, 2001. Reproduced by permission of Taylor & Francis Group, LLC, http://www.taylorandfrancis.com.

The atmospheric residue from Arab light crude oil was reported to contain 27.1 wt% of aromatics, 12.3 wt% of polars, 3 wt% of S, and 0.17 wt% of N. Such a high polar content suggests that the O compounds might be present in an amount as high as 9 wt%. The atmospheric residue from Arab light crude oil was also reported to contain 34 ppm of V, 10 ppm of Ni, 1 ppm of Fe, and 6 wt% of asphaltenes (Ali, 2001). The atmospheric residue from Arab heavy crude oil was reported to have a higher viscosity and higher pour point and contained 42.4 wt% of aromatics. An increased polar content of 17.8 wt%, the S content of 4.2 wt %, and N content of 0.5 wt% indicate that the O compounds might be present in an amount as high as 13 wt%. The atmospheric residue from Arab heavy crude oil was also reported to contain 105 ppm of V, 32 ppm of Ni, 17 ppm of Fe, and 10 wt% of asphaltenes (Ali, 2001). The carbon residue of atmospheric residue significantly increased from 7.5 to 14.2 wt%. The use of heavy oils affects the properties and composition of atmospheric residue. The use of heavy Arab crude oil increases the specific gravity, viscosity, pour point, aromatic and heteroatom contents, metals, asphaltenes, and carbon residue of atmospheric residue.

3.1.2 CATALYTIC CRACKING

An increase in processing of heavy crude oils leads to an increase in heavy fuel oils for which the market is decreasing. There is an increase in demand for gasoline made from the lowest boiling fraction of crude oil, which is naphtha. Crude oils are thermally unstable when exposed to high temperatures. The temperature of distillation is below 350°C to prevent their decomposition. The recent literature reported that as early as 1930, the use of thermal cracking in the presence of silica–alumina catalyst increased the amount of gasoline that could be produced which had higher octane rating (Schwarcz, 2007). Catalytic cracking is used to produce more light fractions than are normally present in crude oil. During catalytic cracking, paraffins, alkyl naphthenes, and alkylaromatics are cracked to form hydrocarbons and olefins (Gary and Handwerk, 2001).

Paraffins   → paraffins and olefins
Alkylnaphthenes → naphthenes and olefins
Alkylaromatics   → aromatics and olefins

The fluid catalytic cracking (FCC) unit uses a fine, powdery catalyst at high temperatures to break long and heavy molecules into shorter ones needed for the production of gasoline and fuel oils. The typical FCC reactor temperature was reported to be in the range of 475°C–510°C and the reactor pressure of 8–30 psig. Commercial cracking catalysts include acid-treated natural alumino–silicates, amorphous synthetic silica–alumina combinations, or crystalline synthetic silica–alumina called zeolites or molecular sieves. According to the literature, most refineries use zeolite catalysts (Gary and Handwerk, 2001). Catalytic cracking of petroleum fractions has been used for a long time and many research papers have been published concerning the kinetics, the mechanism of cracking, and the coke formation (Peixoto and Medeiros, 2001). The FCC products obtained from Alaska North Slope crude oil are shown in Table 3.5.

The light and heavy distillates from atmospheric and vacuum distillation towers can be routed to the cracking units for further processing. The atmospheric gas oil and heavy vacuum gas oil (HVGO) are the main feedstocks to an FCC unit for the production of gasoline and diesel (Ellis and Paul, 2000). The literature reported

TABLE 3.5
FCC Products from Alaska North Slope Crude Oil

Feed	Feed API	Feed, %	Products	Products, %
650°F–850°F	23.2	56.7	Gas	12.4
850°F–1050°F	16.5	43.3	C5 + Naphtha	56.8
			Light gas oil	18.4
			Heavy gas oil	6.6
			Coke	5.8

Source: From Gary, J.H. and Handwerk, G.E., *Petroleum Refining, Technology and Economics*, 4th ed., Marcel Dekker, Inc., New York, 2001. Reproduced by permission of Routledge/Taylor & Francis Group, LLC.

TABLE 3.6
Composition and Olefin Content of Catalytically Cracked Gasoline

Properties	VGO, FCC Gasoline	Atmospheric Residue, FCC Gasoline
Density at 15°C, g/cm^3	0.78	0.778
Saturates, wt%	44.6	41.9
Aromatics, wt%	33.6	27.7
Sulfur, ppm	61	229
1-Olefins, wt%	5.2	7.3
Internal olefins, wt%	16.6	23.1
Research octane number	86.7	87

Source: Reprinted with permission from Hatanaka, S., Yamada, M., and Sadakane, O., *Industrial and Engineering Chemistry Research*, 36, 1519, 1997. Copyright American Chemical Society.

on the composition, S content, and different chemistry olefins present in catalytically cracked gasoline produced from desulfurized vacuum gas oil and low S atmospheric residue (Hatanaka et al., 1997). The composition and olefin content of catalytically cracked gasoline obtained from different feedstocks are shown in Table 3.6.

The FCC unit is a major secondary process for producing fuel products and their quality is affected by the feedstocks. The Merey crude oil residue was used as a feedstock to produce gasoline and the literature reported on relating the feedstock composition to the product composition in catalytic cracking (Sheppard et al., 2003). The composition of different Merey crude oil residues, having a boiling point >650°F, is shown in Table 3.7.

The residue # 1 was obtained, as a whole fraction, while the residues # 2 and # 3 and, many other reported in the paper, were obtained by blending to vary their composition. There are different methods to measure the carbon residue of petroleum oils. Carbon residue measurements which use the Conradson carbon residue (CCR) test procedure are usually reported as carbon residue. The ASTM D 4530 test method is used to measure the microcarbon residue (MCR) and the results are equivalent to the CCR test data. The Merey crude oil residue was reported to contain 3.09 wt% of S, 56 wt% of N, 89 ppm of Ni, 353 ppm of V, and formed 17.3 wt% of MCR. The properties and composition of gasoline obtained by catalytic cracking in an FCC unit of different residues, from Merey crude oil, are shown in Table 3.8.

The FCC gasoline was reported to have a high olefin content ranging from 9.9 to 16.8 wt% which was mostly 1-double bond molecules. The literature reported on different processing options used to manage the sulfur content of feedstocks used in an FCC unit and naphtha (Sloley, 2001). The literature also reported that some feedstocks are more difficult to process. The use of atmospheric residue, vacuum residue, and deasphalted oils, as a feedstock to an FCC unit, was reported

TABLE 3.7
Composition of Different Merey Crude Oil Residues

Composition	Atmospheric Residue # 1	Atmospheric Residue # 2	Atmospheric Residue # 3
C, wt%	84.8	85.1	84.3
H, wt%	10.9	11.6	11.3
S, wt%	3.1	2.6	2.4
N, wt%	0.6	0.1	0.4
Ni, ppm	89	19	17
V, ppm	353	90	79
MCR, wt%	17.3	5.4	5.8

Source: Reprinted with permission from Sheppard, C.M. et al., *Energy and Fuels*, 17, 1360, 2003. Copyright American Chemical Society.

to produce residue FCC naphtha (RFCC) which is more difficult to be sweetened. The chemistry of sulfur compounds was analyzed using gas chromatography, equipped with a flame photometric detector (GC-FPD) which is the specific detector for sulfur compounds (Xia et al., 2002). The boiling point and the chemistry of many different organo sulfur compounds present in RFCC naphtha, before and after sweetening, are shown in Table 3.9.

More than 20 different thiols, including C_3–C_8 normal thiols, iso-thiols, thiophenol, and methylthiophenol, were identified in RFCC naphthas using

TABLE 3.8
Effect of Feed on Properties and Composition of FCC Gasoline

Properties	Residue # 1, FCC Gasoline # 1	Residue # 2, FCC Gasoline # 2	Residue # 3, FCC Gasoline # 3
Specific gravity	0.762	0.736	0.765
Paraffins, wt%	29.1	35	31.8
Naphthenes, wt%	10	13	10.7
Aromatics, wt%	42.5	39.1	43.4
Total olefins, wt%	16.8	11.4	9.9
1-Double bond	16.5	11.4	9.9
2-Double bond	0.3	0.02	0.04
Octane content, wt%			
RON	101	98	100
MON	90	88	89

Source: Reprinted with permission from Sheppard, C.M. et al., *Energy and Fuels*, 17, 1360, 2003. Copyright American Chemical Society.

TABLE 3.9

GC-FPD Analysis of Organo Sulfur Compounds in RFCC Naphtha

Boiling Point, °C	RFCC Naphtha	Sweetened RFCC Naphtha
67–68	1-Propanethiol	
84.2	Thiophene	Thiophene
98.4	1-Butanethiol	
112–115.4	2(3)-Methylthiophene	2(3)-Methylthiophene
126.6	1-Pentanethiol	
136–145	Dimethylthiophene	Dimethylthiophene
151	1-Hexanethiol	
168.7	Thiophenol	
176.2	1-Heptanethiol	1-Heptanethiol
194–195	Methylthiophenol	Methylthiophenol
199.1	1-Octanethiol	1-Octanethiol

Source: From Xia, D. et al., *Petroleum Science and Technology*, 20, 17, 2002. Reproduced by permission of Taylor & Francis Group, LLC, http://www.taylorandfrancis.com.

the GC-FPD technique (Xia et al., 2002). After the sweetening process, the RFCC naphtha did not contain thiols, such as 1-propanethiol, 1-butanethiol, 1-pentanethiol, and 1-hexanethiol, having a boiling point below 170°C but it contained 1-heptanethiol and 1-octanethiol, having a higher boiling point. The literature reported that the oxidation sweetening process has no effect on thiophene derivatives. The thiophene, methylthiophene, and dimethylthiophene having a boiling point below 150°C, were reported present in RFCC naphtha before and after the sweetening process. The sweetening problem of some fuels might be related to the use of residue as feedstocks and a higher content of aromatic compounds including thiophene derivatives. The olefin and heteroatom contents of different light heating fuels are shown in Table 3.10.

Different light heating fuels were reported to have a similar saturate content of 42.7–43 vol% and a similar olefin content of 10.49–10.54 vol%. The light heating fuel, having a boiling range of 178°C–361°C, was reported to contain 0.05 wt% of S and 49 ppm of N. Another light heating fuel, having a higher boiling range of 187°C–394°C, was reported to contain a higher S content of 0.13 wt% and a higher N content of 296 ppm. The literature reported that the presence of olefins formed during the catalytic cracking can affect the product stability. Olefins are easily oxidized and polymerized. Diolefins are more reactive and quickly form high molecular weight polymers leading to sludge formation (Sequeira, 1994). Gasoline, used as a fuel for internal combustion engines, was reported to form deposits called gums. In addition to hydrocarbons, gasoline has a small content of S, O, and a trace of N. The analysis of gums was reported to contain C, H, O,

TABLE 3.10

Olefin and Heteroatom Contents of Different Light Heating Fuels

Properties	Light Heating Fuel #1	Light Heating Fuel #2
Boiling range, °C	178–361	187–394
Saturates, vol%	43	42.7
Olefins, vol%	10.54	10.49
Sulfur, wt%	0.05	0.13
Nitrogen, ppm	49	296

Source: From Severin, D. and David, T., *Petroleum Science and Technology*, 19, 469, 2001. Reproduced by permission of Taylor & Francis Group, LLC, http://www.taylorandfrancis.com.

and N (Pereira and Pasa, 2005). The formation of gum precursors, related to the presence of olefins in catalytically cracked gasoline, was studied in the conversion of VGO and n-C16 using commercial FCC catalysts (Puente and Sedran, 2004). While catalytic cracking is used to produce better gasoline, the literature reported on the formation of polymeric compounds which can affect their storage stability and the engine performance.

3.1.3 HYDRODESULFURIZATION

The reaction of any petroleum fraction with hydrogen, in the presence of a catalyst, is known as hydroprocessing. The main operating variables are temperature, hydrogen partial pressure, and space velocity. Depending on the processing conditions, hydroprocessing is classified as hydrotreating (HT) or hydrocracking (HC). The chemical reactions involve removal of sulfur, nitrogen, oxygen, and metals as well as the partial saturation of unsaturated hydrocarbons. The S compounds, such as mercaptans, disulfides, sulfides, and thiophenes are converted to hydrocarbons and hydrogen sulfide. The N compounds are converted to hydrocarbons and ammonia. The O compounds are converted to hydrocarbons and water. Heavy metals present in feedstocks are usually removed during hydrogen processing (Speight, 1999).

S-compounds $\rightarrow$ RH and H_2S
N-compounds $\rightarrow$ RH and NH_3
O-compounds $\rightarrow$ RH and H_2O

The typical HT operation requires the reactor temperature of 270°C–340°C, the hydrogen partial pressure of 690–20,700 kPag, and the space velocity (LHSV) of 1.5–8.0. Most HT operations use cobalt–molybdate catalysts which are cobalt and molybdenum oxides on alumina (Gary and Handwerk, 2001). Increasing temperature and hydrogen partial pressure increases S and N removal. Increasing pressure also increases the hydrogen saturation and reduces coke formation. Increasing space velocity decreases conversion, hydrogen consumption, and coke formation (Gary and Handwerk, 2001).

HC is used to break down the heavy fractions of crude oils in the presence of catalyst and pressurized hydrogen gas. During HC, the condensed type molecules, such as naphthalene or anthracene, are converted into polynaphthenes and further crack to lower molecular weight 1-ring naphthenes and alkanes. One of the most important reactions in HC is the partial hydrogenation of polycyclic aromatics followed by rupture of the saturated rings to form substituted monocyclic aromatics. The side chain may split off to form iso-paraffins. During the HC of petroleum many chemical reactions can occur including ring-opening, cyclization, and isomerization (Speight, 1999). The typical HC operation requires the reactor temperature of 290°C–400°C and the reactor pressure of 1200–2000 psig. The HC catalysts are crystalline mixtures of silica–alumina containing the rare-earth metals such as platinum, palladium, tungsten, and nickel (Gary and Handwerk, 2001). The effect of hydroprocessing on Jobo crude oil is shown in Table 3.11.

Hydroprocessing is based on feeding the hydrogen and the feed into a reactor, filled with catalyst, at high temperature and pressure. The main chemical reactions are known as hydrodesulfurization (HDS), hydrodenitrogenation (HDN), hydrodeoxygenation (HDO), and hydrodemetallization (HDM). While the principal operating variables are temperature, hydrogen partial pressure, and space velocity, the reactor temperature is the primary means of conversion control. Hydrogen partial pressure is the most important parameter controlling aromatic saturation (Gary and Handwerk, 2001). HC is used to decrease the boiling range of the feed. The use of HC technology requires the use of expensive molecular hydrogen. The literature reported on the use of water–syngas provided in situ from the water–gas

TABLE 3.11
Hydroprocessing of Jobo Crude Oil

Jobo Crude Oil	Crude Oil Feed	Hydroprocessing Product
API gravity	8.5	22.7
Sulfur, wt%	4.0	0.8
Nickel, ppm	89	5
Vanadium, ppm	440	19
Residue, wt%	13.8	2.8

Source: From Gary, J.H. and Handwerk, G.E., *Petroleum Refining, Technology and Economics*, 4th ed., Marcel Dekker, Inc., New York, 2001. Reproduced by permission of Routledge/Taylor & Francis Group, LLC.

TABLE 3.12
Effect of Temperature on the Hydroprocessing of Heavy Maya Crude Oil

Heavy Maya Crude Oil Properties	Crude Oil Feed	Hydroprocessing at 380°C	Hydroprocessing at 440°C
API gravity	20.9	23.5	29.7
Sulfur, wt%	3.4	2.5	1.5
Nitrogen, wt%	0.37	0.34	0.28
Vanadium, ppm	299	201	91
Nickel, ppm	55	49	29
Asphaltenes, wt%	12.4	8.7	4.9

Source: Reprinted with permission from Ancheyta, J. et al., *Energy and Fuels*, 17, 1233, 2003. Copyright American Chemical Society.

shift reaction as an alternative hydrogen (Jian et al., 2005). The effect of temperature on the hydroprocessing of heavy Maya crude oil is shown in Table 3.12.

At the lower temperature of 380°C, the S content decreased from 3.4 to 2.5 wt% and the N content decreased from 0.37 to only 0.34 wt%. The V content decreased from 299 to 201 ppm, and Ni content decreased from 55 to only 49 ppm. The asphaltene content also decreased from 12.4 to 8.7 wt%. At the high temperature of 440°C, the S content further decreased to 1.5 wt% and the N content further decreased to 0.28 wt%. The V content further decreased to 91 ppm and the Ni content further decreased to 29 ppm. The asphaltene content also further decreased to 4.9 wt%.the S, N, Ni, V, and asphaltene contents were significantly reduced after hydroprocessing of heavy Maya crude oil; however, even after the hydroprocessing at the high temperature of 440°C, the crude oil contains some S, N, metals, and asphaltenes. The literature reported on the hydroprocessing of a North Gujarat crude oil, 30 wt% by weight with light gas oil, over a commercial HT catalyst containing 20 wt% of molybdenum oxide, 5 wt% of nickel oxide, traces of KOH, and about 75 wt% of aluminum oxide (Chaudhuri et al., 2001). The effect of hydroprocessing conditions on product yields and composition is shown in Table 3.13.

With an increase in the temperature, from 390°C to 450°C, and an increase in the hydrogen pressure, from 6.8 to 20.4 MPa, about 30%–60% of the atmospheric residue (365°C + cut) was reported to be converted to light distillates (Chaudhuri et al., 2001). The S content was reported to gradually decrease from 0.1 to 0.03 wt%, the N content gradually decreased from 200 to less than 1 ppm, the oxygen content gradually decreased from 54 to less than 1 ppm, and the metal content also gradually decreased from 130 ppm to none. With an increase in the temperature and the severity of HT, the residue yield gradually decreased from 40.1% to 32.3%. Mild HC conditions were required to produce middle distillates.

TABLE 3.13
Effect of Hydroprocessing Conditions on Product Yields and Composition

Temperature, °C	390	390	450
Pressure, MPa	6.8	6.8	20.4
Time, h	0.25	0.5	0.5
Sulfur, wt%	0.1	0.09	0.03
Nitrogen, wppm	200	170	0.1
Oxygen, wppm	54	40	0.1
Metals, wppm	130	41	0
Gases and naphtha, yield %	4.3	8.4	8.6
Kerosene, yield %	19.7	23.6	19.7
Diesel, yield %	35.8	34.6	35.3
Residue, yield %	40.1	33.3	32.3

Source: From Chaudhuri, U.R. et al., *Petroleum Science and Technology*, 19, 535, 2001. Reproduced by permission of Taylor & Francis Group, LLC, http://www.taylorandfrancis.com.

HT is applied to a wide range of feedstocks, ranging from naphtha to reduced crude, for the purpose of saturating olefins, aromatics and decreasing the sulfur and nitrogen contents without changing the boiling point of the feed. When HT is used specifically for sulfur removal, it is called hydrodesulfurization. The reaction variables are reactor temperature and pressure, space velocity, hydrogen consumption, nitrogen content, and hydrogen sulfide content of the gases. The reactor effluent is cooled before separating the oil from the hydrogen-rich gas. The oil is stripped of any remaining hydrogen sulfide and light ends in a stripper. The gas might be treated to remove hydrogen sulfide and recycled to the reactor (Gary and Handwerk, 2001).

According to the literature, the severity of the process increases with an increase in the boiling range of the feedstock and the presence of sulfoaromatic compounds. The literature also reported on installation of a hydrotreater hot separator which had severe foaming and gas carry-under problems (Turner et al., 1999). The principle of hot separation consists of separating a gas product mixture at 210°C–230°C which created the opportunity to increase the plant output and the heat utilization process (Lugovskoi et al., 2000). The effect of boiling range on the chemistry of S molecules found in middle distillates, produced from Astrakhan crude oil, is shown in Table 3.14.

The ease of sulfur removal was reported to vary and lower boiling compounds are desulfurized more easily than higher boiling ones. The difficulty in sulfur removal was reported to increase from paraffins to naphthenes and aromatics (Gary and Handwerk, 2001). With an increase in the boiling range of middle distillates, the total S content increased from 0.23 to 2.8 wt%; however, a decrease in mercaptans and disulfides with an increase in sulfides and thiophenes was

TABLE 3.14
Effect of Boiling Range on the Chemistry of S Molecules Found in Middle Distillates

Astrakhan Crude Oil, Sulfur Content	Middle Distillate, bp < 180°C	Middle Distillate, bp 180°C–350°C	Middle Distillate, bp > 350°C
Mercaptans, wt%	0.09	0.18	0
Sulfides, wt%	0.1	0.36	0.9
Disulfides, wt%	0.02	0	0
Thiophenes, wt%	0.02	0.67	1.9

Source: Modified from Irisova, K.N., Talisman, E.L., and Smirnov, V.K., *Chemistry and Technology of Fuels and Oils*, 39, 20, 2003.

reported (Irisova et al., 2003). During the hydroconversion process, the mercaptans, sulfides, and disulfides were reported as the most reactive while the thiophenes and their derivatives were less reactive. The literature reported on high boiling points of aromatic thiophenes, such as benzothiophene (bp 221°C), dibenzothiophene (bp 334°C), and different alkyldibenzothiophenes (bp 335°C–382°C) (Irisova et al., 2003). The effect of hydrotreatment on the S content of middle distillates, having different boiling ranges, is shown in Table 3.15.

After HT, the middle distillate having a higher boiling range was reported to have a higher sulfide and thiophene content leading to a higher S content. The literature reported that in the diesel cut, the main S compounds are sulfides and thiophenes, and the resistance of thiophenes to hydroconversion increases with an increase in the number of aromatic and naphthene rings in their structure (Irisova et al., 2003). According to the literature, some refinery units have difficulties to produce a low level of S fuels because additional hydrogen is required to hydrogenate the unsaturated and polyaromatic hydrocarbons. For a production of

TABLE 3.15
Effect of HT on S Content of Different Middle Distillates

Astrakhan Crude Oil, Sulfur Content	HT Middle Distillate, bp < 180°C	HT Middle Distillate, bp 180°C–350°C
Mercaptans, wt%	0	0
Sulfides, wt%	0.009	0.09
Disulfides, wt%	0	0
Thiophenes, wt%	0.001	0.22

Source: Modified from Irisova, K.N., Talisman, E.L., and Smirnov, V.K., *Chemistry and Technology of Fuels and Oils*, 39, 20, 2003.

diesel fuel having an S content below 350 ppm, a high hydrogen pressure and a high hydrogen/feedstock ratio is required (Irisova et al., 2003).

The literature reported on the technical and economical aspects of catalytic dehydrodesulfurization process. Depending on the hydrogen supply and the catalyst system, the hydroconversion of thiophenes was reported to follow different chemical reactions. The use of aluminum–cobalt–molybdenum catalysts or aluminum–nickel–molybdenum catalysts leads to different thiophenes' hydroconversion (Irisova et al., 2003). The HDS process was reported to be efficient in removing thiols, sulfides, and disulfides but less effective in removing aromatic thiophenes, such as benzothiophenes, dibenzothiophenes, and their alkyl derivatives (Hernandez-Maldonado et al., 2004). To meet an increased environmental demand for cleaner burning fuels, many refineries use severe HT to decrease their aromatic and S contents.

3.2 HYDROPROCESSING OF BASE STOCKS

While conventional refining uses temperature and solvents to separate the hydrocarbon molecules, nonconventional refining uses temperature and catalysts to break up, combine, or reshape the hydrocarbon molecules. The viscosity index (VI) describes the relationship of viscosity to temperature and higher VI base stocks tend to display less change in viscosity with temperature. To remove unstable and low VI components from mineral base stocks, conventional refining uses solvent extraction. The solvent extraction process selectively dissolves the aromatic compounds in an extract phase while leaving more paraffinic molecules in a raffinate phase. Naphthenes are distributed between the extract and the raffinate phases. By controlling the oil/solvent ratio and the temperature, the degree of the separation between the extract and raffinate phases can be controlled. The increasing demand for automotive engine lubricants to have a lower volatility and improved stability and cold flow properties, led to a growing interest in hydroconversion processes of vacuum distillates and heavy feedstocks. The feed is either distillate from the high vacuum unit, deasphalted oil from the propane deasphalting unit, or the raffinate from the extraction unit. The typical feedstocks to a lube oil hydrocracker are shown in Table 3.16.

The purpose of HC is to convert high boiling petroleum feedstocks to lower boiling products by cracking the hydrocarbons in the feed and hydrogenating the unsaturated materials. During HC, the condensed type molecules, such as naphthalene or anthracene, are converted into lower molecular weight hydrocarbons. The processing scheme for the production of hydrocracked base stocks from vacuum gas oil was reported. After HC of VGO, distillation separates light fuels and additional processing, such as extraction, dewaxing, and finishing, is needed to obtain hydrocracked base stocks (Prince, 1997). Hydrocracked base stocks, similar to solvent refined base stocks, require dewaxing to lower their pour points. According to the literature, the saturation of polyaromatic rings and naphthenic ring opening can increase the VI and decrease the pour point of hydrocarbons. Some 1-ring naphthenic molecules having alkyl side chains were

TABLE 3.16
Typical Feedstocks to Lube Oil Hydrocracker

Unrefined distillates	Solvent refined distillates (raffinates)
Deasphalted oils	Solvent refined deasphalted oils
Mixtures of the above	Slack wax and scale wax

Source: From Sequeira, A. Jr., *Lubricant Base Oil and Wax Processing*, Marcel Dekker, Inc., New York, 1994. Reproduced by permission of Routledge/Taylor & Francis Group, LLC.

reported to have a high VI of 130 and a pour point of $-10°C$ (Mang, 2001). The effect of hydroprocessing severity on VI and dewaxed oil yield of the deasphalted oil (DAO), from Middle East crude oil, is shown in Table 3.17.

With an increase in hydroprocessing severity, a decrease in dewaxed oil yield, viscosity and an increase in VI, from 74 to 103, is observed. Early literature reported on the hydrotreatment of mixtures of distillates and deasphalted oils leading to a VI of 120–125 (Beuther et al., 1964). To produce high VI base stocks, selected crude oils, a severe hydrogen refining process, high wax content feedstocks, or isomerization of wax need to be used (Sequeira, 1994). There are many different combinations of solvent extraction and HC used for the manufacture of base stocks. Instead of solvent extraction, HC can be used to reduce the aromatic content of the raffinate (Sequeira, 1994). The selective HC process, used to crack the wax molecules to light hydrocarbons, is known as isodewaxing. The relative cracking rates of paraffins indicated that the long chain molecules crack easily. The cracking rates of hydrocarbons were reported to significantly decrease with an increase in the degree of branching (Sequeira, 1994). The literature reported on

TABLE 3.17
Effect of Hydroprocessing Severity on VI and Dewaxed Oil Yield

Properties	Deasphalted Oil, DAO Feed	Hydroprocessing, Low Severity Product	Hydroprocessing, High Severity Product
Dewaxed oil yield, vol%		77.7	64.4
API gravity	19.8	24.9	28.6
Viscosity SUS at 210°F	231	149.9	75.5
VI	74	84	103
Dewaxed pour point, °C	54	−23	−26

Source: From Sequeira, A. Jr., *Lubricant Base Oil and Wax Processing*, Marcel Dekker, Inc., New York, 1994. Reproduced by permission of Routledge/Taylor & Francis Group, LLC.

TABLE 3.18

Lube Oil VI and Yields of Products Produced during
Isodewaxing of Slack Wax

Isodewaxing Process	Slack Wax Feed	Lube Oil Product	Product Yields, wt%
Gravity API	35.9	38	Fuel gas (4.5)
Viscosity, cSt at 100°C		6.28	Naphtha (13.6)
Viscosity, cSt at 50°C	6.8		Diesel (22.5)
Viscosity, cSt at 40°C	13.5	32.8	Lube oil (60.2)
VI	172	145	
Cloud point, °C		−11	
Pour point, °C		−15	

Source: From Sequeira, A. Jr., *Lubricant Base Oil and Wax Processing*, Marcel
Dekker, Inc., New York, 1994. Reproduced by permission of
Routledge/Taylor & Francis Group, LLC.

isodewaxing of slack wax leading to a lube oil base stock, having a very high VI
(VHVI), above 140. The VI of lube oil and yields of different products, obtained
during isodewaxing of slack wax, are shown in Table 3.18.

The isodewaxing of slack wax, having a VI of 172, leads to a lube oil base
stock, having a VI of 145. The unconverted wax needs to be removed, usually by
solvent dewaxing, to achieve the pour point of −15°C or lower. Another variant
of hydroprocessing of base stocks is the process of slack wax hydroisomerization
introduced to meet the ever increasing demand for high VI and low volatility base
stocks. The term hydroisomerization indicates that the isomerization process
predominates over HC. Both reactions are taking place and, depending on the
feed and processing conditions, one will predominate. The literature reported that
the slack wax feed can vary from high paraffin medium and heavy slack waxes to
soft hydrocarbon waxes (Cody et al., 1992).

Over the past two decades, the HC and hydroisomerization of long chain
paraffins over bifunctional catalysts have been studied. The bifunctional catalysts
are characterized by the presence of acidic sites which provide an isomerization/
cracking function and metal sites which provide a hydrogenation/dehydrogena-
tion function (Calemma et al., 2000). Catalysts containing a noble metal, such as
Pt or Pd, were reported to show a higher selectivity for hydroisomerization.
A high surface area catalyst useful in the wax hydroisomerization process was
reported to contain 0.01–5 wt% of Pt, 0.1–10.5 of F, and over 50% of alumina
used as a refractory metal oxide support (Cody et al., 1992). More recently, high
selectivity for the isomerization of long chain paraffins and wax feeds was
reported for the catalysts composed of silicoaluminophosphate molecular sieve
containing Pt (Calemma et al., 2000). The effect of hydroprocessing on VI of
different lube oil base stocks, having a similar viscosity, is shown in Table 3.19.

TABLE 3.19
Effect of Hydroprocessing on VI of Different Lube Oil Base Stocks

Processing	Solvent Refining	Hydrocracking #1	Hydrocracking #2	Hydroisomerization
Viscosity at 40°C, cSt	—	35.4	—	—
Viscosity at 100°C, cSt	5.2	—	5.6	5.0
VI	98	97	125	146
Pour point, °C	−15	−18	−15	−18

Source: From Pillon, L.Z., Asselin, A.E., and MacAlpine, G.A., *Lubricating Oil Having an Average Ring Number of Less Than 1.5 Per Mole Containing Succinic Anhydride Amine Rust Inhibitor*, US Patent 5,225,094, 1993; Prince, R.J., *Chemistry and Technology of Lubricants*, 2nd ed., Mortier, R.M. and Orszulik, S.T., Eds., Blackie Academic & Professional, London, 1997.

The literature reported on the properties of solvent refined versus hydroprocessed base stocks. The typical solvent refined base stock, having a viscosity of 5.2 cSt at 100°C and dewaxed to a pour point of −15°C, has a VI of 98. The patent literature reported on the properties of hydrocracked base stock, having a VI of only 97 and solvent dewaxed to a pour point of −18°C (Pillon et al., 1993). Depending on the feed, another hydrocracked base stock was reported to have a high VI of 125 and dewaxed to a pour point of −15°C. The slack wax isomerate base stock was reported to have an extra high VI (XHVI) of 146 and solvent dewaxed to −18°C (Prince, 1997).

In recent years, solvent extraction and solvent dewaxing of lube oil distillates has been replaced by catalytic processing to meet the ever increasing demand for higher quality mineral base stocks. There are two types of dewaxing processes in use today. The wax is removed by solvent dewaxing which is a crystallization–filtration process. The process uses refrigeration to crystallize the wax and solvent to dilute the oil to allow for a rapid filtration and wax separation from the oil. In catalytic dewaxing, also known as isodewaxing, the selective HC process is used to crack the wax molecules to light hydrocarbons and no wax is produced. The effect of solvent and catalytic dewaxing on VI and composition of lube oil base stock is shown in Table 3.20.

In conventional solvent refining and catalytic dewaxing, the n-paraffin content of waxy raffinates is removed and a decrease in viscosity and VI is reported. After solvent dewaxing, the VI decreases from 116 to 100. After catalytic dewaxing, the VI decreases further to 89 and a lower content of iso-paraffins is reported. During catalytic dewaxing, the cracking of some iso-paraffins decreases the VI of dewaxed base stock. No wax is produced from catalytic dewaxing unless the wax is removed by solvent dewaxing prior to catalytic dewaxing. The use of hydroprocessing technology leads to lube oil base stocks which, depending on the feed and processing conditions, have a different composition and VI. Based on their VI,

TABLE 3.20
Effect of Solvent and Catalytic Dewaxing on VI and Composition

Dewaxing Process Properties	Waxy Raffinate, 100N	Solvent Dewaxing, 100N	Catalytic Dewaxing, 100N
Viscosity SUS at 100°F	93	106	117
VI	116	100	89
Pour point, °C	27	−18	−15
Saturate fraction			
n-Paraffins	21	—	—
Iso-paraffins	13	24	21
Naphthenes	33	37	39
Other saturates	12	13	13
Aromatic fraction			
Alkylbenzenes	6	8	8
Other aromatics	15	18	19

Source: From Taylor, R.J., McCormack, A.J., and Nero, V.P., *Preprints Division of Petroleum Chemistry*, ACS, 37, 1337, 1992. Reproduced by permission of Chevron Corporation.

aromatic and S contents, lube oil base stocks are classified into four API groups. For the same viscosity base stocks, an increase in the VI leads to lower volatility. The volatility of base stocks is measured as the percent loss of material, usually at 250°C, in the standard Noack volatility test (ASTM D 5800). Base stocks having VI above 140 are known as extra high viscosity index (XHVI) base stocks. The VI, composition, and volatility of different nonconventional base stocks, having a similar viscosity of 4 cSt at 100°C, are shown in Table 3.21.

There are different methods to produce nonconventional base stocks. According to literature, some refineries produce nonconventional base stocks by HC of fuel hydrocracker bottoms, HC plus isodewaxing, or slack wax isomerization (Kline & Company, 2002). The patent literature reported on the method for producing high quality lube oil feedstocks from unconverted oil from a fuel hydrocracker operating in the recycle mode. After HC of VGO, a series of fractional distillations separates light hydrocarbons and unconverted oil which is fed to another vacuum distillation unit to produce feedstocks of high quality lube oil base stocks (Kwon et al., 1997). The early patent literature indicates that different processing conditions and catalysts may be used to hydroisomerize the slack wax (Cody and Brown, 1990). The more recent literature reported that lubricating oils can be produced by hydroisomerization of medium and heavy slack waxes, derived from solvent refined oils, over Pt supported on silicone–alumina oxides in a continuous down flow trickle bed (Calemma et al., 1999). The effect of viscosity on the volatility of different nonconventional base stocks, having a similar viscosity of 5–6 cSt at 100°C, is shown in Table 3.22.

TABLE 3.21
VI, Composition, and Volatility of Different Nonconventional Base Stocks

Base Stock Properties	HVI, Group I	HVI, Group II	VHVI, BP HC4	VHVI, Texaco HVI	VHVI, Shell XHVI
Kin. viscosity at 100°C, cSt	3.81	4.01	3.99	4.44	4
VI	92	96	128	127	144
Pour point, °C	−18	−12	−27	−15	−18
Saturates, wt%	81.8	98.9	92.3	93.7	100
Aromatics, wt%	18.2	1.1	7.7	6.3	0
S, wt%	0.46	0.003	0.08	0.01	0.002
N, ppm	17	5	2	2	2
Volatility, % off	32	31	15.8	17	16.4

Source: From Phillipps, R.A., *Synthetic Lubricants and High-Performance Functional Fluids*, Rudnick, L.R. and Shubkin, R.L., Eds., 2nd ed., Marcel Dekker, Inc., New York, 1999. Reproduced by permission of Routledge/Taylor & Francis Group, LLC.

TABLE 3.22
Effect of Viscosity on Volatility of Different Nonconventional Base Stocks

Base Stock Properties	HVI, Group I	HVI, Group II	VHVI, BP HC6	XHVI, Shell	XHVI, Exxsyn
Kin. viscosity at 100°C, cSt	5.3	5.66	5.59	5.14	5.8
VI	96	97	130	146	142
Pour point, °C	−9	−12	−12	−18	−21
Saturates, wt%	65.1	100	98	100	100
Aromatics, wt%	34.9	0	2	0	0
S, wt%	0.4	0.006	0.001	0.002	0.001
N, ppm	53	5	5	2	0
Volatility, % off	16	17.5	8.9	11	9

Source: From Phillipps, R.A., *Synthetic Lubricants and High-Performance Functional Fluids*, Rudnick, L.R. and Shubkin, R.L., Eds., 2nd ed., Marcel Dekker, Inc., New York, 1999. Reproduced by permission of Routledge/Taylor & Francis Group, LLC.

HVI base stocks, having a conventional VI of 96–97 and classified as group I, can have a varying aromatic content, from very low to as high as 35 wt%, and their S and N contents can vary. VHVI base stock, having a VI of 130, was reported to have 2 wt% of aromatics and very low S and N contents. XHVI base stocks, having a VI of 142–146, contain practically no aromatics, S, and N and a significant decrease in their volatility is observed. XHVI base stock, having a higher VI of 146, was found to contain no aromatics and only 20 ppm of S and 2 ppm of N. XHVI 5 base stock, having a lower viscosity, was found to have a lower volatility of 11% off. Exxsyn 6 base stock, having a higher kinematic viscosity of 5.8 cSt at 100°C and a lower VI of 142, was found to contain no aromatics and have a lower S content of 10 ppm and no N. With an increase in viscosity of Exxsyn 6 base stock, a further decrease in volatility to 9% off is observed. The literature reported on properties and potential product application of lube oil base stocks produced by hydroisomerization of slack wax derived as a by-product of solvent dewaxing of lubricating oils (Patel et al., 1997). There is a continuous interest in high VI base stocks obtained by catalytic hydroprocessing (Kazakova et al., 1999). Lower oil consumption demands lower volatility base stocks. Many engine oils require the use of base stocks having a high VI to meet the volatility and the low temperature specifications.

3.3 THERMAL CRACKING

The literature reported on thermal cracking of pure n-hexadecane at 330°C–375°C in liquid and gas phases (Wu et al., 1996). Since the thermal decomposition of crude oil is significant above 350°C, to separate the heavier portion of crude oil the distillation is carried under vacuum. The vacuum still is used to prevent thermal cracking, discoloration of the product, and equipment fouling due to coke formation (Gary and Handwerk, 2001). Under severe thermal conditions, all petroleum products undergo thermal cracking and have a tendency to form carbon residue. The literature reported on kinetics and mechanisms of petroleum hydrocarbon thermal cracking (Xiao et al., 1997). There are three general descriptions of petroleum cracking which are called catalytic cracking, HC, and thermal cracking, also known as pyrolysis. The literature reported on gas oil, having a boiling range of 350°C–390°C and a pour point of 28°C, which was thermally cracked in an autoclave in the presence of polymeric surfactant (Nabih, 2002). The effect of temperature and time on the pour point of the gas oil fraction is shown in Table 3.23.

At 375°C, no significant effect of temperature on the pour point of gas oil was reported. After 20 min the pour point decreased only to 27°C. However, after 20 min at 425°C it decreased to 19°C and after 20 min at 475°C it decreased to 20°C. With an increase in temperature, above 375°C, a significant increase in thermal cracking is observed. The recent literature reported that paraffinic oil was stable at 360°C for about 300 h and then thermal breakdown and cracking was observed. Degradation led to the formation of gases, liquids, and 5 wt% of coke. The distillation of liquids indicated the composition varied from light naphtha to heavy material (Oyekunie and Susu, 2005).

TABLE 3.23
Effect of Temperature on the Pour Point of Gas Oil Fraction

Temperature, Time (min)	375°C, Pour Point, °C	425°C, Pour Point, °C	475°C, Pour Point, °C
0	28	28	28
5	28	26	20
10	28	23	16
15	28	21	13
20	27	19	10

Source: From Nabih, H.I., *Petroleum Science and Technology*, 20, 77, 2002. Reproduced by permission of Taylor & Francis Group, LLC, http://www.taylorandfrancis.com.

Long paraffinic side chains attached to the aromatic rings are the primary cause of high pour points of paraffinic base residue. Visbreaking is considered to be a mild thermal cracking process used to reduce the viscosity and pour point of residue by breaking off and cracking the long side chains. The heavy nonvaporized portion of crude oil is broken down into smaller components to be used in gasolines and heating oils. Atmospheric residue is also known as long residue and vacuum residue is known as short residue (Gary and Handwerk, 2001). Thermal cracking (visbreaker) uses heat to break down the long, heavy molecules remaining at the bottom of the distillation towers. The visbreaking products from different residues are shown in Table 3.24.

The gasoline from a visbreaker can go to a hydrotreater, catalytic cracker, or even back to a crude oil unit. Visbreaker gas oil goes to a catalytic cracker. The bunker oil, also called the residual fuel oil, is the residue left after naphtha–gasoline

TABLE 3.24
Visbreaking Products from Different Residues

Residue Feed, Visbreaking Products	Kuwait Long Residue Yields, %	Agha-Jari Short Residue Yields, %
Gas	2.5	2.4
Naphtha	5.9	4.6
Gas oil	13.5	14.5
Tar	78.1	78.5

Source: Modified From Gary, J.H. and Handwerk, G.E., *Petroleum Refining, Technology and Economics*, 4th ed., Marcel Dekker, Inc., New York, 2001.

and other fuel oils have been removed (Speight, 2006). The literature reported that when a residue is obtained from crude oil and the thermal decomposition has commenced, the product is known as pitch. The naphthenic acids, found in higher molecular weight nonvolatile residue, are claimed to be products of the air blowing of the residue (Speight, 1999). The bottom from a visbreaker, called pitch, goes to bunker and when mixed with other light oils can be used as ship fuel. The literature reported on the optimization of the visbreaking unit and other processing requirements (Musienko et al., 2000).

Coking is a severe thermal cracking process and there are several coking processes such as delayed coking, fluid coking, flexicoking, and their variations. Delayed coking is the oldest and most widely used process which converts petroleum residue into gas products, liquid, and coke. The literature reported on the delayed coking process. The heavy feedstock or residue is heated to the cracking/coking temperature, above 350°C, and the hot liquid is charged, usually up flow, to a coke drum where the coking reactions occur. The coke drums, arranged in pairs, are used alternatively to provide the residence time for the coking reactions to take place and to accumulate the coke. Liquid and gaseous products pass to a fractionator for separation and coke deposits in the drum (Speight, 1999). The delayed coker products, from Alaska North Slope crude oil, are shown in Table 3.25.

Coking units convert heavy feedstocks into a solid coke and lower boiling hydrocarbon products which can be converted into higher value fuels or suitable for use by other refinery units. The literature reported that the coker naphthas have boiling ranges up to 220°C, are olefinic, and must be upgraded by hydroprocessing to saturate olefins and remove sulfur and nitrogen. Middle distillates, boiling in a range of 220°C–360°C, also require a hydrogen treatment to improve storage stability, remove sulfur, and decrease nitrogen. The coker gas oil fraction, boiling up to 510°C, is usually low in metals and may be used as the feedstock for FCC (Speight, 2006).

TABLE 3.25
Delayed Coking Products

Delayed Coking Products	Vol%	API	Sulfur, wt%
Gas			5.9
Light naphtha	7.6	65	0.4
Heavy naphtha	14.1	50.1	0.7
Light coker gas oil (LCGO)	37.3	30	0.9
Heavy coker gas oil (HCGO)	20.1	14.3	2
Coke			2.6

Source: From Gary, J.H. and Handwerk, G.E., *Petroleum Refining, Technology and Economics*, 4th ed., Marcel Dekker, Inc., New York, 2001. Reproduced by permission of Routledge/Taylor & Francis Group, LLC.

The naphtha or gasoline fraction may be split into light and heavy cuts. After HT for sulfur removal and olefin saturation, the light cut can be isomerized to improve octane or blended directly into finished gasoline. The heavy cut is hydrotreated and reformed. The gas oil fraction is usually split into light coker gas oil (LCGO) and heavy coker gas oil (HCGO) before further processing. The LCGO can be hydrotreated and fed to an FCC unit (Gary and Handwerk, 2001). The HCGO was reported to be used as heavy fuel or sent to the vacuum distillation unit. An oversized fractionator can be used to maximize the diesel product and minimize the HCGO to an FCC unit (Ellis and Paul, 2000). The coking process is also used to produce a high quality coke from heavy catalytic gas oils and decanted oils from an FCC unit (Gary and Handwerk, 2001). Different feeds, processing options, and blends are used to manufacture various petroleum products. FCC naphtha makes up 30%–40% gasoline in a typical refinery pool (Lavanyja et al., 2002). The source of aromatics and olefins in gasoline is shown in Table 3.26.

Processing of crude oils involves many other processes including the processing of petrochemical feedstocks and production of aromatic solvents. Catalytic reforming uses heat and pressure, along with a special platinum catalyst in the presence of hydrogen, to change the molecular structure of naphtha. Light fractions produced during the distillation of crude oil, such as naphtha, are converted into higher octane gasoline components. Through subsequent separation, distillation, and extraction of the reformed product, aromatic solvents such as benzene, toluene, and xylene are produced (Gary and Handwerk, 2001). Alkylation is a polymerization process uniting olefins and iso-paraffins, such as butylene and isobutane, to produce a high octane blend agent for gasoline. Some of the unsaturates are produced by the FCC unit but most are produced by steam cracking or low molecular weight polymerization of low molecular weight

---

TABLE 3.26

Source of Aromatics and Olefins in Gasoline

Blend Stock	Pool, %	Aromatics, %	Olefins, %
LSR	3.1	10	2
FCC naphtha	38	30	29
Lt. HC naphtha	2.4	3	0
Lt. coker naphtha	0.7	5	35
Reformate	27.2	63	1
Alkylate	12.3	0.4	0.5
Isomerate	3.7	1	0
Polymer	0.4	0.5	96
n-Butane	3.1	0	2.6

Source: From Gary, J.H. and Handwerk, G.E., *Petroleum Refining, Technology and Economics*, 4th ed., Marcel Dekker, Inc., New York, 2001. Reproduced by permission of Routledge/Taylor & Francis Group, LLC.

TABLE 3.27

Effect of Different Feedstocks on Desulfurization of Diesel

Astrakhan Crude Oil Composition	Mixed Diesel Cuts	Diesel Cut and Visbreaking Naphtha
Feed		
Olefins, wt%	1.2–7	8.4–17.5
Sulfur, wt%	0.9–1.2	0.9–1.2
After HDS process		
Sulfur, wt%	0.04–0.08	0.05–0.1
S conversion, %	94–97	92–96

Source: Modified From Irisova, K.N., Talisman, E.L., and Smirnov, V.K., *Chemistry and Technology of Fuels and Oils*, 39, 20, 2003.

components. The technology and economics of petroleum refining, including the environmental aspects of refinery fuels and processing of feedstocks for petrochemical manufacture, were published (Gary and Handwerk, 2001). The effect of different feedstocks on the desulfurization (HDS) of diesel, under the same processing conditions, is shown in Table 3.27.

According to the literature, deep desulfurization of diesel blending streams has gained considerable importance due to strict environmental regulations which limit the S content to very low levels. Using the same reactors and processing conditions, the literature reported on the higher conversion of S molecules in feedstocks containing straight-run diesel cuts (Irisova et al., 2003). The literature also reported on HDS of straight-run gas oil, coker gas oil, and their blends. Despite a higher sulfur content, straight-run gas oil was easier to desulfurize to below 350 ppm and at lower temperature when compared to coker gas oil. More severe operating conditions, such as higher temperature and lower LHSV, were required to desulfurize coker gas oil. The difference in S and N compounds was reported to affect the HDS process (Al-Barood et al., 2005). The refining processes vary from refinery to refinery, depending upon the properties of the crude oil being processed and the specific petroleum products being produced. The use of asphaltic and heavy crude oils affects the processing options and the processing equipment required in modern refinery.

3.4 BITUMEN UPGRADING

Conventional refining is based on pumping of crude oil trapped in porous rocks and their distillation. Oil sands, which contain sand, water, clays, and bitumen, are another source of crude oils. Bitumen, from oil sands, can be recovered by mining and water extraction to separate from sand and clay, or by steam injection of underground wells and collecting bitumen released by heat. The term "heavy oil" is used to describe bitumen in tar sand from which the heavy bituminous material is recovered by a mining operation. Bitumen is a thick crude oil and includes a variety of dark to black materials of semisolid, viscous to brittle character that can contain a high mineral impurity, above 50 wt% (Speight, 1999). The literature reported that with a growing need for titanium and zirconium products, the pilot plant was built to recover titanium and zirconium from oil sands (*ACCN Canadian Chemical News*, 2004). Syncrude Canada Ltd. operates a surface mining oil sands plant and produces synthetic crude oil from the extracted bitumen. Heavy oils and bitumens can be upgraded to produce products which resemble conventional crude oil and are known as "synthetic crude oil" or "syncrude." Crude oils which have not been cracked or subjected to other treatments causing significant chemical changes are known as virgin stocks. Synthetic crude, having a wide boiling range, is a product of catalytic cracking, HC, or coking. The typical processing steps of producing synthetic crude from bitumen are shown in Table 3.28.

Model water-in-hydrocarbon emulsions were studied to assess the effect of solids on the emulsion stability of bitumen. Solids adsorbed at the interface prevent bridging between the water droplets while the particles in the continuous phase increase the emulsion stability by preventing the close contact between the water droplets (Sztukowski and Yarranton, 2004). The solids were reported to act as emulsifying agents but only in the presence of surfactants.

In the presence of oil-soluble surfactant, both oil-wet and water-wet solids were capable of stabilizing the water-in-oil emulsions of bitumen. In the presence of water-soluble surfactant, only water-wet solids were found to stabilize the oil-in-water emulsions (Yaghi et al., 2001). The flocculation of aqueous bitumen emulsion during transportation and processing mostly depends on the resin/asphaltene ratio of the bitumen. It was reported that stable bitumen dispersions are obtained when values of resin/asphaltene ratios are high (Jada et al., 2001). Bitumen, in the form of a bitumen-in-water emulsion, can be transported using a pipeline and recovered by lowering pH.

TABLE 3.28
Processing of Synthetic Crude from Bitumen

Steps	Processing
Mining	Mining shovels dig oil sand ore and load it onto huge trucks
Ore preparation	Trucks take oil sand ore to crushers where it is broken down
Extraction	Warm water is added to the oil sand and mixed to form slurry
	Dirty bitumen, water, and sand are separated
Froth treatment	Froth settlers use countercurrent decantation process
	Bitumen is washed with a hydrocarbon solvent
Upgrading	Catalytic and thermal cracking of heavy bitumen molecules
	Sulfur and nitrogen are removed from synthetic crude

Froth, which is a product of water extraction of bitumen from oil sands, has a high viscosity and cannot be transported using a conventional pipeline. There are different technologies for transporting froth; however, it is usually emulsified which decreases its viscosity. Froth can be emulsified using NaOH and the concentration of NaOH, salt concentration, temperature, and mechanical energy was reported to affect the emulsion stability and viscosity. The concentration of sodium hydroxide, temperature, and mechanical energy input was reported to affect the formation of an emulsion and its viscosity. The literature reported on the effect of water pH on the stability of a bitumen-in-water emulsion (Xu et al., 2001). The recent literature reported on high maintenance costs and production losses in oil sands mining from material degradation caused by wear and corrosion. The wear-resistant coatings and overlays, made of polymers, metal alloys, ceramics, and composites, are being developed and tested. The properties and composition of different bitumen samples are shown in Table 3.29.

The definition of heavy oil is usually based on the API gravity of less than 20. Very heavy crude oils and tar sand bitumens have API of less than 10. With a decrease in API of different bitumen oils, from 14.4 to 10.3, their S content was reported to vary from as low as 0.6 to 4.4 wt%, their N was reported to vary from 0.5 to 1 wt%, and their metal content was reported to vary. The Ni content varied in a range of 53–120 ppm and V in a range of 25–108 ppm. With a decrease in API of different bitumen oils, their carbon residue, measured as Ramsbottom carbon residue (RCR), was reported to vary in a range of 3.5–21.6 wt%.

The asphaltene precipitation is known to affect the crude oil recovery leading to well bore plugging and the sediment formation. The literature reported that heavy crude oils, residue, and bitumens are characterized by a large asphaltene content which are described as condensed polyaromatic rings containing a high heteroatom content and having a high boiling point above 500°C (Trasobares et al., 1999). The literature reported on the colloidal nature of Athabasca

---

TABLE 3.29
Properties and Composition of Different Bitumen Oils

Properties	Bitumen # 1	Bitumen # 2	Bitumen # 3
Gravity API	14.4	11.1	10.3
Sulfur, wt%	0.6	4.4	0.8
Nitrogen, wt%	1	0.5	1
Nickel, ppm	120	53	98
Vanadium, ppm	25	108	25
Carbon residue, wt%	3.5	21.	12.5

Source: From Speight, J.G., *The Chemistry and Technology of Petroleum*, 3rd
          ed., Marcel Dekker, Inc., New York, 1999. Reproduced by permission
          of Routledge/Taylor & Francis Group, LLC.

TABLE 3.30
Other Compounds Coprecipitating with Athabasca C5
Insolubles

Other Compounds	Chemistry	Number of Species
Hydrocarbons	Terpanes	
	Hopanes	
	Steroids	
Sulfur molecules	Sulfoxides	>40
Nitrogen molecules	Carbazoles	
Oxygen molecules	Ketones	>50
	Alcohols	>10
	Carboxylic acids	>90
Metals	Vanadyl porphyrins	

Source: Reprinted with permission from Strausz, O.P., Peng, P., and Murgich, J., *Energy and Fuels*, 16, 809, 2002. Copyright American Chemical Society.

asphaltenes indicating that it is affected by the selected paraffinic solvent. Precipitation with n-pentane (C5) solvent increases the asphaltene content by 15–98 wt% which was reported to be caused by resins and low molecular weight asphaltene components (Strausz et al., 2002). Acetone extraction, successive precipitations, or gel permeation chromatography (GPC) can be used to separate asphaltenes from other components. The chemistry of other compounds, coprecipitating with Athabasca C5 insolubles, is shown in Table 3.30.

While the color of Athabasca C5 insolubles was brown, the other compounds were reddish-brown materials and contained sulfur, nitrogen, and a wide range of oxygen containing compounds (Strausz et al., 2002). The literature reported on the measurement of Athabasca and Cold Lake asphaltenes, in the presence of inorganic solids, using ultraviolet at wavelengths of 288 and 800 nm (Alboudwarej et al., 2004). The bitumen extraction and upgrading processes pose a unique combination of wear and corrosive conditions. Particles of quartz are the main abrasive media in the oil sands. The main corrosive species were reported to be chloride compounds in the oil sand feeds and dissolved oxygen in water (Fisher, 2006).

The Syncrude upgrading process was reviewed in the literature. Refining of bitumen involves upgrading to remove coke, fractionation to remove naphtha, kerosene, and gas oil and HT to remove sulfur (Yui and Chung, 2004). The synthetic crude oils are refined by the usual refinery system and used as a refinery feedstock to produce different petroleum products. MCR is affected by the boiling range of molecules and thus can vary depending on the processing temperature. Cracking of molecules at lower temperatures might lead to some changes in MCR due to a decrease in their molecular weight and their boiling

TABLE 3.31
Properties and Composition of Oil Sand Derived
Heavy Gas Oil

Oil Sand Derived Heavy Gas Oil	Properties and Composition
Density at 20°C, g/cm^3	0.9859
Boiling range, °C	210–655
Sulfur, wt%	4.1
Nitrogen, wt%	0.39
Asphaltenes, wt%	1.55
Carbon residue (MCR), wt%	1.98

Source: Reprinted with permission from Bej, S.K., Dalai, A.K., and Adjaye,
J., *Energy and Fuels*, 15, 1103, 2001. Copyright American
Chemical Society.

point. The hydrotreated heavy gas oil, produced from bitumen, was reported to be used mainly as synthetic crude for blending with conventional crude oil (Bej et al., 2001). The properties and composition of the oil sand derived heavy gas oil are shown in Table 3.31.

The liquid products obtained by processing of bitumen in a fluid coker or LC-Fining hydrocracker were reported to contain high contents of S and N and required HT over Ni–Mo alumina catalysts at high temperatures and pressures (Bej et al., 2001). During the hydrotreatment of heavy gas oil, a change in the boiling range of the feed and a mild HC process taking place was reported. The effect of different HT conditions was studied and during removal of S and N, some reduction in MCR was observed (Bej et al., 2001). Processing of bitumen, having a higher viscosity, higher aromatic, heteroatom, and metal contents and higher carbon residue, is more difficult and conventional catalysts are reported less effective. The commercial catalytic HC process removed the sulfides only from the front-cut fractions and a relatively small decrease in the thiophenic sulfur was observed. The literature reported on the effect of different support on the activity of sulfided molybdenum catalysts used to desulfurize dibenzothiophene and the use of mixed oxides (Zhao et al., 2001). While the hydrocarbon types were reported similar in conventional distillates and distillates from bitumen, the bitumen-derived distillates have higher boiling points. Feeds from oil sands bitumen were reported to affect the refining of petroleum products and the gas oil conversion.

3.5 CATALYST ACTIVITY

Deactivation of catalysts is the main concern during hydroprocessing of petroleum fractions and the profitability of the process depends on the lifetime of the catalyst. Loss of active sites was reported to be responsible for catalyst deactivation mainly due to active site poisoning by deposits of coke and metals. Performance evaluation of hydroprocessing catalysts and a review of experimental techniques were published (Bej, 2002). The recent literature reported on the characterization of sulfated nickel zeolite catalyst and its activity by studying the cumene conversion at the temperature of 300°C–425°C (Saad and Mikhail, 2005). A small quantity of dienes in FCC gasoline was reported to degrade its quality and make a further processing difficult. The literature reported on the selective hydrogenation of dienes in FCC gasoline using alumina supported nickel catalysts (Yi et al., 2005). Olefins are known to polymerize and form sediments.

It is important that the metal content of desalted crude oils is low. The accumulation of nickel on the zeolite catalysts, used to crack vacuum gas oils, was reported to reduce the volume of gasoline and diesel fractions (Ali, 2001). Nickel interferes with upgrading processes severely deactivating many heterogeneous catalysts and catalyzing undesirable side reactions (Reynolds, 2001). The antimony containing deactivators were reported to decrease the poisoning effect of nickel on the catalyst (Baranova et al., 2002). The literature reported that the presence of Fe in feed can lead to reaction with hydrogen sulfide during hydroprocessing and the formation of FeS adsorbing onto catalyst leads to its deactivation (Liu et al., 2004). For light fractions such as naphtha and middle distillates, catalyst deactivation is minimal but for heavy oils and residues, catalyst deactivation can be severe leading to short periods of operation (Ancheyta et al., 2002). The effect of heavy Maya crude oil on deactivation of commercial Mo–Ni catalyst was studied and the carbon content of catalyst was used to measure the coke-like sediment formation on spent catalyst (Ancheyta et al., 2002). The analysis of fresh and spent commercial Mo–Ni catalyst, used for hydroprocessing of heavy Maya crude oil, is shown in Table 3.32.

After 490 h time on stream at 400°C, the C content of spent catalyst increased, from 0 to 18.3 wt%, and the metal deposition was about 10 wt% based on the

TABLE 3.32
Composition of Fresh and Spent Mo–Ni Hydrotreating Catalyst

Composition	Fresh Catalyst	Spent Catalyst
Molybdenum, wt%	10.66	5.56
Nickel, wt%	2.88	2.64
Sodium, wt%	0.04	0.13
Carbon, wt%	0	18.3
Sulfur, wt%	0	8.56
Vanadium, wt%	0	4.41
Iron, wt%	0	0.08

Source: Reprinted with permission from Ancheyta, J. et al., *Energy and Fuels*, 16, 1438, 2002. Copyright American Chemical Society.

fresh catalyst. According to the literature, the coke formation shown as carbon content is mainly due to the presence of asphaltenes. The asphaltenes which are polynuclear aromatic systems containing heteroatoms and metals are too big to penetrate into the catalyst and deposit on the surface causing deactivation (Ancheyta et al., 2002). After HT of heavy Maya crude oil, using different Ni–Mo catalysts, the asphaltenes were precipitated and their heteroatom content was analyzed. With an increase in the temperature of the HT process, the nitrogen and metal content of asphaltenes were reported to increase while the sulfur content was reported to decrease (Ancheyta et al., 2004). The literature reported on VGO, derived from Escravos crude oil, having a high basic nitrogen content and causing catalyst poisoning in an FCC unit and having a negative effect on yields (Bhaskar et al., 2003). The properties and N content of VGO, from Escravos crude oil, are shown in Table 3.33.

Hydrotreatment of feedstocks for use in an FCC unit is required to reduce the sulfur and nitrogen contents of fuel products. The bottom product from mild HC of HVGO was reported to be a good feedstock for use in an FCC unit because of their low aromatic, sulfur, and nitrogen contents leading to better quality fuels (Bhaskar et al., 2002). According to the literature, the ratio of basic type nitrogen compounds (BN) to total nitrogen (TN) of crude oils is approximately constant, in the range of $0.3 +/- 0.05$, irrespective of the source of crudes (Speight, 1999). The trend in recent years toward cutting deeper into the crude to obtain stocks for catalytic cracking can increase the presence of the N-containing molecules which are mostly concentrated in the higher boiling fractions (Speight, 1999). Their chemistry might also vary and can lead to an increase in the basic type N compounds. The literature reported on heavy Russian crude oil, from the Van-Eganskoe field, containing 0.33 wt% of N which was reported to have a

TABLE 3.33
Properties and N Content of VGO from Escravos Crude Oil

VGO from Escravos Crude Oil	Properties and Composition
API gravity	22.4
Kin. viscosity at 100°C, cSt	8.06
Pour point, °C	39
Saturates, wt%	58.1
Aromatics, wt%	41.9
S, wt%	0.29
N, wt%	0.17

Source: From Bhaskar, M., Valavarasu, G., and Balaraman, K.S., *Petroleum Science and Technology*, 21, 1439, 2003. Reproduced by permission of Taylor & Francis Group, LLC, http://www.taylorandfrancis.com.

significant negative effect on catalytic refining and the quality of fuels and lube oil base stocks (Kovalenko et al., 2001). The chemistry and the relative content of nitrogen molecules identified in heavy Russian crude oils, from the Van-Eganskoe field, are shown in Table 3.34.

The crude oil was separated into different fractions, using silica gel, which were desorbed using different polarity solvents. The chemistry and content of different nitrogen molecules were analyzed by infrared (IR) spectroscopy and mass spectrometry (MS). The IR analysis indicated the presence of pyridine benzologs ($1598-1567 \ cm^{-1}$), a carbonyl functional group of amides (1680, 1660, and $1640 \ cm^{-1}$), a carbonyl functional group of aromatic acids ($1700 \ cm^{-1}$), and a carbonyl functional group of esters and their main structure was identified as naphthene-substituted (Kovalenko et al., 2001). The most significant N compounds were identified to be quinolines, benzoquinolines, and S-containing compounds represented by thiophenoquinolines. The O containing N compounds were mostly represented by cyclic amides of the pyridone type, their hydrogenated analogs, such as lactams, quinoline carboxylic acids, and esters of quinoline carboxylic acids (Kovalenko et al., 2001). The different N compounds were extracted with varying concentrations of sulfuric acid in acetic acid and it was

TABLE 3.34
Chemistry of Nitrogen Compounds Identified in Heavy Russian Crude Oil

Heavy Russian Crude Oil	Heteroatom Molecules	Relative Content, %
N-compounds	Pyridines	8.9
	Quinolines	15.2
	Benzoquinolines	17.6
	Dibenzoquinolines	12.5
	Azapyrenes	8.5
NS-compounds	Thiazole	11.3
	Thiophenoquinolines	13.6
	Benzothiophenoquinolines	11.7
	Dibenzothiophenoquinolines	0.8
NO-compounds	Pyridones	16.4
	Quinolones	10
	Benzoquinolones	0.8
	Lactams	54.8
NO_2-compounds	Quinoline carboxylic acids	11.1
	Esters of quinoline carboxylic acids	7

Source: Modified From Kovalenko, E.Yu. et al., *Chemistry and Technology of Fuels and Oils*, 37, 265, 2001.

reported that heavy Russian crude oils, from the Van-Eganskoe field, contained 0.33 wt% of N including 0.023 wt% of weak basic N and 0.077 wt% of strong basic N (Kovalenko et al., 2001). The type and chemistry of nitrogen molecules found in crude oils are shown in Table 3.35.

The N compounds in heavy Russian crude oils, from the Van-Eganskoe field, were represented by 48% weak bases and 53% strong bases. The NO compounds, represented by cyclic amides of the pyridone type, their hydrogenated analogs, such as lactams, quinoline carboxylic acids, and esters of quinoline carboxylic acids, were identified to have weak basic properties. The N compounds, such as quinolines, benzoquinolines, and their derivatives, and NS compounds, represented by thiophenoquinolines and their derivatives, were identified to have strong basic properties (Kovalenko et al., 2001). Nitrogen is usually more difficult to remove than sulfur from hydrocarbon streams and any treatment effective in reducing nitrogen will reduce the sulfur. HDN of pyrrole over different catalysts was reported to produce ammonia, C_1-C_2 hydrocarbons and other nitrogen compounds (Zhao et al., 2001). The literature reported on the effect of temperature and the pressure, with and without the catalyst, on the hydroconversion of different chemistry N-containing molecules (Bressler and Gray, 2002).

TABLE 3.35
Type and Chemistry of Nitrogen Molecules Found in Crude Oils

Nitrogen Type	Name	Formula
Nonbasic N	Pyrrole	C_4H_5N
	Indole	C_8H_7N
	Carbazole	$C_{12}H_9N$
	Benzocarbazole	$C_{16}H_{11}N$
Weak basic	Cyclic amides of pyridones	C_nH_{2n-z} NO ($_{z\ 9-27}$)
	Lactams	C_nH_{2n-z} NO$_2$ ($_{z\ 9-27}$)
	Quinoline carboxylic acids	
	Esters of quinoline carboxylic acids	
Strong basic	Pyridine	C_5H_5N
	Quinoline	C_9H_7N
	Indoline	C_8H_9N
	Benzoquinoline	$C_{13}H_9N$
	Dibenzoquinolines	C_nH_{2n-z} N ($_{z\ 9-27}$)
	Thiophenoquinolines	C_nH_{2n-z} NS ($_{z\ 9-27}$)
	Benzothiophenoquinolines	
	Dibenzothiophenoquinolines	

Source: From Kovalenko, E.Yu. et al., *Chemistry and Technology of Fuels and Oils*, 37, 265, 2001; Speight, J.G., *The Chemistry and Technology of Petroleum*, 4th ed., CRC Press, Boca Raton, 2006.

TABLE 3.36
Effect of Temperature and Pressure on Hydroconversion of 2-Amino-Biphenyl

Processing	330°C/600 psi/No Catalyst	430°C/600 psi/No Catalyst
Conversion	No reaction	10%
Products		Carbazole
		1,1-Biphenyl
		Aniline

Source: From Bressler, D.C. and Gray, M.R., *Energy and Fuels*, 16, 1076, 2002.

The effect of temperature and pressure, without catalyst, on hydroconversion of 2-amino-biphenyl is shown in Table 3.36.

At the temperature of 330°C, no reaction takes place. With an increase in the temperature to 430°C and no catalyst, the 10% conversion of 2-amino-biphenyl leads to three products which are carbazole, 1,1-biphenyl, and aniline (Bressler and Gray, 2002). Two of these products, carbazole and aniline, still contain N; however, their chemistry is different. The effect of temperature and pressure, in the presence of catalyst, on hydroconversion of 2-amino-biphenyl is shown in Table 3.37.

At the temperature of 330°C, and in the presence of catalyst, it was reported the 10% conversion of 2-amino-biphenyl leads mainly to one product which is cyclohexylbenzene. With an increase in the temperature to 380°C, and in the presence of catalyst, the 20% conversion of 2-amino-biphenyl leads to three products which are cyclohexylbenzene, 1,1-bicyclohexyl, and 1,1-biphenyl (Bressler and Gray, 2002). With an increase in the temperature and the presence of catalyst, only a small increase in the conversion of 2-amino-biphenyl is observed. The effect of temperature and pressure, in the presence of catalyst, on hydroconversion of basic quinoline is shown in Table 3.38.

HT at 330°C of basic type N compounds, such as quinoline, leads to one main product which was reported to be 1,2,3,4-tetrahydroquinoline (Bressler and

TABLE 3.37
Effect of Temperature and Catalyst on Hydroconversion of 2-Amino-Biphenyl

HT Conditions	330°C/600 psi/Catalyst	380°C/600 psi/Catalyst
Conversion	10%	20%
Products	Cyclohexylbenzene	Cyclohexylbenzene
		1,1-Bicyclohexyl
		1,1-Biphenyl

Source: From Bressler, D.C. and Gray, M.R., *Energy and Fuels*, 16, 1076, 2002.

TABLE 3.38

Effect of Temperature and Catalyst on Hydroconversion
of Basic Quinoline

HT Conditions	330°C/600 psi/Catalyst	380°C/600 psi/Catalyst
Conversion	90%	92%
Products	1,2,3,4-Tetrahydroquinoline	1,2,3,4-Tetrahydroquinoline
		5,6,7,8-Tetrahydroquinoline
		Propylcyclohexane
		1-Propylcyclohexene
		4-Nonane

Source: From Bressler, D.C. and Gray, M.R., *Energy and Fuels*, 16, 1076, 2002.

Gray, 2002). Despite a high conversion of quinoline, in the range of 90%, the main product is 1,2,3,4-tetrahydroquinoline, which is still basic type N molecule. Hydroconversion of quinoline at a higher temperature of 380°C was reported to lead to many different products, such as 1,2,3,4-tetrahydroquinoline, 5,6,7,8-tetrahydroquinoline, propylcyclohexane, propylbenzene, 4-nonane, and 1-propylcyclohexene including the formation of olefins (Bressler and Gray, 2002). Despite a high conversion of quinoline, in the range of 92%, the main products are basic type tetrahydroquinolines and some olefins. The current HDN process uses sulfided Ni–Mo on alumina catalysts and nonbasic carbazoles were reported to be the most difficult to remove by a conventional HT process due to their low basicity and a high stability of carbon–nitrogen bond. One alternative approach was reported to use bacterial cells or enzymes as catalysts (Bressler and Gray, 2002). The presence of heteroatoms, such as S and N, increases with an increase in the boiling range of petroleum fractions. The effect of boiling range on heteroatom content of different middle distillates is shown in Table 3.39.

With an increase in boiling range, an increase in aromatic and heteroatom content is observed. The middle distillate, having a low boiling range of 143°C–347°C, was reported to contain 0.05 wt% of S, 125 ppm of N, and 1 vol% of olefins. Another middle distillate, having a boiling range of 155°C–374°C, was reported to contain 0.09 wt% of S, 575 ppm of N, and no olefins. Middle distillate, having a higher boiling range of 191°C–362°C, was reported to contain a higher S content of 0.21 wt%, 463 ppm of N, and also no olefins (Severin and David, 2001). With an increase in boiling range, an increase in viscosity leads to a higher flash point. The chemistry of heteroatom compounds, present at the trace level, is difficult to isolate and identify. The literature described different analytical approaches to sample preparations containing low contents of S, N, and O containing compounds and the actual content and chemistry of some heteroarom containing compounds found in different petroleum fractions (Severin and David, 2001). The content and chemistry of heteroatom molecules found in different petroleum fractions are shown in Table 3.40.

TABLE 3.39
Effect of Boiling Range on Heteroatom Content of Different Middle Distillates

Composition	Middle Distillate #1	Middle Distillate #2	Middle Distillate #3
Boiling range, °C	143–347	155–374	191–362
Density at 20°C, g/mL	0.8337	0.8598	0.8494
Viscosity, mm²/s	2.11	4.05	3.56
Saturates, wt%	71.9	47.6	57.5
Aromatics, wt%	28.1	52.4	42.5
Olefins, wt%	1	0	0
Sulfur, wt%	0.05	0.09	0.21
Nitrogen, ppm	125	575	463
Flash point, °C	58	63	88

Source: From Severin, D. and David, T., *Petroleum Science and Technology*, 19, 469, 2001. Reproduced by permission of Taylor & Francis Group, LLC, http://www.taylorandfrancis.com.

Light heating fuel was reported to contain 76 ppm of dibenzothiophene, 39 ppm of dibenzofuran, 21 ppm of carbazole, and 3 ppm of quinoline. Middle distillate was reported to contain a higher content of dibenzothiophene and a higher content of N compounds, such as indoles, carbazole, and a higher content of basic quinoline. Some middle distillates were also reported to contain phenols and dibenzofurane. VGO, having a higher boiling point, was reported to contain

TABLE 3.40
Content and Chemistry of Heteroatom Molecules Found in Petroleum Fractions

Content, ppm	Light Heating Fuel	Middle Distillate	VGO
Benzothiophene			
Dibenzothiophene	76	121	73
			679
Dibenzofuran	39		
Indole		74	75
1-Methylindole			21
Carbazole	21	116	4
8-Methylquinoline (BN)	3	35	30

Source: From Severin, D. and David, T., *Petroleum Science and Technology*, 19, 469, 2001.

benzothiophene, a significantly higher content of dibenzothiophene, and a significantly higher content of N compounds, including indoles, carbazole, and quinoline (Severin and David, 2001). HT is used to remove undesirable heteroatom molecules, mostly containing sulfur, and HC is used to reduce the boiling point of feed. The typical HC operation requires the reactor temperature of 290°C–400°C and the reactor pressure of 1200–2000 psig (Gary and Handwerk, 2001). The removal of S compounds is currently achieved by the HDS process at the temperatures of 300°C–340°C and pressures of 20–100 atm of H_2 in the presence of Co–Mo/Al_2O_3 or Ni–Mo/Al_2O_3 catalysts (Hernandez-Maldonado et al., 2004). The typical operating variables of hydroprocessing reactions are shown in Table 3.41.

Hydrotreatment was reported to be highly effective in removing thiols, sulfides, and disulfides but less effective in removing aromatic thiophenes, such as benzothiophenes, dibenzothiophenes, and their alkyl derivatives. With an increase in boiling point and heteroatom content, HDS and HDN of fuels become more difficult.

With an increased demand for cleaner burning fuels, the use of severe HT has become a common process in many refineries. The literature reported that at the present time, it is impossible to achieve a good quality product and long life on stream of the catalyst with a single catalyst. Dibenzothiophene was identified to be the most abundant sulfur compound, the content of which was found to

TABLE 3.41
Typical Operating Variables of Hydroprocessing Reactions

Hydroprocessing Variables	Typical Range
HT process variables	
Reactor temperature	270°C–340°C
Reactor pressure	690–20,700 kPag
Space velocity (LHSV)	1.5–8.0
HC process variables	
Reactor temperature	290°C–400°C
Reactor pressure	1200–2000 psig
HDS process variables	
Reactor temperature	300°C–340°C
Reactor pressure	20–100 atm of H_2

Source: From Gary, J.H. and Handwerk, G.E., *Petroleum Refining, Technology and Economics*, 4th ed., Marcel Dekker, Inc., New York, 2001; Hernandez-Maldonado, A.J., Yang, R.T., and Cannella, W., *Industrial and Engineering Chemistry Research*, 43, 6142, 2004.

TABLE 3.42
Effect of Lube Oil Hydrocracking Process Conditions
on Catalyst Life

HC Lube Oil Variables	Range	Typical
Temperature, °C	330–450	385–440
Pressure, psig	1500–4000+	2500–3000
Space velocity, Vo/Vc/hr	0.25–1.25	0.5–1.0
Lube oil yield, vol%	30–90	40–80
Catalyst life, years	1–3	1–2

Source: From Sequeira, A. Jr., *Lubricant Base Oil and Wax Processing*, Marcel Dekker, Inc., New York, 1994. Reproduced by permission of Routledge/Taylor & Francis Group, LLC.

gradually increase with an increase in the boiling range of petroleum fractions, from middle distillates to VGO. HC is used to crack heavy molecules in the feed and the commercial HC catalysts are cobalt–molybdenum, nickel–molybdenum, iron–cobalt–molybdenum, or nickel–tungsten on alumina or silica–alumina support (Sequeira, 1994). The lube oil HC process conditions and their effect on the catalyst life are shown in Table 3.42.

The literature reported on the two-stage HC of gas oils produced from conventional, coker, and oil sands which involves hydrogenation, heteroatom removal, and cracking. During the first HT step, the conversion of gas oils to middle distillates was correlated to the S and polyaromatic content of the feed. During the HC step, the conversion of gas oils was reported to be affected by the nitrogen content (Aoyagi et al., 2003). Although an increase in temperature improves S and N removal, excessive temperature must be avoided because of the increased coke formation leading to catalyst deactivation. The coke deposit on the catalyst was reported to vary depending on the temperature due to different mechanisms of coke formation below 400°C and above 450°C. The processes of polymerization and alkylation of slightly condensed coke were reported to compete with the HC process (Shakun et al., 2001). The effect of catalyst on the coke deposit and the HC activity, measured at 360°C, is shown in Table 3.43.

Under the same HC conditions, the coke deposit was found to be similar in a range of 4.1%–4.6% of the weight of catalyst; however, a significant difference in activity was observed. The degree of decrease in the HC activity was found to vary from 36% to 66% indicating that other factors affect the catalyst activity. Alumina is widely used as hydroprocessing catalyst support and the acidic content of the feed might also lead to alumina deactivation. According to the literature, strong mineral acids react with aluminum oxide leading to hydrolysis of aluminum–oxygen bond. Dissolution of aluminum leads to changes in the crystalline structure affecting the activity of the catalyst (Dolbear, 1998). With an

TABLE 3.43
Effect of Catalyst on Coke Deposit and Hydrocracking Activity

Catalyst-Al$_2$O$_3$	Coke, wt%	Decrease in Activity, %
5% MoO$_3$	4.1	66
12% MoO$_3$	4.6	36

Source: Modified From Shakun, A.N., Yasyan, Yu.P., and Litvinova, S.M., *Chemistry and Technology of Fuels and Oils*, 37, 123, 2001.

increase in the boiling point of different petroleum fractions, their total acid number (TAN) was reported to increase. The asphaltic and heavy crude oils contain a higher content of heteroatoms, metals, asphaltenes, resins, have a higher carbon residue and, in many cases, have a higher TAN. The literature on hydro-processing of heavy Maya crude oil reported that the use of different catalyst support, such as alumina, alumina–silica, or alumina–titania, also affects the catalyst deactivation (Maity et al., 2005). The carbon and vanadium contents of spent Co–Mo catalysts, having different support, are shown in Table 3.44.

After 60 h of operation at 380°C, the alumina supported Co–Mo catalyst contained a carbon content of 19 wt% and a vanadium content of 0.87 wt%. The alumina–silica supported Co–Mo catalyst contained a carbon content of 9.6 wt% and a vanadium content of 0.45 wt%. The alumina–titania supported Co–Mo catalyst contained the least carbon content of 6 wt% and the least vanadium content of 0.16 wt%. Alumina support was reported to be used to prepare the HT catalysts; alumina–silica is considered to be acidic and used to prepare the HC catalysts while alumina–titania was used to prepare HDS catalysts (Maity et al., 2005). The literature indicated that the most coke deposition was related to the largest pore diameter and an increase in the adsorption of big asphaltene

TABLE 3.44
Effect of Support on Composition of Spent Co–Mo Catalysts

Spent Co–Mo Catalyst Composition	Al$_2$O$_3$ (100%)	Al$_2$O$_3$–SiO$_2$ (95:5%)	Al$_2$O$_3$–TiO$_2$ (95:5%)
Carbon, wt%	19	9.6	6
Vanadium, wt%	0.87	0.45	0.16

Source: Reprinted with permission from Maity, S.K., Ancheyta, J., and Rana, M.S., *Energy and Fuels*, 19, 343, 2005. Copyright American Chemical Society.

molecules. Also, it was related to support acidity and an increase in HC of asphaltenes leading to more coke deposit (Maity et al., 2005). The typical catalysts used in processing of conventional crude oils were reported less effective when processing bitumen-derived distillates. X-ray photoemission spectroscopy (XPS) was used to determine sulfur molecules and their distribution in fractions of Athabasca fluid coking residue and HC residue. The sulfur compounds, identified by XPS, were sulfides, mercaptans, and thiophenes (Zhao et al., 2002). While the hydrocarbon types were reported to be similar in conventional distillates and distillates from bitumen, the bitumen-derived distillates have higher boiling points. The literature reported on the effect of HT on heteroatoms present in the gas oil fraction boiling in a range of 433°C–483°C. After HT, the thiophenic and sulfide sulfur was significantly reduced. While the overall content of nitrogen was reduced, its concentration in the polyaromatic fraction was reported to increase and no significant effect of HT on the oxygen content was observed (Woods et al., 2004). The nitrogen species in bitumen were reported to be more difficult to remove when compared to other crude oils (Yui and Chung, 2004). The recent literature reported on processing schemes for production of sulfur-free fuels which contain less than 10 ppm of sulfur (Iki, 2007). The processing schemes for production of sulfur-free fuels are shown in Table 3.45.

It was reported that in 2004, the specification for S content of diesel oil in Japan was reduced from 500 to 50 ppm. At the present, the market is supplied with sulfur-free fuels which contain less than 10 ppm of S. The sulfur-free production of fuels is achieved by a combination of hydrodesulfurization (HDS), hydrocracking (HC), and the proprietary ROK-Finer process for selective catalytic cracked gasoline (Iki, 2007). A review of FCC coke yields, sources, and formation affecting catalyst and the unit performance was recently published (Jawad, 2007). The recent literature also reported on the catalyst industry trends which can lead to higher product yields, longer operating cycles, and better quality products (Gonzales, 2007). The literature reported on the Swedish refinery meeting the diesel specification for domestic and export markets with flexible

TABLE 3.45

Scheme of Sulfur-Free Fuel Production

Atmospheric Distillation	Processing	Products
Naphtha	HDS/Reformer	Gasoline
Kerosene	HDS	Kerosene
Gas oil	HDS	Diesel
Vacuum gas oil	HC	Diesel
Vacuum gas oil	HDS/FCC/ROK-Finer	Gasoline

Source: From Iki, H., *PTQ Catalysis*, 41, 2007.

process strategies and effective catalysts (Egby and Larsson, 2007). In recent years strict regulations against environmental pollution further decreased the sulfur content of gasoline and diesel indicating the need for optimizing the catalyst formulations and processing conditions. An increase in hydroprocessing of bitumen-derived distillates was reported to decrease the catalyst activity. To reduce the S and N, and decrease the coke deposition on the catalyst, further improvements in hydroconversion technology are needed.

REFERENCES

ACCN Canadian Chemical News, 56(8), 6, 2004.

Al-Barood, A., Qabazard, H., and Stanislaus, A., *Petroleum Science and Technology*, 23 (7&8), 749, 2005.

Alboudwarej, H. et al., *Petroleum Science and Technology*, 22(5&6), 647, 2004.

Ali, M.A., *Petroleum Science and Technology*, 19(9&10), 1063, 2001.

Ancheyta, J. et al., *Energy and Fuels*, 16, 1438, 2002.

Ancheyta, J. et al., *Energy and Fuels*, 17, 1233, 2003.

Ancheyta, J., Centeno, G., and Trejo, F., *Petroleum Science and Technology*, 22(1&2), 219, 2004.

Aoyagi, K., McCaffrey, W.C., and Gray, M.R., *Petroleum Science and Technology*, 21 (5&6), 997, 2003.

Baranova, S.V. et al., *Petroleum Chemistry*, 42(5), 306, 2002.

Behbehani, H., Al-Qallaf, M.A., and El-Dusouqui, O.M.E., *Petroleum Science and Technology*, 23(3&4), 219, 2005.

Bej, S.K., Dalai, A.K., and Adjaye, J., *Energy and Fuels*, 15, 1103, 2001.

Bej, S.K., *Energy and Fuels*, 16, 774, 2002.

Beuther, H., Donaldson, R.E., and Henke, A.M., *I&EC Product Research and Development*, 3(3), 174, 1964.

Bhaskar, M., Valavarasu, G., and Balaraman, K.S., *Petroleum Science and Technology*, 20 (7&8), 879, 2002.

Bhaskar, M., Valavarasu, G., and Balaraman, K.S., *Petroleum Science and Technology*, 21 (9&10), 1439, 2003.

Bressler, D.C. and Gray, M.R., *Energy and Fuels*, 16, 1076, 2002.

Calemma, V. et al., *Preprints Division of Petroleum Chemistry*, ACS, 44(3), 241, 1999.

Calemma, V., Peratello, S., and Perego, C., *Applied Catalysis A*: General 190, 207, 2000.

Chaudhuri, U.R. et al., *Petroleum Science and Technology*, 19(5&6), 535, 2001.

Cody, I.A. and Brown, D.L., *Wax Isomerization Using Small Particle Low Fluoride Content Catalysts*, US Patent 4,923,588, 1990.

Cody, I.A. et al., *High Porosity, High Surface Area Isomerization Catalyst and Its Use*, PCT Int. Patent Appl., 1992.

Dolbear, G.E., Hydrocracking: reactions, catalysts, and processes, in *Petroleum Chemistry and Refining*, Speight, J.G., Ed., Taylor and Francis, Washington, 1998.

Egby, C. and Larsson, R., *PTQ Catalysis*, 12(2), 25, 2007.

Ellis, P.J. and Paul, C.A., *Delayed Coking Fundamentals*, AIChE 2000 Spring National Meeting, Atlanta, May 5–9, 2000.

Fisher, G., *Canadian Chemical News*, 58(6), 13, 2006.

Gary, J.H. and Handwerk, G.E., *Petroleum Refining, Technology and Economics*, 4th ed., Marcel Dekker, Inc., New York, 2001.

Gonzales, R.G., *PTQ Catalysis*, 12(2), 7, 2007.

Hatanaka, S., Yamada, M., and Sadakane, O., *Industrial and Engineering Chemistry Research*, 36, 1519, 1997.

Hernandez-Maldonado, A.J., Yang, R.T., and Cannella, W., *Industrial and Engineering Chemistry Research*, 43, 6142, 2004.

Iki, H., *PTQ Catalysis*, 12(2), 41, 2007.

Irisova, K.N., Talisman, E.L., and Smirnov, V.K., *Chemistry and Technology of Fuels and Oils*, 39(1&2), 20, 2003.

Jada, A., Salou, M., and Siffert, B., *Petroleum Science and Technology*, 19(1&2), 119, 2001.

Jawad, Z., *PTQ Catalysis*, 12(2), 29, 2007.

Jian, C. et al., *Petroleum Science and Technology*, 23(11&12), 1453, 2005.

Kazakova, L.P., Esipko, E.A., and Boldinov, V.A., *Petroleum Chemistry*, 39, 433, 1999.

Kline & Company, Inc., Report on GLT Specialties: High Value Opportunity or Threat, 2002.

Kovalenko, E.Yu. et al., *Chemistry and Technology of Fuels and Oils*, 37(4), 265, 2001.

Kwon, S.-H., Min, W.-S., and Lee, Y.-K., *Journal of Cleaner Production*, 5(1&2), 174, 1997.

Lavanyja, M. et al., *Petroleum Science and Technology*, 20(7&8), 713, 2002.

Liu, G., Xu, X., and Gao, J., *Energy and Fuels*, 18, 918, 2004.

Liu, R. et al., *Petroleum Science and Technology*, 23(5&6), 711, 2005.

Lugovskoi, A.I. et al., *Chemistry and Technology of Fuels and Oils*, 36(5), 337, 2000.

Maity, S.K., Ancheyta, J., and Rana, M.S., *Energy and Fuels*, 19, 343, 2005.

Mang, T., Base oils, in *Lubricants and Lubrications*, Mang, T. and Dresel, W., Eds., Wiley-VCH, Weinheim, 2001.

Musienko, G.G., Ermakov, V.P., and Solovkin, V.G., *Chemistry and Technology of Fuels and Oils*, 36(5), 342, 2000.

Nabih, H.I., *Petroleum Science and Technology*, 20(1&2), 77, 2002.

Oyekunie, L.O. and Susu, A.A., *Petroleum Science and Technology*, 23(2), 199, 2005.

Patel, J.A. et al., *Preprints Division of Petroleum Chemistry*, ACS, 42(1), 200, 1997.

Peixoto, F.C. and Medeiros, J.L., *AIChE Journal*, 47(4), 935, 2001.

Pereira, R.C.C. and Pasa, V.M.D., *Energy and Fuels*, 19, 426, 2005.

Phillipps, R.A., Highly refined mineral oils, in *Synthetic Lubricants and High-Performance Functional Fluids*, Rudnick, L.R. and Shubkin, R.L., Eds., 2nd ed., Marcel Dekker, Inc., New York, 1999.

Pillon, L.Z., Asselin, A.E., and MacAlpine, G.A., *Lubricating Oil Having an Average Ring Number of Less Than 1.5 Per Mole Containing Succinic Anhydride Amine Rust Inhibitor*, US Patent 5,225,094, 1993.

Prince, R.J., Base oils from petroleum, in *Chemistry and Technology of Lubricants*, 2nd ed., Mortier, R.M. and Orszulik, S.T., Eds., Blackie Academic & Professional, London, 1997.

Puente, G. and Sedran, U., *Energy and Fuels*, 18, 460, 2004.

Reynolds, J.G., *Petroleum Science and Technology*, 19(7&8), 979, 2001.

Saad, L. and Mikhail, S., *Petroleum Science and Technology*, 23(11&12), 1463, 2005.

Schwarcz, J., *ACCN Canadian Chemical News*, 59(1), 6, 2007.

Sequeira, A. Jr., *Lubricant Base Oil and Wax Processing*, Marcel Dekker, Inc., New York, 1994.

Severin, D. and David, T., *Petroleum Science and Technology*, 19(5&6), 469, 2001.

Shakun, A.N., Vasyan, Yu.P., and Litvinova, S.M., *Chemistry and Technology of Fuels and Oils*, 37(2), 123, 2001.

Sheppard, C.M. et al., *Energy and Fuels*, 17, 1360, 2003.

Sloley, A.W., *Hydrocarbon Processing* (International Edition), 80(2), 75, 2001.

Speight, J.G., *The Chemistry and Technology of Petroleum*, 3rd ed., Marcel Dekker, Inc., New York, 1999.

Speight, J.G., *The Chemistry and Technology of Petroleum*, 4th ed., CRC Press, Boca Raton, 2006.

Strausz, O.P., Peng, P., and Murgich, J., *Energy and Fuels*, 16, 809, 2002.

Sztukowski, D. and Yarranton, H.W., *Journal of Dispersion Science and Technology*, 25(3), 299, 2004.

Taylor, R.J., McCormack, A.J., and Nero, V.P., *Preprints, Division of Petroleum Chemistry*, ACS, 37(4), 1337, 1992.

Trasobares, S. et al., *Industrial and Engineering Chemistry Research*, 38, 938, 1999.

Turner, J., Asquith, R.J., and Atkinson, R., *Hydrocarbon Processing* (International Edition), 78, 119, 1999.

Woods, J.R. et al., *Petroleum Science and Technology*, 22(3&4), 347, 2004.

Wu, G. et al., *Industrial and Engineering Chemistry Research*, 35, 4747, 1996.

Xia, D. et al., *Petroleum Science and Technology*, 20(1&2), 17, 2002.

Xiao, Y. et al., *Industrial and Engineering Chemistry Research*, 36, 4033, 1997.

Xu, Y., Dabros, T., and Czarnecki, J., *Petroleum Science and Technology*, 19(5&6), 623, 2001.

Yaghi, B., Benayoune, M., and Al-Bemani, A., *Petroleum Science and Technology*, 19(3&4), 373, 2001.

Yi, J. et al., *Petroleum Science and Technology*, 23(2), 109, 2005.

Yui, S. and Chung, K.H., *ACCN Canadian Chemical News*, 56(8), 16, 2004.

Zhao, R. et al., *Petroleum Science and Technology*, 19(5&6), 495, 2001.

Zhao, S. et al., *Petroleum Science and Technology*, 20(9&10), 1071, 2002.

4 Asphaltene Stability

4.1 CONVENTIONAL PROCESSING

In the area of petroleum science, asphaltenes are the most studied yet the least understood components of crude oils. An extensive book review of asphaltenes and asphalts was published (Chilingarian and Yen, 2000). The term asphaltenes is used to describe the insoluble content precipitated from petroleum oils using paraffinic solvents. Crude oils are usually characterized by API gravity, distillation range, pour point, sulfur content, metals, salt content, and carbon residue. The definition of "heavy oil" is usually based on the API gravity of <20. Very heavy crude oils and tar sand bitumens have an API <10. Usually, the higher the density, or lower the API gravity of the crude oil, the higher the sulfur content. The total sulfur in the crude oil can vary from 0.04 wt% in a light paraffin oil to 5 wt% in a heavy crude oil. Heavy crude oils contain a higher content of heteroatom content, metals, and asphaltenes and have a higher carbon residue. The carbon residue is roughly related to the asphalt content of the crude oil and the quantity of the lubricating oil fraction that can be recovered (Gary and Handwerk, 2001). Only some crude oils are suitable to produce lube oil base stocks. The literature reported that correlations have been developed to characterize the crude oils in terms of typical assay parameters and their effect on increased processing cost related to desalting and corrosion formation (Van den Berg et al., 2003). The effect of crude oil properties and composition on their processing is shown in Table 4.1.

The composition of asphaltenes found in crude oils is important in predicting yields and the operating parameters for different processes. The early literature reported that asphaltenes present in crude oils typically have 79–85 wt% of carbon, 7–9 wt% of hydrogen, 0.3–10 wt% of sulfur, and 0.6–3 wt% of nitrogen (Moschopedis and Speight, 1974). The literature reported on asphaltene stability in crude oils and the influence of oil composition (Wang and Buckley, 2003). The asphaltenes are usually isolated from crude oils as n-pentane, n-hexane, or n-heptane insolubles and the use of different paraffinic solvents was reported to affect the asphaltene composition. The H/C ratios of the n-heptane insolubles precipitated from crude oils were reported to be lower indicating a higher degree of aromaticity when compared to H/C ratios of the n-pentane insolubles

TABLE 4.1
Effect of Crude Oil Properties and Composition on Their Processing

Crude Oil Properties	Typical Range	Crude Oil Processing
API gravity	10–50	High gravity is more valuable
Distillation range	Vary	Indicates product quantity
Pour point	Vary	Related to wax content
Sulfur, wt%	0.1–5	Low content is more valuable
Nitrogen, wt%	0.1–2	Catalyst poison
Oxygen, wt%	0.1–0.5	Corrosion problem
Metals	Vary	Catalyst poison
Salt	Vary	Low content is more valuable
Asphaltenes	Vary	Low content is more valuable
Carbon residue	Vary	Low content is more valuable

Source: From Sequeira, A. Jr., *Lubricant Base Oil and Wax Processing*, Marcel Dekker, Inc., New York, 1994. Reproduced by permission of Routledge/Taylor & Francis Group, LLC.

(Speight, 2006). The effect of API gravity on the heteroatom content of *n*-pentane insolubles, precipitated from Arab crude oils, is shown in Table 4.2.

With a decrease in API gravity of Arab crude oils, and an increase in their viscosity, their *n*-pentane insolubles were found to have higher S and N contents but lower O content. The use of Arab heavy crude oil also leads to *n*-pentane insolubles, having a higher H/C and N/C ratios, indicating a lower polyaromatic content. The literature reported that asphaltenes contain S compounds as highly condensed thiophene derivatives. Their N content was reported to be mostly in the form of pyrroles while their O content was reported to contain carboxylic,

TABLE 4.2
Effect of API Gravity on Heteroatom Content of *n*-Pentane Insolubles

C5 Insolubles	Arab Light, API 33.8	Arab Medium, API 30.4	Arab Heavy, API 28.8
Carbon, wt%	84.23	83.65	83.17
Hydrogen, wt%	7.76	8.31	8.28
Sulfur, wt%	6.3	6.41	7.18
Nitrogen, wt%	0.75	0.65	0.84
Oxygen, wt%	0.96	0.98	0.53
H/C ratio	1.1	1.18	1.19
N/C ratio	0.009	0.008	0.01

Source: From Siddiqui, M.N., *Petroleum Science and Technology*, 21, 1601, 2003. Reproduced by permission of Taylor & Francis Group, LLC, http://www.taylorandfrancis.com.

ketones, and phenolic functional groups (Siddiqui, 2003). The free phenolic group in asphaltenes appears as a sharp peak at 3610 cm^{-1} and a hydrogen bonded —OH appears as a broad peak at 3100–3350 cm^{-1} (Siddiqui, 2003). According to the literature, n-pentane insolubles, precipitated from Arab heavy crude oil, have less acidic but more basic properties (Siddiqui, 2003). Intermolecular forces in aggregates of asphaltenes and resins other than H-bonding, such as van der Waals, electrostatic, and charge transfer, were also reported (Murgich, 2002). The effect of API gravity on the metal content of n-pentane insolubles, precipitated from Arab crude oils, is shown in Table 4.3.

The use of Arab heavy crude oils also leads to n-pentane insolubles having a higher metal content, mostly V and Ni, but also a higher content of Ca, Mg, and Al indicating a decrease in the efficiency of the desalter to remove salts from heavy oils. The nonporphyrin V and Ni were reported to be associated with heteroatoms, such as S, N, and O, or are associated with the aromatics in metalloporphyrins (Siddiqui, 2003). The n-pentane insolubles precipitated from Arab heavy crude oil were reported to have the highest content of V and Ni. The use of gel permeation chromatography (GPC) indicated that the n-insolubles, precipitated from Arab heavy crude oil, have the highest molecular weight (Siddiqui, 2003).

The first step in refining involves the fractionation of the crude oil at atmospheric pressure. The next step is the fractionation of the high boiling fraction under vacuum. The conventional processing of crude oils leads to the atmospheric and vacuum residues. There are different methods to measure the carbon residue of petroleum oils. Carbon residue measurements which use the Conradson carbon

TABLE 4.3
Effect of Heavy Crude Oil on Metal Content of n-Pentane Insolubles

C5 Insolubles	Arab Light, API 33.8	Arab Medium, API 30.4	Arab Heavy, API 28.8
Vanadium, ppm	267	181	366
Nickel, ppm	62	121	124
Iron, ppm	143	14	18
Copper, ppm	2	2	1
Sodium, ppm	80	115	73
Potassium, ppm	11	11	7
Calcium, ppm	48	59	106
Magnesium, ppm	19	18	24
Aluminum, ppm	22	29	36
Titanium, ppm	0	0	1
Zinc, ppm	6	4	4
Phosphorus, ppm	3	3	1

Source: From Siddiqui, M.N., *Petroleum Science and Technology*, 21, 1601, 2003. Reproduced by permission of Taylor & Francis Group, LLC, http://www.taylorandfrancis.com.

TABLE 4.4
Properties and Composition of Different Residue and Vacuum Gas Oil

Properties	Atmospheric Residue	Vacuum Gas Oil	Vacuum Residue
API gravity	13.9	22.4	5.5
Viscosity at 100°C, cSt	55	—	1900
Pour point, °C	18	—	—
Sulfur, wt%	4.4	2.97	5.45
Nitrogen, wt%	0.26	0.12	0.39
Vanadium, ppm	50	<1	102
Nickel, ppm	14	<1	32
C7 insolubles, wt%	2.4	0	7.1
Carbon residue (CCR), wt%	12.2	0.09	23.1
Carbon residue (RCR), wt%	9.8	<0.1	—

Source: From Speight, J.G., *The Chemistry and Technology of Petroleum*, 3rd ed., Marcel Dekker, Inc., New York, 1999. Reproduced by permission of Routledge/Taylor & Francis Group, LLC.

residue (CCR) test procedure are usually reported as carbon residue. The literature reported that carbon residue of an oil can be also measured by the Ramsbottom carbon residue (RCR) test (Shell, 2001). Despite a wide spread use of carbon residue data as quality control parameter for petroleum feedstocks and products, its chemical significance is not known. The early literature reported on the additivity of CCR data and its dependence on elemental composition for residual oils in relation to optimization of thermal upgrading (Roberts, 1989). The properties and composition of different residue and vacuum gas oil, produced from Kuwaiti crude oil, are shown in Table 4.4.

The atmospheric residue, from Kuwaiti crude oil, was reported to have an API gravity of 13.9 and a pour point of 18°C. The composition indicated the presence of 4.4 wt% of sulfur, 0.26 wt% of nitrogen, and the presence of metals, including 50 ppm of V and 14 ppm of Ni. The asphaltene content, measured as *n*-heptane (C7) insolubles, was reported to be 2.4 wt%. The atmospheric residue, from Kuwaiti crude oil, was also reported to have carbon residue of 12.2 wt%, measured as CCR, and 9.8 wt%, measured as RCR (Speight, 1999).

The vacuum gas oil (VGO), from Kuwaiti crude oil, indicated a higher API gravity of 22.4, a lower S content of 2.97 wt%, a lower N content of 0.12 wt%, and practically no metals and no asphaltenes. VGO was reported to have a low carbon residue of 0.09 wt%, measured as CCR, and a low carbon residue of <0.1 wt%, measured as RCR. During crude oil distillation, asphaltenes are not volatilized and remain in the vacuum reduced crude along with the metals and no asphaltenes and metals are present in VGOs.

The vacuum residue, from the same Kuwaiti crude oil, indicated a lower API gravity of 5.5, a higher sulfur content of 5.45 wt%, a higher N content of 0.39 wt%, and a high metal content, including 102 ppm of V and 32 ppm of Ni.

TABLE 4.5
Chemistry of Volatile Aromatic Products from the Thermal
Decomposition of Asphaltenes

Molecular Type	Chemistry
1-Ring aromatics	Alkylbenzenes
Polynuclear aromatics	
2-Ring condensed aromatics	Alkylnaphthalenes
3-Ring condensed aromatics	Alkylphenanthrenes
4-Ring condensed aromatics	Alkylchrysenes
Aromatic S-containing compounds	Alkylbenzothiophenes
	Alkyldibenzothiophenes

Source: Reprinted with permission from Speight, J.G., *Polynuclear Aromatic Compounds*, Ebert, L.B., ed., American Chemical Society, Washington, DC, 1988. Copyright American Chemical Society.

The asphaltene content, measured as C7 insolubles, increased to 7.1 wt% and a CCR carbon residue was also reported to significantly increase to 23.1 wt% (Speight, 1999). The chemistry of volatile aromatic products from the thermal decomposition of asphaltenes is shown in Table 4.5.

The presence of highly condensed aromatic structures in asphaltenes, precipitated from crude oils, leads to formation of coke precursors during their thermal decomposition. The literature reported that thermal decomposition of asphaltenes leads to a high content of nonvolatile residue, in a range of 38–63 wt% (Speight, 1988). More recent literature reported that at the temperature of 350°C–800°C, the pyrolysis of asphaltenes produced a substantial amount of alkanes, having up to 40 carbon atoms per molecule, and O-containing compounds identified to be carboxylic, phenolic, and ketonic type molecules. The pyrolysis of asphaltenes indicated that the oxygen-containing molecules were found to be more volatile (Speight, 1999). The use of scanning electron microscopy (SEM) identified the presence of S, V, Fe, Mg, and Si in asphaltenes precipitated from vacuum residue using different n-alkane solvents. Depending on the precipitating solvent, some asphaltenes were also reported to contain Zn and Sn (Bragado et al., 2001). The presence of polyaromatic hydrocarbons and metals might lead to a decrease in volatility of some heteroatom containing molecules and an increase in nonvolatile residue.

4.2 HYDROPROCESSING

The use of catalytic and thermal cracking processing, under the high temperature conditions, affects the composition of petroleum products and their asphaltene content. Hydrotreating is used to remove sulfur by catalytically reacting the product stream with hydrogen. The recovered sulfur is sold as a product to other industries. While hydrotreatment (HT) and hydrocracking (HC) convert low grade feedstocks to gasoline and distillate fuels, the variability in the feed composition affects the severity of processing conditions. Sulfur compounds, such as mercaptans and sulfides, found in light cuts are easy to remove (Sanchezllanes and Ancheyta, 2004). The literature reported on the analysis of heavy ends of petroleum needed to predict the effect of conversion on the volume of products in fluid catalytic cracker (FCC) unit and HC. In a modern refinery, the catalytic cracker is used to crack paraffinic atmospheric and VGOs while the hydrocracker is used to crack more aromatic cycle oils and coker distillates (Gary and Handwerk, 2001). Coking of vacuum residue was used primarily to produce coker gas oil suitable for use as a feed to a catalytic cracker. This reduced the coke formation on the cracker catalyst (Gary and Handwerk, 2001). More aromatic coker distillate feeds resist catalytic cracking but are easily cracked in the presence of pressure and hydrogen. In recent years, the coking process is also used to prepare hydrocracker feedstocks (Gary and Handwerk, 2001). The typical HC feedstocks and HC petroleum products are shown in Table 4.6.

Distillation separates crude oil into useful components which may be end products or feedstocks for other refinery processes, such as catalytic cracking, hydrocracking, or thermal cracking. Delayed and fluid coking produces a signifi-cant amount of light olefins, including ethylene, propylene, and butane. The naphtha fraction was reported to contain olefins and diolefins and the presence of diolefins requires special treatment to prevent polymerization in hydrotreaters (Gray and McCaffrey, 2002).

Atmospheric residue desulfurization process is widely used for residue upgrading and different catalysts are used to decrease the S, N, and metal contents of atmospheric residue. The straight-run atmospheric residue, used as feed, is high

---

TABLE 4.6
Typical Hydrocracking Feedstocks and HC
Petroleum Products

HC Feedstocks	HC Petroleum Products
Kerosene	Naphtha
Straight-run diesel	Naphtha and/or jet fuel
Atmospheric gas oil	Naphtha, jet fuel and/or diesel
Vacuum gas oil	Naphtha, jet fuel, diesel, lube oil
FCC LCO	Naphtha
FCC HCO	Naphtha and/or distillates
Coker LCGO	Naphtha and/or distillates
Coker HCGO	Naphtha and/or distillates

in S, N, metals, asphaltenes, and carbon residue. The partially hydrotreated atmospheric residue was reported to contain a lower S content of 0.53 wt%, a lower N content of 0.21 wt%, a lower V content of 10 ppm, and a lower Ni content of 6 ppm. After HT, the atmospheric residue was also reported to contain a lower asphaltene content of 1.3 wt% and have a lower carbon residue of 5.2 wt% (Marafi and Stanislaus, 2001). A rapid coke build up on the catalyst was observed during the HT of atmospheric residue. The catalyst deactivation by a coke deposition was reported to be affected by the feedstock composition (Marafi and Stanislaus, 2001). The effect of partially hydrotreated atmospheric residue on the composition of distilled VGO is shown in Table 4.7.

VGO, distilled from the partially hydrotreated atmospheric residue, was reported to have a lower S content of 0.4 wt%, N content of 0.11 wt%, and have no metals and no asphaltenes. VGO, distilled from the partially hydrotreated atmospheric residue, was also reported to have a lower carbon residue of 0.02 wt% (Marafi and Stanislaus, 2001). When using the alumina supported hydrotreating catalysts, the alumina support provides the required porosity, surface area, and the acidic and basic sites. The acidity of catalysts can be modified by using sodium and fluoride ions as acidity modifying agents (Marafi and Stanislaus, 2001). An increase in the catalyst acidity by the use of fluoride, at low level of 2 wt%, increased the HDS and HDN activity, however, the use of higher fluoride content of 5 wt% decreased the catalyst activity. The use of catalyst containing sodium increased the coke formation leading to catalyst deactivation (Marafi and Stanislaus, 2001). Although coke deposition on the catalyst cannot be totally eliminated, it can be minimized. The literature reported that to protect the catalyst, it is important to reduce the nitrogen content of the feed to below 10 ppm (Speight, 1999). The literature reported that the straight-run atmospheric residue, used as

TABLE 4.7
Effect of Partially HT Atmospheric Residue on Composition of VGO

Properties	Partially HT Atmospheric Residue	Distilled VGO
API gravity	20.49	24.94
Density at 15°C, g/mL	0.9304	0.904
Sulfur, wt%	0.53	0.4
Nitrogen, wt%	0.21	0.11
Vanadium, ppm	10	0
Nickel, ppm	6	0
Asphaltenes, wt%	1.3	0
Carbon residue, wt%	5.2	0.02

Source: From Marafi, M. and Stanislaus, A., *Petroleum Science and Technology*, 19, 697, 2001. Reproduced by permission of Taylor & Francis Group, LLC, http://www.taylorandfrancis.com.

TABLE 4.8

Effect of Different Catalyst Beds on Properties and Composition
of Residue

Properties	Atmospheric, Residue Feed	Catalyst Bed # 1, HDM Process	Catalyst Bed # 2, HDS Process
Viscosity at 50°C, cSt	765.1	208.5	112.9
Carbon, wt%	83.4	85.6	86.5
Hydrogen, wt%	11.1	12.0	12.7
Sulfur, wt%	4.3	2.3	0.7
Nitrogen, wt%	0.27	0.23	0.15
Oxygen, wt% (calc.)	0.9	0	0
Vanadium, ppm	69	21	15
Nickel, ppm	21	12	8
Asphaltenes, wt%	3.8	2.5	0.9
Carbon residue, wt%	12.2	8.5	5

Source: Reprinted with permission from Hauser, A. et al., *Energy and Fuels,* 19, 544, 2005.
Copyright American Chemical Society.

feed, needs to pass through different catalyst beds to achieve the required product
quality (Hauser et al., 2005). The effect of different catalyst beds on the properties
and composition of straight-run atmospheric residue is shown in Table 4.8.

When atmospheric residue passed through catalyst bed # 1, mostly designed
to remove metals, a decrease in viscosity, S, metals, asphaltenes, and carbon
residue is observed. However, no significant reduction in N content was reported.
After hydrotreated atmospheric residue was passed through an additional catalyst
bed # 2, mostly designed to remove S, a further decrease in viscosity, S, metals,
asphaltenes, and carbon residue is observed. Only after passing through the
second catalyst bed, a reduction in N was reported. According to literature,
asphaltenes are considered to be coke precursors because they have a high
molecular weight, high aromaticity, and are the least reactive components of the
feed (Hauser et al., 2005). The effect of different catalyst beds on the content and
composition of asphaltenes is shown in Table 4.9.

After passing through two different catalyst beds, a significant decrease in the
asphaltene content is observed. The presence of oxygen is difficult to measure
and, while the literature reports on the presence of heteroatom content, such as S
and N in asphaltenes, a high content of calculated oxygen is also observed.
Despite a decrease in heteroatoms, metals, asphaltenes, and carbon residue, the
hydrotreated feeds are not "easier feedstocks" in terms of preventing the coke
formation and catalyst deactivation (Hauser et al., 2005). The propensity to form
coke on different catalysts was studied and the coking tendency was higher for
hydrotreated atmospheric residue, having a higher saturate content and depleted
of asphaltenes, when compared to straight-run atmospheric residue. According to
the literature, the carbon content of spent catalyst beds # 1 increased to about

TABLE 4.9
Effect of Different Catalyst Beds on Content and Composition of Asphaltenes

Composition	Atmospheric, Residue Feed	Catalyst Bed # 1, HDM Process	Catalyst Bed # 2, HDS Process
Asphaltenes, wt%	3.8	2.5	0.9
Carbon, wt%	77.2	76.5	83
Hydrogen, wt%	7.1	6.9	11.7
Sulfur, wt%	8.8	5.2	1.2
Nitrogen, wt%	0.9	0.9	0.2
Oxygen, wt% (calc.)	6	10.5	3.9
Vanadium, ppm	561	8	7
Nickel, ppm	172	5	3

Source: Reprinted with permission from Hauser, A. et al., *Energy and Fuels*, 19, 544, 2005. Copyright American Chemical Society.

18 wt% while the carbon content of spent catalyst bed # 2 increased to about 23 wt% (Hauser et al., 2005). Under industrial conditions, the initial catalyst deactivation is caused by rapid coke deposition followed by metal accumulation and very rapid final deactivation by pore mouth clocking. The literature reported on the use of an NMR technique to analyze the chemistry of coke deposited on the catalysts which indicated a less aromatic character of coke generated from hydrotreated feed. According to the literature, another route to coking which is formation and polymerization of olefins formed during the HC step needs to be considered (Hauser et al., 2005). The effect of hydrotreating temperature on S, metals, and asphaltene contents of Kuwaiti product oil, obtained from atmospheric residue, is shown in Table 4.10.

TABLE 4.10
Effect of HT Temperature on Composition of Kuwaiti Product Oil

Arabian Heavy Crude Oil, Atmospheric Residue	Product Oil, Sulfur, wt%	Product Oil, (Ni + V), ppm	Product Oil, Asphaltenes, wt%
Residue feed	4.4	86	4.07
HT temperature			
360°C	0.34	17	1.18
370°C	0.33	14	1.47
380°C	0.22	10	1.01
390°C	0.12	5	0.72

Source: Reprinted with permission from Bartholdy, J. et al., *Energy and Fuels*, 15, 1059, 2001. Copyright American Chemical Society.

TABLE 4.11

Effect of HT Temperature on Asphaltenes from Kuwaiti Product Oil

Arabian Heavy Crude Oil, Atmospheric Residue	Product Oil, Asphaltenes, wt%	Asphaltenes, (S + N + O), wt%	Asphaltenes, MW
Residue feed	5.91	10.4	2400
HT temperature			
360°C	2.87	10.1	2375
370°C	1.77	9.9	2330
380°C	2.04	9	2140
390°C	1.35	8.4	2270
405°C	0.86	6.3	1325

Source: Reprinted with permission from Bartholdy, J. et al., *Energy and Fuels*, 15, 1059, 2001. Copyright American Chemical Society.

The literature reported that at low hydrotreating temperature, the main reaction is hydrogenation leading to stable products. At higher temperatures of 370°C–390°C, the process becomes cracking dominated and the sludge formation is observed (Bartholdy et al., 2001). The effect of hydrotreating temperature on the composition and molecular weight of asphaltenes, separated from Kuwaiti product oils, is shown in Table 4.11.

An increase in the HT temperature leads to a decrease in asphaltene content, having a lower heteroatom content and a lower molecular weight. With a decrease in asphaltene content, a decrease in product stability was reported. Kuwaiti product oils, produced from atmospheric residue below 370°C, were reported to have good stability during their transportation and storage. With an increase in the HT temperature, despite a lower asphaltene content, a decrease in product stability leads to sludge. Below 370°C, the main reaction was reported to be dominated by hydrogenation leading to stable products while above 370°C, it was reported to be mainly dominated by HC leading to less stability and sludge (Bartholdy et al., 2001).

To increase the production of fuels, cracked stocks are blended with straight-run products which can lead to an increase in their instability. The literature reported on the use of methanol extraction and hydrostabilization to improve the stability of blends containing cracked light cycle oil (LCO) and straight-run gas oil (SRGO) (Sharma et al., 2003). The accelerated storage stability test indicated that some HCGO, produced under the same processing conditions, were found to have a different storage stability. The literature reported that the furfural extraction followed by ethanol extraction was required to improve the storage stability of cracked diesel fuel. After the double extraction process, the sulfur content was reduced by 75% and the nitrogen content was reduced by 94%. A reduction in sediment formation, from 27.56 to 0.43 mg/100 mL, was reported (Liu and Yan, 2003).

4.3 THERMAL CRACKING

4.3.1 COKING

It is generally recognized that vacuum residue has a colloidal structure where asphaltenes and resins form micelles and the remaining fractions exist as a dispersing medium. The literature reported that the coke formation during the thermal or catalytic cracking of vacuum residue is affected by the chemical reactions and also by the collapse of asphaltenes–heavy resins micelle structures (Li et al., 2001). A number of studies reported that an important variable in determining the coke yield is CCR in the feed. The literature reported on kinetics of CCR conversion in the catalytic hydroprocessing of a residue from Maya crude oil (Trasobares et al., 1998). The composition of coke varies with the source of the crude oil but it contains high molecular weight hydrocarbons rich in carbon and poor in hydrogen (Speight, 1999).

Coking is a severe thermal cracking process which converts heavy oils and residue to lower boiling products and coke. The various coking processes vary from refinery to refinery. The gas–oil fraction is usually split into light coker gas oil (LCGO) and heavy coker gas oil (HCGO) before further processing. HCGO, used as a feed to a hydrocracker, was reported to increase coke formation leading to catalyst deactivation. The formation of coke, caused by the presence of asphaltenes, was reported to be the main reason for catalyst deactivation. The composition and BN/TN ratio of HCGO, having good storage stability and containing no visible sediment, are shown in Table 4.12.

HPLC-2 analysis of HCGO indicated the presence of 58 wt% of aromatics, including 1-ring, 2-ring, 3-ring, and 4+ ring aromatics, and a significant polar content, above 5 wt%. The HCGO was found to contain a sulfur content of 2.8 wt% and nitrogen content of 0.16–0.18 wt%. The metal content of HCGO was found to

TABLE 4.12
Composition and BN/TN Ratio of HCGO

Composition	HCGO #1	HCGO #1
Density at 15°C, g/cc	0.9414	0.9438
Saturates, wt%	41.86	41.76
Aromatics, wt%	58.14	58.24
Sulfur, wt%	2.8	2.8
Nitrogen, wt%	0.18	0.16
Metals, ppm	<1	<1
BN/TN ratio	0.3	0.3

Source: From Pillon, L.Z., *Petroleum Science and Technology*, 19, 673, 2001. Reproduced by permission of Taylor & Francis Group, LLC, http://www.taylorandfrancis.com.

be below 1 ppm. The ASTM D 2896 test method for Base Number of Petroleum Products by Potentiometric Perchloric Acid Titration is used to determine the basic nitrogen (BN) content of petroleum oils by titration with perchloric acid. Basic constituents of petroleum include BN compounds but also other compounds, such as salts of weak acids (soaps), basic salts of polyacidic bases, and salts of heavy metals. The BN/TN ratio of 0.3 indicated the typical presence of basic N compounds without any increase in polar basic type contaminants.

The ASTM D 3279 test method for "n-heptane insolubles" is used to determine the asphaltene content of crude oils, gas oils, heavy fuel oils, and asphalts. The sample containing n-heptane solvent is placed on the hot plate and secured under a reflux condenser. In the case of HCGO samples, the reflux time is usually 30 min. The dispersed mixture is cooled at RT for a period of 1 h and filtered, while warm. The sediment is dried in the oven at 107°C and weighted. HCGO was found to contain a low, below 100 ppm, C7 insolubles content having a black color.

Petroleum coke is the residue left by the destructive distillation of petroleum residue and it was reported to contain some volatiles and high boiling hydrocarbons. To eliminate all volatiles, petroleum coke is calcined at high temperatures, above 1000°C. Even after calcination, coke was reported to contain some volatiles (Gary and Handwerk, 2001). The metals present in crude oils that the desalter does not remove will end up in the coker feed. Petroleum coke was reported to contain vanadium and nickel. Iron, silicon, and ash are in the coke as particulates (Ellis and Paul, 2000). The FTIR analysis of petroleum coke and black color C7 insolubles, precipitated from HCGO, is shown in Table 4.13.

The FTIR analysis of petroleum coke indicated the presence of only a trace amount of aromatic hydrocarbons and the presence of carbon black particles. The XRF analysis of petroleum coke indicated the presence of S and many metals, including V, Ni, Fe, Ca, Al, and Si. The FTIR analysis of the black color C7 insolubles, precipitated from HCGO, indicated the presence of multiring aromatics, sulfates, CH, CH_3CH_3, and CH_3 groups. The presence of a broad peak at 3100–3300 cm^{-1} indicated the presence of hydrogen bonding through —OH groups. The ASTM D 4530 test method for the determination of carbon

---

TABLE 4.13
FTIR Analysis of Petroleum Coke and Black Color C7 Insolubles from HCGO

Petroleum Coke	Color Black C7 Insolubles
Carbon black	Multiring aromatics
Trace aromatics	Sulfates
	—OH-bonding

Source: From Pillon, L.Z., *Petroleum Science and Technology*, 19, 673, 2001. With permission.

TABLE 4.14

C7 Insolubles and Carbon Residue of Different HCGO

Properties	HCGO # 1	HCGO # 2
Density at 15°C, g/cc	0.9436	0.9497
Carbon, wt%	80.2	85.1
Hydrogen, wt%	11.5	12.4
Sulfur, wt%	2.63	2.84
Nitrogen, wt%	0.18	0.18
C7 insolubles, ppm	46	171
Carbon residue (MCR), wt%	0.18	0.29

Source: From Pillon, L.Z., *Petroleum Science and Technology*, 19, 863, 2001. Reproduced by permission of Taylor & Francis Group, LLC, http://www.taylorandfrancis.com.

residue (micromethod) is used to determine the relative coke forming tendency of petroleum products. The petroleum oil is heated at 500°C and under a nitrogen blanket. The ASTM D 4530 test method is used to measure the microcarbon residue (MCR) and the results are equivalent to the CCR test data. The C7 insoluble content and carbon residue of different HCGO samples are shown in Table 4.14.

HCGO # 1, containing 2.6 wt% of S and 0.18 wt% of N, was found to precipitate only 46 ppm of C7 insolubles and had an MCR of 0.18 wt%. Another HCGO # 2, containing 2.8 wt% of S and 0.18 wt% of N, was found to precipitate 171 ppm of C7 insolubles and had an MCR of 0.29 wt%. Petroleum oil samples might contain wax or other precipitates and they need to be effectively mixed to assure their uniformity. The solubility of some molecules present in HCGO will vary depending on the temperature and severity of mixing. The use of reflux, as recommended by the ASTM D 3279 procedure, increases the temperature of mixing to 98°C. After standing for 1 h, the filtration temperature described as "warm" indicates the temperature of about 40°C. HCGO samples, stored at RT, were found to have a temperature of 21°C–22°C and were mixed with *n*-heptane solvent under different conditions ranging from slow to vigorous. The cooling time was varied in the range from 30 to 60 min and the drying time was varied in the range from 10 to 20 min at 107°C. No significant effect of the HCGO storage temperature, mixing, cooling, and drying time on the C7 insoluble precipitation was observed. However, without any reflux, only mixing and stirring of *n*-heptane solvent with HCGO # 1 was found to affect the C7 insoluble content and their color indicating differences in their chemistry. The FTIR analysis of C7 insolubles, precipitated under modified conditions and having a different color, is shown in Table 4.15.

Under standard test conditions, reflux of HCGO with *n*-heptane at 98°C leads to a lower content of C7 insolubles and having black color. The FTIR analysis of the black color C7 insolubles indicated the presence of multiring aromatics, sulfates, CH, CH_3CH_3, and CH_3 groups. The presence of a broad peak at

TABLE 4.15
FTIR Analysis of Different C7 Insolubles Precipitated from HCGO

HCGO, C7 Insolubles	Standard Test, ASTM D 3279	Modified Test (Mixing at RT)
Content	Lower	Higher
Color	Black	Brown
FTIR	Multiring aromatics	Multiring aromatics
	Sulfates	Sulfates
	—OH-bonding	Carboxylates
		—OH-bonding

Source: From Pillon, L.Z., *Petroleum Science and Technology*, 19, 673, 2001. With permission.

3100–3300 cm^{-1} indicated the presence of hydrogen bonding through —OH groups. Under modified testing conditions, mixing of the HCGO with n-heptane at RT leads to a higher content of C7 insolubles and having colors ranging from beige to brown. The FTIR analysis of the brown color C7 insolubles indicated the presence of multiring aromatics, sulfates, carboxylates, CH, CH$_3$CH$_3$, and CH$_3$ groups. The presence of a broad peak at 3100–3300 cm^{-1} indicated the presence of hydrogen bonding through —OH groups.

The brown colored C7 insolubles, produced under modified conditions, were found to contain carboxylates. The black colored C7 insolubles, produced under standard conditions, were found to have a higher content of multiring aromatics. No presence of carboxylate groups would explain a lower content of C7 insolubles precipitated under standard test conditions. While the thermal degradation during coking decreases the n-heptane insoluble content of HCGO, the presence of —OH bonding, similar to the asphaltene structure present in crude oils, was identified by FTIR analysis. The effect of different paraffinic solvents on the insoluble content of HCGO is shown in Table 4.16.

TABLE 4.16
Effect of Different Paraffinic Solvents on Insoluble Precipitation

Standard Test, ASTM D 3279	Solvent, n-Heptane	Solvent, n-Pentane
Reflux time, min	30	30
Reflux temp., °C	98	40
Standing time, h	1	1
Filtration temp., °C	40	RT
Insolubles, ppm	375	502
Color	Black	Black

Source: From Pillon, L.Z., *Petroleum Science and Technology*, 19, 863, 2001. Reproduced by permission of Taylor & Francis Group, LLC, http://www.taylorandfrancis.com.

Following the standard ASTM D 3279 test conditions, another HCGO was found to contain 375 ppm of *n*-heptane insolubles. The use of *n*-pentane, as another paraffinic solvent, increased the insolubles precipitation from 375 to 502 ppm. The use of a different paraffinic solvent, often reported in the literature, as producing a higher asphaltene content of crude oils and petroleum fractions, affects the actual testing conditions. The reflux temperature of *n*-heptane is 98°C while the reflux temperature of *n*-pentane is only 40°C. Such a significant difference in the reflux temperature will not only affect dissolution of some molecules but also the filtration temperature. After the standard 1 h standing time, the use of *n*-heptane will lead to higher filtration temperature resulting in a lower *n*-heptane insoluble content. The use of *n*-pentane solvent leads to an increase in the insoluble content of HCGO and also affects their chemistry by precipitating additional molecules, not present in *n*-heptane insolubles.

4.3.2 "True" Asphaltenes

The "true" asphaltenes are defined as the crude oil constituents that are insoluble in an alkane but soluble in toluene. Aromatic type solvents, such as toluene, have the highest solvency to dissolve organic compounds. Asphaltenes are, by definition, insoluble in the lower molecular weight saturates and soluble in toluene. When asphaltenes are precipitated from a crude oil with paraffinic solvents, some other material also precipitates. This other material is called "solids" and contains ash, fine clays, and some adsorbed hydrocarbons (Yarranton and Musliyah, 1996). The literature reports on the asphaltenes, precipitated from crude oils, as having a complex chemical structure in terms of bridged aromatics versus large condensed aromatic cores. Their structure and molecular weight will be affected by the processing and asphaltenes were reported to require high boiling points, above 600°C, to correlate with carbon residue data (Gray, 2003). The literature reported on the chemistry and reactivity of asphaltenes in terms of cracking, coking, and tendency to aggregate. The bonding between the asphaltene molecules was studied using x-ray diffraction and small angle x-ray scattering techniques (Tanaka et al., 2004). To measure and characterize the true asphaltene properties of insolubles present in HCGO, the toluene insoluble solids need to be removed. The effect of different paraffinic solvents and toluene on the true asphaltene content of HCGO is shown in Table 4.17.

Following the standard ASTM D 3279 test method and, after 1 h standing time, the filtration temperature described as warm was about 40°C. The HCGO was found to contain 400 ppm of *n*-heptane insolubles and their color was black. The *n*-heptane insolubles, precipitated from the HCGO, were found to contain 176 ppm of toluene insoluble solids and only 224 ppm of true toluene soluble asphaltenes. Only 56 wt% of the *n*-heptane insolubles were actually soluble in toluene. Following the same test method and using *n*-pentane as solvent, the insoluble content of the HCGO increased to 530 ppm and their color was also black. The *n*-pentane insolubles, precipitated from the same HCGO, were found to contain 159 ppm of toluene insoluble solids and 371 ppm of true asphaltenes.

TABLE 4.17

Effect of Paraffinic Solvents and Toluene on the True Asphaltene Content

HCGO, Insolubles	Solvent, n-Heptane	Solvent, n-Pentane
Insolubles (total), ppm	400	530
Toluene insoluble solids, ppm	176	159
Toluene soluble asphaltenes, ppm	224	371
Toluene soluble yield, %	56	70

Source: From Pillon, L.Z., *Petroleum Science and Technology*, 19, 673, 2001. With permission.

Only 70 wt% of the *n*-pentane insolubles were soluble in toluene. The FTIR analysis of toluene insoluble solids and toluene soluble asphaltenes, precipitated using different paraffinic solvents from HCGO, is shown in Table 4.18.

The FTIR analysis of solids, toluene insoluble fraction of *n*-heptane insolubles, indicated the presence of multiring aromatics, sulfates, carboxylates, and some —CH, —CH$_2$CH$_3$, —CH$_3$ groups. The presence of a broad peak at 3100–3300 cm^{-1} indicated the presence of hydrogen bonded —OH groups which might include S, N, and O. The FTIR analysis of solids indicated that the presence of carboxylates is significantly lower than the presence of sulfates.

The FTIR analysis of true asphaltenes, toluene soluble fraction of *n*-heptane insolubles, indicated the presence of multiring aromatics and —CH, —CH$_2$CH$_3$, and —CH$_3$ groups. A small presence of —OH group was also observed. The FTIR analysis of the toluene soluble fraction of *n*-pentane insolubles also indicated the presence of multiring aromatics, —CH, —CH$_2$CH$_3$, —CH$_3$ groups, and a small presence of —OH group. The presence of an additional peak at 1730 cm^{-1} indicated the presence of esters. The use of *n*-pentane solvent, as a precipitating solvent, increased the insoluble precipitation and also affected the composition of

TABLE 4.18

FTIR of Solids and Toluene Soluble Asphaltenes Precipitated from HCGO

HCGO C7 Insolubles, Toluene Insoluble Solids	HCGO C7 Insolubles, Toluene Soluble	HCGO C5 Insolubles, Toluene Soluble
FTIR	FTIR	FTIR
Multiring aromatics	Multiring aromatics	Multiring aromatics
Sulfates	—OH-bonding	Esters
Carboxylates		—OH-bonding
—OH-bonding		

Source: From Pillon, L.Z., *Petroleum Science and Technology*, 19, 673, 2001. With permission.

TABLE 4.19
Composition of Solids and True Asphaltenes
Precipitated from HCGO

HCGO, C7 Insolubles	Toluene Insoluble, Solids	Toluene Soluble, Asphaltenes
Carbon, wt%	76.8	82.6
Hydrogen, wt%	5.62	8.84
Sulfur, wt%	10.2	6.5
Nitrogen, wt%	2.59	1.41
Oxygen, wt% (calc.)	4.79	0.65
H/C ratio	0.87	1.28

Source: From Pillon, L.Z., *Petroleum Science and Technology*, 19, 863, 2001. Reproduced by permission of Taylor & Francis Group, LLC, http:// www.taylorandfrancis.com.

asphaltenes by precipitation of esters. The composition of toluene insoluble solids and toluene soluble asphaltenes, precipitated from HCGO, is shown in Table 4.19.

The toluene insoluble solids were found to contain 76.8 wt% of C, 5.62 wt% of H, 10.2 wt% of S, 2.59 wt% of N, and 4.79 wt% of calculated O. The toluene soluble true asphaltenes were found to contain 82.6 wt% of C, 8.84 wt% of H, 6.5 wt% of S, 1.41 wt% of N content, and 0.65 wt% of calculated O. In terms of heteroatom content, the toluene insoluble solids, precipitated from HCGO as C7 insolubles, were found to contain a higher S, N, and a significantly higher O content. A lower H/C indicates an increase in polynuclear aromatic systems while a higher H/C ratio indicates the presence of long aliphatic chains. The H/C atomic ratio of solids was found to decrease, from 1.28 in toluene soluble asphaltenes to 0.87, indicating a decrease in polynuclear aromatic hydrocarbons. While the presence of metals can vary depending on the crude oils, their content can further vary depending on the efficiency of the desalter. Metals, such as Fe, K, and Mg, as well as Cl not removed during the desalting step will concentrate in the residue. The XRF analysis of toluene insoluble solids and toluene soluble asphaltenes, precipitated from HCGO, is shown in Table 4.20.

The toluene insoluble solids were found to contain 1.8 wt% of sulfate, 0.47 wt% of chloride, and a high amount of alkali metal such as Na. A significant presence of other metals such as Fe, Al, and Si is observed. Silicones are used to prevent foaming. While no presence of V and Ni was found in toluene soluble asphaltenes, the presence of Ni was found in toluene insoluble solids. A high content of chlorides, sulfates, and the presence of carboxylates, identified by FTIR, indicates that the solids fraction of C7 insolubles is a mixture of metal salts. The formation of salts would explain the reason for their solubility in HCGO and their tendency to precipitate in the presence of paraffinic solvents, such as *n*-heptane and *n*-pentane, and also their insolubility in organic solvents, such as toluene.

TABLE 4.20

XRF Analysis of Solids and True Asphaltenes

Precipitated from HCGO

HCGO C7 Insolubles, XRF Analysis, wt%	Toluene Insoluble, Solids	Toluene Soluble, Asphaltenes
Sulfates	1.8	0
Aluminum	1.41	0.05
Chloride	0.47	0.15
Sodium	0.41	0
Silicone	0.22	0.16
Calcium	0.18	0.01
Iron	0.13	0.05
Magnesium	0.05	0.02
Potassium	0.03	0.01
Nickel	0.02	0
Copper	0.01	0
Titanium	0.01	0
Zinc	0.01	0
Phosphorus	0.01	0
Cobalt	0	0.01
Vanadium	0	0

Source: From Pillon, L.Z., *Petroleum Science and Technology*, 19, 673, 2001. Reproduced by permission of Taylor & Francis Group, LLC, http://www.taylorandfrancis.com.

The XRF analysis of toluene soluble asphaltenes indicated the presence of 0.15 wt% of chloride and a small amount of K, Ca, and Mg indicating the presence of alkali chloride salts not removed during the desalting step. A significant presence of other metals, such as Fe, Al, and Si, was also identified. While the metal content of HCGO was found to be below 1 ppm, many different metals were found present in C7 insolubles precipitated from HCGO. The silicone presence in petroleum feedstocks is of concern due to its high surface activity. Silicones are known as excellent antistick agents; however, once silicone adheres to a metal surface, it is difficult to remove. For long-term storage, silicone dispersions require thickeners to prevent coalescence of silicone droplets (ProQuest, 1996).

The GC/MS analysis of toluene soluble asphaltenes was carried out in an He purge gas in four steps to allow for the selective desorption of organic molecules. In the first step, the asphaltene sample was heated at temperatures from 50°C to 200°C. In the second step, the same asphaltene sample was heated from 50°C to 400°C. In the third step, the same asphaltene sample was heated from 50°C to 600°C and in the fourth step, the same asphaltene sample was heated from 50°C to 700°C. The GC/MS analysis of toluene soluble asphaltenes, precipitated from HCGO, is shown in Table 4.21.

TABLE 4.21
GC/MS Analysis of Toluene Soluble Asphaltenes Precipitated from HCGO

50°C–200°C	200°C–400°C	400°C–600°C	600°C–700°C
Water	Water	Water	Water
Toluene	CO_2	CO_2	CO_2
Xylene	Toluene	Benzene	
Alkylbenzenes	Styrene	Toluene	
	Hexadecanoic acid	Other aromatics	
	Aromatic acids	Unsaturated hydrocarbons	
	Ester phthalates		
	Other O compounds		

Source: From Pillon, L.Z., *Petroleum Science and Technology*, 19, 673, 2001. With permission.

In the first step, at temperatures from 50°C to 200°C, GC/MS analysis showed the presence of water and desorption of toluenes, xylenes, and alkylbenzenes. In the second step, from 200°C to 400°C, GC/MS analysis indicated the presence of CO_2, water, aromatic hydrocarbons and only desorption of O compounds, such as phthalic acid, hexadecanoic acid, and phthalates. In the third step, from 400°C to 600°C, GC/MS analysis indicated the presence of CO_2, water, aromatic hydrocarbons, and unsaturated hydrocarbons indicating the presence of thermal cracking and olefin formation. In the higher temperature range of 600°C–700°C, GC/MS analysis indicated only the presence of CO_2 which indicates the presence of thermal decomposition taking place. The major drawback to the application of GC/MS to study organic compounds is the amount of material that remains as nonvolatile residue. The effect of temperature on the volatile content of toluene soluble asphaltenes, precipitated from HCGO, is shown in Table 4.22.

Polar molecules adsorb at low temperatures and desorb at high temperatures. Desorption of organic molecules usually takes place, below 400°C, and at higher

TABLE 4.22
Volatile Content of Toluene Soluble Asphaltenes

GC/MS Analysis, Temperature, °C	Toluene Soluble Asphaltenes, Volatiles, %
50–200	65.7
200–400	1.6
400–600	5.8
600–700	0.8

Source: From Pillon, L.Z., *Petroleum Science and Technology*, 19, 673, 2001. With permission.

temperatures, above 500°C, their thermal decomposition occurs. In the case of toluene soluble asphaltenes, precipitated from HCGO, in the temperature range of 50°C–400°C only 67 wt% of the sample was volatile. A relatively high volatility of 1-ring aromatics and some volatile O compounds with no presence of volatile S and N compounds, in the temperature range up to 400°C and above, was observed.

A sample studied by mass spectrometry (MS) may be a gas, liquid, or solid. Enough of the sample must be converted to the vapor state to obtain the stream of molecules that must flow into the ionization chamber (Pavia et al., 1996). Different techniques are used to monitor the amount and the chemistry of different molecules concentrating in the gas phase. Temperature programmed desorption of molecules is a fundamental tool used in catalysis (ACCN, 2005). Despite the fact that the S and N were present in toluene soluble asphaltenes as indicated by the elemental analysis, during GC/MS analysis, the S and N molecules remained in the nonvolatile residue and only the presence of O containing volatile molecules was observed.

4.4 SEDIMENT FORMATION

4.4.1 REACTIVITY

It is accepted that the asphaltenes present in crude oils form micelles which are stabilized by resins kept in solution by aromatics. Resins are considered to be natural asphaltene dispersants. The effect of different surfactants on the dispersion of asphaltenes, in crude oils, was reported.

Alkyl phenols were found to show good peptizing properties but were found to partially lose their capacity when oxyethylenic groups were incorporated into the molecule. Long chain aliphatic alcohols and alkylbenzenes were found to be inefficient as dispersing agents. Primary aliphatic amines were found to show some ability to disperse asphaltenes (Gonzales and Middea, 1991). The effect of different molecules on the colloidal stability of asphaltenes, dispersed in pentane, was also reported. Nonylphenol was found to act as a dispersing agent. Stearic acid was found not to affect the colloidal stability of asphaltenes while hexadecylamine was found to act as a flocculant (Lian et al., 1994). The literature reported on the use of dodecylbenzenesulfonic acid (DBSA) as a dispersant to study the electrophoretic mobility of asphaltene particles in ethanol. DBSA was found to decrease the electrophoretic mobility of asphaltene particles until a constant positive value was reached and the neutralization of the positive charges of the asphaltene particles was achieved by the adsorption of DBSA (Leon et al., 2000). The more recent literature reported that synthetic dispersants can increase the solubility of asphaltenes in crude oils and are more effective than resins when used at high concentrations. Synthetic dispersants were reported to make asphaltenes soluble in n-heptane and thereby convert them to resins (Wiehe and Jermansen, 2003). The effect of different commercial dispersants on the stability of HCGO and the C7 insolubles precipitation is shown in Table 4.23.

TABLE 4.23
Effect of Dispersants on Stability of HCGO and C7 Insolubles
Precipitation

HCGO	Sediment	C7 Insolubles, ppm	C7 Insolubles, ppm (repeated experiments)
Neat HCGO	0	171	172
Dispersants (1 wt%)			
Dodecylphenol	0	165	128
Nonylphenol	0	161	171
DBSA, Na salt	0	118	145

Source: From Pillon, L.Z., *Petroleum Science and Technology*, 19, 863, 2001.
Reproduced by permission of Taylor & Francis Group, LLC, http://www.
taylorandfrancis.com.

Before the addition of any dispersants, HCGO was found to contain no sediment and have C7 insoluble content of 171 ppm. The repeated C7 insoluble content of 172 ppm indicated a good repeatability of the ASTM D 3279 test procedure. After the addition of 1 wt% of dodecylphenol, no sediment was observed and the C7 insoluble content was found to vary in a range of 128–165 ppm. After the addition of 1 wt% of nonylphenol, no sediment was observed and the C7 content was found to vary in a range of 161–171 ppm. After the addition of 1 wt% of DBSA Na salt also, no sediment was observed and C7 insolubles content was found to vary in a similar range of 118–145 ppm. The literature reported that two types of polymers, dodecylphenolic resin and poly(octadecene maleic anhydride) prevented asphaltenes, obtained from crude oils, from flocculating in *n*-heptane through the acid–base interaction. At low polymer-to-asphaltene weight ratios, asphaltenes flocculate with themselves and with the polymer while at higher polymer-to-asphaltene weight ratios, asphaltene–polymer aggregates are peptized by the extra polymer and become more stable (Chang and Fogler, 1996). The poly(octadecene maleic anhydride), under the name poly(maleic anhydride-1-octadecene), is produced by reacting maleic anhydride with an olefin, such as 1-octadecene, and is commercially available. The effect of poly(maleic anhydride-1-octadecene), which is polymer, on the stability of HCGO and the C7 insolubles precipitation, is shown in Table 4.24.

While the commercial poly(maleic anhydride-1-octadecene) is advertised as an oil-soluble polymer, some heating and stirring was required to dissolve it in HCGO. Before any addition of a polymer, the HCGO was found to contain no sediment and have a low C7 insoluble content of only 46 ppm. After the addition of 0.1–1 wt% of polymer, no sediment was formed, however, some changes in C7 insolubles precipitation were observed. After the addition of 0.1–0.5 wt% of the polymer, the C7 insolubles precipitation was found to vary in a range of 141–164 ppm, indicating some precipitation increase and its color changed from black to beige. After the addition of 1 wt% of the polymer, the C7 insolubles precipitation

TABLE 4.24

Effect of Poly(Maleic Anhydride-1-Octadecene) Polymer on C7 Insolubles

HCGO	Sediment	C7 Insolubles, ppm	C7 Insolubles Color
Neat HCGO	0	46	Black
Polymer (0.1 wt%)	0	146	Beige
Polymer (0.3 wt%)	0	164	Beige
Polymer (0.5 wt%)	0	141	Beige
Polymer (1 wt%)	0	89	Beige

Source: From Pillon, L.Z., *Petroleum Science and Technology*, 19, 863, 2001. Reproduced by permission of Taylor & Francis Group, LLC, http://www.taylorandfrancis.com.

did not increase any further but actually decreased to 89 ppm and it was beige. FTIR spectrum of C7 insolubles, precipitated in the presence of 0.3 wt% of poly (maleic anhydride-1-octadecene), indicated the presence of sulfates and an increase in the presence of C=O peaks characteristic of an anhydride group. The presence of C=O peaks characteristic of an acid type functional group and the presence of a broad peak at 3500 cm^{-1} characteristic of H-bonded —OH groups were also observed indicating the presence of unreacted poly(maleic anhydride-1-octadecene) polymer.

The literature reported that in crude oils, the multiple polar anhydride groups on a poly(maleic anhydride-1-octadecene) polymer molecule were found to associate with more than one asphaltene molecule resulting in the hetero-coagulation (Chang and Fogler, 1996). According to the literature on asphaltenes present in crude oils, unsaturated bonding of maleic anhydride can react with unsaturation of multiring aromatic structures and a subsequent hydrolysis can yield a product bearing carboxylic acid function (Speight, 1999). The effect of maleic anhydride on the stability of HCGO and C7 insolubles precipitation is shown in Table 4.25.

After an addition of only 0.1 wt% of maleic anhydride, while no sediment formation was observed, the C7 insolubles content of HCGO drastically increased, from 52 to 635 ppm, and its color changed from black to brown. After an addition of 0.3 wt% of maleic anhydride, while no sediment formation was observed, the C7 insolubles content of HCGO further increased to 926 ppm and its color was brown. After an addition of 0.5 wt% of maleic anhydride, while no further increase in C7 insolubles was observed, its color changed from brown to beige. With an additional increase in the maleic anhydride content to 1 wt%, while no sediment formation was observed, the C7 insolubles content of HCGO further increased to 1028 ppm and its color was beige. FTIR spectrum of C7 insolubles, precipitated in the presence of 0.1 wt% of maleic anhydride, indicated the presence of aromatics, multiring aromatics and only one C=O peak at 1720 cm^{-1}, characteristic of an acid type functional group. The presence of a broad peak at 3100–3300 cm^{-1} indicated the presence of H-bonded —OH groups. No presence of C=O peaks

TABLE 4.25
Effect of Maleic Anhydride on Stability of HCGO and C7 Insolubles Precipitation

HCGO	Sediment	C7 Insolubles, ppm	C7 Insolubles Color
Neat HCGO	0	52	Black
Maleic anhydride (0.1 wt%)	0	635	Brown
Maleic anhydride (0.3 wt%)	0	926	Brown
Maleic anhydride (0.5 wt%)	0	817	Beige
Maleic anhydride (1 wt%)	0	1028	Beige

Source: From Pillon, L.Z., *Petroleum Science and Technology*, 19, 863, 2001. Reproduced by permission of Taylor & Francis Group, LLC, http://www.taylorandfrancis.com.

characteristic of anhydride and C=C peaks characteristic of unsaturation indicated no presence of any unreacted maleic anhydride left in HCGO. The effect of temperature and the presence of air on the sediment formation and C7 insoluble precipitation from HCGO is shown in Table 4.26.

A fresh sample of HCGO, containing no sediment and having a relatively low C7 insoluble content of 46 ppm, was heated without a nitrogen blanket. After heating for 2 h at 82°C, no sediment was formed but some increase in C7 insolubles precipitation was observed. After heating for 3 h at 82°C, the sediment was formed and a drastic increase in C7 insolubles precipitation was observed. With an increase in C7 insolubles precipitation, no increase in MCR was observed. Heteroatom content provides polarity and is responsible for oil incompatibility. Its interaction with oxygen changes functional group composition which means that oxidation will affect the oil stability. The literature reported that an increase in the process temperature and the use of chemical additives can increase the oil instability leading to solid precipitation (Wiehe, 2003). Additives are used during the coking process and their thermal stability and boiling point can vary which might lead to an increase in coke-like deposits affecting the catalyst activity or increase the carbon residue. Silicone antifoaming agents are used during the coking process. Demulsification of the oil-in-water emulsion from the column top of a

TABLE 4.26
Effect of Temperature and Air on the Sediment Formation and C7 Insolubles Content

HCGO Stability	Sediment	C7 Insolubles, ppm	C7 Insolubles Color
Fresh HCGO	0	46	Black
2 h at 82°C/air	0	Increase	Black
3 h at 82°C/air	Yes	Drastic increase	Black

delayed coking unit was studied and some new types of cationic demulsifiers were reported more effective in breaking the emulsion (Liu et al., 2004). The use of additives, which is more economical than the optimization of the processing conditions, might further affect the oil stability and even increase the carbon residue.

4.4.2 STORAGE STABILITY

The literature reported that olefins are formed during the catalytic and thermal cracking of petroleum oils and many petroleum products become unstable during their storage or handling. Without the use of hydrogen, the catalytically and thermally cracked petroleum products have a tendency to form a sediment (Gary and Handwerk, 2001). One of the most common operating problems when using HC technology is the formation of coke-like sediment. The carbonaceous sediment was reported to deposit on the reactor, and downstream vessels as well as on the catalyst surface leading to rapid catalyst deactivation and equipment fouling. The addition of aromatic-rich diluents was reported to decrease the sediment formation (Marafi et al., 2005). Popcorn coke was reported to be produced by polymerization of olefins present in thermally cracked oils. The polymerization rate increases with the temperature; however, it was reported to be the greatest in the temperature range of 500°F–580°F (260°C–304°C). The olefins conjugated to aromatics were reported to be the most reactive (Wiehe, 2003). The literature reported that the self-incompatible crude oils, hydrotreated oils, and thermally cracked oils, produced from converted residue, contain insoluble asphaltenes which cause fouling. The oil compatibility model and testing method was developed for detecting self-incompatible oils (Wiehe, 2004). The composition and the storage stability of HCGO, produced under the same processing conditions, are shown in Table 4.27.

TABLE 4.27
Composition and Storage Stability of HCGO

Composition and Stability	HCGO #1	HCGO #2
Density at 15°C, g/cc	0.9438	0.9414
Carbon, wt%	84.0	82.5
Hydrogen, wt%	11.6	11.8
Sulfur, wt%	2.8	2.8
Nitrogen, wt%	0.16	0.18
Visible sediment	No	Yes
C7 insolubles, ppm	35	310
MCR, wt%	0.21	0.2

Source: From Pillon, L.Z., *Petroleum Science and Technology*, 19, 673, 2001. Reproduced by permission of Taylor & Francis Group, LLC, http://www.taylorandfrancis.com.

TABLE 4.28

FTIR Analysis of Toluene Washed Sediment Found in HCGO

FTIR of Sediment from HCGO (Toluene Insoluble Fraction)	FTIR of Sediment from HCGO (Toluene Soluble Fraction)
Aromatics	Aromatics
Multiring aromatics	Multiring aromatics
Carbon black	Aromatic ethers
Sulfates	Carboxylates

Source: From Pillon, L.Z., *Petroleum Science and Technology*, 19, 673, 2001. With permission.

Only some HCGO samples were found to contain suspended particles and required filtration to remove the sediment. HCGO samples, stored for 2 years, were found to contain some visible sediment. The presence of sediment, which could be separated using a 0.8 μm filter, was found to be below 100–150 ppm. HCGO, containing sediment, was found to have an increased C7 insoluble content with no significant increase in MCR. Olefins, formed during the thermal cracking, can polymerize leading to sediment formation but also many other molecules. Heteroatom content provides polarity and interaction with other functional groups can lead to agglomeration and sediment precipitation. The sediment was separated from HCGO, using a 0.8 μm filter, washed with toluene and analyzed. The FTIR analysis of toluene washed sediment, found in HCGO, is shown in Table 4.28.

The FTIR analysis of the toluene insoluble fraction of sediment indicated the presence of aromatics, multiring aromatics, sulfates, —CH, —CH$_2$CH$_3$, and —CH$_3$ groups and the presence of carbon black particles. A broad peak at 3100–3300 cm^{-1} indicated the presence of hydrogen bonding through the —OH group which might involve —OH—S, —OH—N, and —OH—O interactions. The FTIR analysis of the toluene soluble fraction of sediment indicated the presence of aromatics, multiring aromatics, aromatic ethers, carboxylates, —CH, —CH$_2$CH$_3$, and —CH$_3$ groups. No presence of a hydrogen bonded —OH group was observed. The toluene soluble fraction of sediment was found to contain aromatic ethers and carboxylates, not present in C7 insolubles, indicating the presence of oxidation by-products. The optical microscopy analysis of sediment and C7 insolubles, precipitated from HCGO, is shown in Table 4.29.

The optical microscopy of the sediment, separated as filter residue, indicated the presence of a heterogeneous mixture containing large black, rust, and brown particles. The optical microscopy of the *n*-heptane insolubles, precipitated from HCGO without filtration and containing sediment, indicated the presence of some small particles and several large black and rust colored particles. After the filtration and the sediment removal, the optical microscopy of the *n*-heptane insolubles precipitated from HCGO, indicated the presence of a homogeneous and uniform mixture. The color of C7 insolubles was dark and described as

TABLE 4.29

Optical Microscopy Analysis of Sediment and C7 Insolubles

Sediment from HCGO (Filter Residue)	C7 Insolubles from HCGO (Containing Sediment)	C7 Insolubles from HCGO (after Sediment Removal)
Heterogeneous mixture	Heterogeneous mixture	Homogeneous mixture
Several large particles:	Several large particles:	Small dark particles:
black, brown, and rust	black, brown, and rust	very uniform

Source: From Pillon, L.Z., *Petroleum Science and Technology*, 19, 673, 2001. With permission.

between plum and aubergine. SEM, equipped with energy dispersive x-ray spectroscopy (EDS), can be used to determine the particle size and their composition. The SEM–EDS analysis of sediment and C7 insolubles, precipitated from HCGO, is shown in Table 4.30.

The SEM–EDS analysis of sediment confirmed the presence of large particles "chained" together forming agglomerates. The main components of sediment were carbon, sulfur, oxygen, and iron. The SEM–EDS analysis of C7 insolubles, precipitated in the presence of sediment, indicated also the presence of some large particle, >300 μm and chained together forming agglomerates. The main components of C7 insolubles were carbon, sulfur, and iron. The large black particles were primarily carbon and found to resemble a thermal coke. The large rust colored particles were primarily iron indicating the presence of rust particles. The SEM–EDS analysis of C7 insolubles, precipitated from HCGO filtrate and after the sediment removal, indicated the presence of a homogeneous mixture containing small particles, <1 μm, also chained together forming

TABLE 4.30

SEM–EDS Analysis of Sediment and C7 Insolubles from HCGO

SEM–EDS of Sediment (Filter Residue)	SEM–EDS of C7 Insolubles (Containing Sediment)	SEM–EDS of C7 Insolubles (after Sediment Removal)
Majority particles >300 μm	Majority particles >300 μm	Majority particles <10 μm
Forming agglomerates	Forming agglomerates	Forming agglomerates
Main components:		
C, S, O, Fe	C, S, Fe	C, S, O
	Some large black particles	
	Main component: C	
	Some large rust particles	
	Main component: Fe	

Source: From Pillon, L.Z., *Petroleum Science and Technology*, 19, 673, 2001. With permission.

agglomerates. The main components were carbon, sulfur, and oxygen. The presence of heteroatoms, metals, and asphaltenes in the feed was reported to cause sludge and coke formation leading to the catalyst deactivation. Even small amounts of iron, copper, and particularly nickel and vanadium can affect the activity of the catalyst (Speight, 1999). The refining of crude oils leads to some products which are unstable and form sediments. The amount of cracking needs to be limited to prevent polymerization and sludge formation during the storage of petroleum products (Gary and Handwerk, 2001). The sediment from HCGO was also found to contain many other components, such as oxidation by-products, particles of coke and rust.

REFERENCES

ACCN, *Canadian Chemical News*, 57(1), 48, 2005.
Bartholdy, J. et al., *Energy and Fuels*, 15, 1059, 2001.
Bragado, G.A.C., Guzman, E.T., and Yacaman, M.J., *Petroleum Science and Technology*, 19(1&2), 45, 2001.
Chang, C.L. and Fogler, H.S., *Fuel Science and Technology International*, 14, 75, 1996.
Chilingarian, G.V. and Yen, T.F., *Developments in Petroleum Science 40B*, Asphaltenes and Asphalts, Elsevier Science, Amsterdam, 2000.
Ellis, P.J. and Paul, C.A., *Delayed Coking Fundamentals*, AIChE 2000 Spring National Meeting, Atlanta, May 5–9, 2000.
Gary, J.H. and Handwerk, G.E., *Petroleum Refining, Technology and Economics*, 4th ed., Marcel Dekker, Inc., New York, 2001.
Gonzales, G. and Middea, A., *Colloids and Surfaces*, 52, 207, 1991.
Gray, M.R., *Energy and Fuels*, 17, 1566, 2003.
Gray, M.R. and McCaffrey, W.C., *Energy and Fuels*, 16, 756, 2002.
Hauser, A. et al., *Energy and Fuels*, 19, 544, 2005.
Leon, O. et al., *Polymer Science and Technology*, 18, 913, 2000.
Li, S., Liu, C., and Liang, W., *American Chemical Society, Abstracts of Papers*, 222nd ACS National Meeting, Chicago, August 26–30, 2001.
Lian, H., Lin, J.R., and Yen, T.F., *Fuels*, 73, 423, 1994.
Liu, G., Xu, X., and Gao, J., *Petroleum Science and Technology*, 22(3&4), 233, 2004.
Liu, Z. and Yan, F., *Petroleum Science and Technology*, 21(11&12), 1887, 2003.
Marafi, M. and Stanislaus, A., *Petroleum Science and Technology*, 19(5&6), 697, 2001.
Marafi, M., Al-Barood, A., and Stanislaus, A., *Petroleum Science and Technology*, 23 (7&8), 899, 2005.
Moschopedis, S.E. and Speight, J.G., *Fuel*, 53, 222, 1974.
Murgich, J., *Petroleum Science and Technology*, 20(9&10), 938, 2002.
Pavia, D.L., Lampman, G.M., and Kriz, G.S., *Introduction to Spectroscopy*, Harcourt Brace College Publishers, Orlando, 1996.
Pillon, L.Z., *Petroleum Science and Technology*, 19(5&6), 673, 2001.
Pillon, L.Z., *Petroleum Science and Technology*, 19(7&8), 863, 2001.
ProQuest Document ID 10552585, *Chemical Engineering*, 103(12), 99, 1996.
Roberts, I., *Preprints-ACS, Division of Petroleum Chemistry*, 34(2), 251, 1989.
SanchezIlanes, M.T. and Ancheyta, J., *Petroleum Science and Technology*, 22(1&2), 73, 2004.

Sequeira, A. Jr., *Lubricant Base Oil and Wax Processing*, Marcel Dekker, Inc., New York, 1994.

Sharma, Y.K., Reimert, R., and Singh, I.D., *Petroleum Science and Technology*, 21(7&8), 1055, 2003.

Shell, http://www.shell-lubricants.com (accessed May 2001).

Siddiqui, M.N., *Petroleum Science and Technology*, 21(9&10), 1601, 2003.

Speight, J.G., Evidence for the types of polynuclear aromatic systems in nonvolatile fractions of petroleum, in *Polynuclear Aromatic Compounds*, Ebert, L.B., Ed., American Chemical Society, Washington, DC, 1988.

Speight, J.G., *The Chemistry and Technology of Petroleum*, 3rd ed., Marcel Dekker, Inc., New York, 1999.

Speight, J.G., *The Chemistry and Technology of Petroleum*, 4th ed., CRC Press, Boca Raton, 2006.

Tanaka, R. et al., *Energy and Fuels*, 18, 1118, 2004.

Trasobares, S. et al., *Industrial and Engineering Chemistry Research*, 37, 11, 1998.

Van den Berg, F.G.A. et al., *Petroleum Science and Technology*, 21(3&4), 557, 2003.

Wang, J. and Buckley, J.S., *Energy and Fuels*, 17, 1445, 2003.

Wiehe, I., *Petroleum Science and Technology*, 21(3&4), 673, 2003.

Wiehe, I., *Petroleum Science and Technology*, 25(3&4), 333, 2004.

Wiehe, I. and Jermansen, T.G., *Petroleum Science and Technology*, 21(3&4), 527, 2003.

Yarranton, H.W. and Musliyah, J.H., *Thermodynamics*, 42, 3533, 1996.

5 Daylight Stability

5.1 EFFECT OF HYDROPROCESSING SEVERITY ON SOLVENCY AND STABILITY

The preferred molecules for lube oil manufacture are iso-paraffins having a high viscosity index (VI) and low pour point. The viscosity of base stocks is defined by the boiling range of hydrocarbons while their VI is defined by their chemistry. The typical VI of solvent refined paraffinic base stocks is in a range of 85–95 and many crude oils are not suitable for the production of lube oil base stocks. Both paraffinic and iso-paraffinic hydrocarbons are high VI components of petroleum while hydrocarbons containing ring structures, such as aromatic and some poly-naphthenic hydrocarbons, are low VI components. The processes of hydrotreating, hydrocracking, and hydroisomerization were introduced to meet the ever increasing demand for higher quality mineral base stocks having a higher VI. Hydrogenation processes for the conversion of petroleum fractions can be classified as nondestructive and destructive. Nondestructive hydrogenation is generally used for the purpose of improving product quality without changing the boiling range. Destructive hydrogenation, such as hydrocracking, is characterized by the cleavage of carbon–carbon bonds (Speight, 2006).

Some petroleum products require the hydrogen treatment to saturate olefins. Hydrogen finishing processes, known as hydrofining, are mild hydrogenation processes. The commercial hydrofining processes are based on heating the feedstock in a furnace and passing it with hydrogen through a reactor filled with catalyst. After passing through the reactor, the treated oil is cooled, separated from the excess hydrogen and pumped to a stripper tower, where hydrogen sulfide is removed by steam, vacuum, or flue gas. The finished product leaves the bottom of the stripper tower and has improved color, odor, and lower sulfur content (Speight, 2006). Under mild hydrofining conditions, nitrogen, sulfur, and oxygen compounds undergo hydrogenolysis to split out ammonia, hydrogen sulfide, and water, respectively. Olefins are saturated and the aromatic contents of finished products are usually not affected (Speight, 2006). The hydrofining process is used to finish naphthas, gas oils, and lube oil base stocks.

The typical hydrotreating operation requires the reactor temperature of 270°C–340°C and the pressure of 100–3000 psig (Gary and Handwerk, 2001). Hydrocracking is used to reduce the boiling point of the feed. The typical

TABLE 5.1

Effect of Hydroprocessing Severity on VI and Solvency Properties of Base Stocks

Processing Severity	Hydrofinishing, Low	Hydrotreatment, Moderate	Hydrocracking, High	Hydroisomerization, High
Feedstock	Solvent refined	Solvent refined	Distillates	Slack wax
Purpose	Saturate olefins	Saturate olefins	Saturate olefins	Saturate olefins
	Remove S and N	Remove S and N	Remove S and N	Remove S and N
		Saturate aromatic	Ring opening	Ring opening
VI	90–105	>105	95–130	>140
Solvency	Excellent	Very good	Moderate	Poor

Source: From Pirro, D.M. and Wessol, A.A., *Lubrication Fundamentals*, 2nd ed., Marcel Dekker, Inc., New York, 2001. Reproduced by permission of Routledge/Taylor & Francis Group, LLC.

hydrocracking operation requires the reactor temperature of 290°C–400°C and the reactor pressure of 1200–2000 psig. During hydrocracking, heat cracking leads to olefins which are saturated in the presence of hydrogen to form paraffins. Without the use of hydrogen, the catalytically and thermally cracked petroleum products have a tendency to form a sediment (Gary and Handwerk, 2001). The typical hydrocracking operation requires the circulation of a large amount of hydrogen to prevent the excessive catalyst fouling. Polynuclear aromatics, such as coronenes, were reported present in some hydrocracked base stocks (Gary and Handwerk, 2001). Commercially available coronene is described as an unsubstituted 6-ring polynuclear aromatic structure (Aldrich, 2007). The effect of hydroprocessing severity on VI and solvency properties of base stocks is shown in Table 5.1.

During the hydrocracking process, the aromatic content can be reduced to low levels through many different reactions such as heteroatom removal, aromatic ring saturation, dealkylation of the aromatic rings, ring opening, straight chain and side chain cracking, and wax isomerization (Cody et al., 2002). The typically low quality feedstocks used in hydrocracking, and the consequent severe conditions required to achieve the VI and volatility might result in the formation of toxic polynuclear aromatic molecules (Cody et al., 2002). Heavy polynuclear aromatics were reported to form in small amounts from hydrocracking reactions and, when the fractionator bottoms are recycled, can build up to concentrations that cause fouling of heat exchanger surfaces and equipment (Gary and Handwerk, 2001).

The main objective of the hydroisomerization reactions is to convert the naphthenic and aromatic ring structures into straight chain and branched chain compounds. It is necessary to crack the side chains on these ring structures to reduce the chain length and saturate the molecules thus formed. Not all ring

TABLE 5.2

Advantages and Disadvantages of Lube Oil Hydrocracking Process

HC Process Advantage	HC Process Disadvantage
Use of poor quality crudes	Catalytically dewaxed bright stocks are hazy
Higher yields	Tendency to darken on exposure to light
Conversion of residual oils to distillate oils	Tendency to form sludge
Higher VI	Exhibit additive solubility problems

Source: From Sequeira, A. Jr., *Lubricant Base Oil and Wax Processing*, Marcel Dekker, Inc., New York, 1994. Reproduced by permission of Routledge/Taylor & Francis Group, LLC.

structures will open (Phillips, 1999). In the case of Shell extra high viscosity index (XHVI) and Exxon Exxsyn base stocks hydroisomerization processes, the feedstock is slack wax. The unconverted wax is removed by solvent dewaxing and it is recycled with the hydrotreated slack wax feed to the reactor (Phillips, 1999). With an increase in VI of petroleum derived base stocks, a decrease in aromatic and naphthenic hydrocarbons with an increase in paraffinic and iso-paraffinic contents will decrease their solvency properties. The use of high VI base stocks in lubricating oils and a decrease in their solvency properties will lead to a decrease in the solubility of additives affecting the lubricant perform-ance. The advantages and disadvantages of the hydrocracking process are shown in Table 5.2.

Olefins are formed during the catalytic and thermal cracking of crude oils and were reported to affect the products stability. Olefins are easily oxidized and polymerize. Diolefins are more reactive and quickly form high molecular weight polymers leading to sludge formation (Sequeira, 1994). Hydrocracked base stocks were reported to darken and form sediment on exposure to light. The severely hydrotreated base stocks, produced from vacuum distillate or deas-phalted vacuum residue, were reported to form a flocculant precipitate upon prolonged exposure to ultraviolet (UV) light known as daylight stability (Bijwaard and Morcus, 1985). The oxidation stability and daylight stability of severely hydrotreated base stocks are shown in Table 5.3.

Oxidation of base stocks, leads to an increase in acidity, viscosity, and sludge. Oxidized base stocks, having a tendency to form sludge, were reported to have daylight stability of 2–9 days. Other oxidized base stocks, having no tendency to form sludge, were reported to have a daylight stability of over 15 days. The use of mild catalytic hydrotreatment and solvent extraction was reported effective in improving their oxidation stability and daylight stability (Bijwaard and Morcus, 1985). The literature reported that in lubricating oils, under low temperature oxidation conditions, peroxides, alcohols, aldehydes, ketones, and water are formed. Under high temperature oxidation conditions, acids are formed (Bardasz and Lamb, 2003). Hydroconversion of under-extracted raffinate was reported to produce base stocks, having a high VI and low volatility for a given viscosity,

TABLE 5.3

Oxidation Stability and Daylight Stability
of Severely Hydrotreated Base Stocks

Acidity, mg eq/100 g	Viscosity Increase, %	Sludge, wt%	Daylight Stability, Days
2	9	0	>15
3	19	0.4	2
3	15	0.2	6
3	20	0.4	5
3	17	0.2	9
4	22	0.5	6
7	35	0	>15

Source: From Bijwaard, H.M.J. and Morcus, A., *Lubricating Base Oil Compositions*, CA Patent 1,185,962, 1985.

with improved oxidation stability and solvency properties (Cody et al., 2002). Different processing techniques are used to meet the volatility targets of base stocks at the max yields. The effect of different processing on volatility and yields of 100N waxy raffinate is shown in Table 5.4.

The volatility can be improved by removing the low boiling front end, known as topping, which increases the viscosity of the oils. Another route to improving the volatility is to remove the high boiling and low boiling ends, known as heart-cut, which maintains a constant viscosity (Cody et al., 2002). Topping of waxy raffinate can lead to a decrease in volatility from 27.8% to 26.2% off, with a decrease in yield from 100% to 95.2%. Heart-cut distillation can lead to waxy raffinates having a lower volatility of 21.7%–22.7% off with a lower yield of only 38%–58% (Cody et al., 2002). The distillate feeds to the extraction zone can be of poor quality and contain

TABLE 5.4

Effect of Processing on Volatility and Yields of 100N
Waxy Raffinate

Kinematic Viscosity, 3.9 cSt at 100°C	Waxy Raffinate, Volatility, % off	Waxy Raffinate, Yield, %
None	27.8	100
Topping	26.2	95.2
Heart-cut distillation	22.7	58
Heart-cut distillation	22.4	50.8
Heart-cut distillation	21.7	38

Source: From Cody, I.A. et al., *Raffinate Hydroconversion Process*, CA Patent 2,429,500, 2002.

TABLE 5.5

Effect of Hydrocracking and Raffinate Hydroconversion
on the Yield of Lube Oil

Properties	Hydrocracking, Two-Stage Process	Raffinate Hydroconversion
Viscosity at 100°C, cSt	6.5	6.5
Volatility, % off	3.3	3.6
Yield, %	30.5	69.7

Source: From Cody, I.A. et al., *Raffinate Hydroconversion Process*, CA Patent 2,429,500, 2002.

over 1 wt% of sulfur and nitrogen. The use of raffinate hydroconversion (RHC) technology is based on hydrotreating, redistilling, and solvent refining. The raffinate from the solvent extraction unit is stripped of solvent and sent to the hydroconversion unit. For the same viscosity base stocks, the use of different processing affects the volatility and yields of base stocks. The effect of a two-stage hydrocracking process and the RHC process on the yield of lube oil fraction is shown in Table 5.5.

The use of the RHC was found to significantly increase the yield of the lube oil fraction, having the same viscosity of 6.5 cSt at 100°C and a similar volatility of 3.3%–3.6% off from 30% to 70%. The use of hydrocracking technology suffers from yield debits and was reported to require high capital investments (Cody et al., 2002). The patent literature reported that hydroconversion of under-extracted raffinate was found to remove multiring aromatics which are known to have an adverse effect on the viscosity, VI, toxicity, and color of base stocks. The dewaxed oil was reported to be suitable for use as a lubricant base stock (Cody et al., 2002). The properties, VI, and the aromatic contents of RHC lube oil base stocks are shown in Table 5.6.

TABLE 5.6

VI, Aromatic Content, and Volatility of RHC
Base Stocks

Properties	RHC Oil	RHC Oil
Viscosity at 100°C, cSt	4.5	5.9
VI	116	114
Pour point, °C	−18	−18
Saturates, wt%	98	97
Aromatics, wt%	2	3
Volatility	14	8

Source: From Cody, I.A. et al., *Raffinate Hydroconversion Process*, CA Patent 2,429,500, 2002.

RHC oil, having a viscosity of 4.5 cSt at 100°C, was reported to have a VI of 116, 2 wt% of aromatics to assure good solvency properties, and a volatility of 14% off. Another RHC oil, having a higher viscosity of 5.9 cSt at 100°C, was reported to have a VI of 114, 3 wt% of aromatics, and a lower volatility of 8% off. The literature reported that the quality of RHC oils, in terms of VI, was better than that of mineral base stocks, however, not as good as the quality of XHVI base stocks produced by hydroisomerization of slack wax which have a higher VI, above 140, and have a lower Noack volatility when compared to other hydro-crackates of the same viscosity (Phillips, 1999). The use of hydroprocessing can involve the hydrotreatment upgrade of solvent raffinate followed by dewaxing and hydrofinishing. The toxicity of the base stock, for a given VI, is controlled during the cold hydrofinishing step by adjusting the temperature and the pressure. The hydroconverted raffinate is subjected to a cold finishing step and sent to a vacuum stripper to separate the low boiling components (Cody et al., 2002). During the RHC step, the processes of hydrocracking and hydroisomerization are minimized.

5.2 EFFECT OF HYDROPROCESSING SEVERITY ON UV ABSORBANCE

The severely hydrotreated and hydrocracked base stocks form sediment after being exposed to UV light known as daylight or UV stability, depending on test conditions. Compounds which are highly colored have adsorption in the visible (VIS) region. Most organic molecules and functional groups are transparent in the portions of the electromagnetic spectrum which we call the UV region, however some molecules containing O, N, S, and aromatics can adsorb, depending on their chemistry (Pavia et al., 1996). When continuous radiation passes through a transparent material, a portion of the radiation may be absorbed. If that occurs, the residual radiation, when it is passed through a prism, yields a spectrum with gaps in it, called an absorption spectrum. The chemistry and UV absorbance of different chemistry aliphatic and aromatic organic molecules were reported (Pavia et al., 1996). The chemistry and UV absorbance of some aliphatic and aromatic organic compounds are shown in Table 5.7.

Aliphatic ketones, aromatic esters, and aromatic acids, such as benzoic acid, were reported to adsorb in a similar UV range of 270–280 nm. Naphthalene, which is a 2-ring condensed aromatic compound, was reported to adsorb in a UV range of 220–320 nm. In the UV range of 280–350 nm, some aliphatic ketones and many different chemistry aromatic molecules can adsorb. Anthracene, which is a 3-ring condensed aromatic compound, was reported to adsorb in a UV range of 255–380 nm (Pavia et al., 1996). Polynuclear aromatics include two or more aromatic rings and can be fused, such as naphthalene and phenantrene, or separate as biphenyl. With an increase in condensed ring number, the UV absorbance increases to 250–500 nm. Coronene, containing six condensed aromatic rings, was found to adsorb in a wide range of 250–400 nm. The literature reported on variation in the UV absorbance of large polynuclear hydrocarbons

TABLE 5.7
Chemistry and UV Absorbance of Aliphatic and Aromatic Organic Molecules

Organic Molecules	Structure	UV Absorbance (nm)
Olefin	$R_2C=CR_2$	175
Aliphatic alcohol	R-OH	180
Aliphatic aldehyde	R-CHO	190, 290
Aliphatic ketone	R_2CO	180, 280
Aliphatic acid	R-COOH	205
Aliphatic ester	R-COOR	205
Benzene	Ar-H	203, 254
Aromatic alcohol	Ar-OH	211
Aromatic aldehyde	Ar-CHO	250
Aromatic ketone	$Ar-COCH_3$	246
Aromatic acid	Ar-COOH	230, 272, 282
Aromatic ester	Ar-COOR	224, 268
Polynuclear aromatic	Naphhtalene	220, 240–320
Polynuclear aromatic	Antracene	255, 290–380

Source: From Pavia, D.L., Lampman G.M., and Kriz, G.S., *Introduction to Spectroscopy*, Harcourt Brace College Publishers, Orlando, 1996.

containing 9–11 condensed aromatic rings (Fetzer, 1988). The UV absorbance of large polynuclear hydrocarbons, containing condensed aromatic rings, is shown in Table 5.8.

According to the literature, some aromatic compounds present in base stocks can form complexes with oxygen even during the storage and develop an UV absorption peak in the region of 260–300 nm (Cooney and Hazlett, 1984). The literature reported on the fuel stability studies where nitrogen heterocycles, such as alkypyrroles, were reported to promote formation of sediment indicating that an autoxidation process was taking place. The molecular association of oxygen

TABLE 5.8
UV Absorbance of Large Polynuclear Hydrocarbons

UV Absorbance	UV Range (nm)	UV Max (nm)
6-Ring condensed (coronene)	250–400	293, 305, 341
9-Ring condensed	>250–500	290, 310, 340, 350, 440, 470
10-Ring condensed	>250–500	280, 350, 470, 520
11-Ring condensed	>250–500	275, 310, 360, 380, 450, 500

Source: From Fetzer, J.C., *Polynuclear Aromatic Compounds*, Ebert, L.B., Ed., American Chemical Society, Washington, DC, 1988.

TABLE 5.9
UV Absorbance of Oxygenated and Deoxygenated 2,5-Dimethylpyrrole

2,5-Dimethylpyrrole	UV Range (nm)	UV Max (nm)
Deoxygenated	>200–250	209, 225
Oxygenated	>200–290	270, 282

Source: From Cooney, J.V. and Hazlett, R.N., *Heterocycles*, 22, 1513, 1984.

with 2,5-dimethylpyrrole was reported to increase the UV absorbance in the range of 285 nm (Cooney and Hazlett, 1984). The UV absorbance of oxygenated and deoxygenated 2,5-dimethylpyrrole is shown in Table 5.9.

The analysis of different liquid phase oxidation products of synthetic and mineral oils claims the broad absorbance peak at about 285 nm to be distinct in the UV spectra of all oxidation products. The literature reported that unsaturated ketones adsorb in this region (Naidu et al., 1984). Naphthenic base stocks are known to have excellent solvency properties due to a high content of naphthenic and aromatic hydrocarbons. The recent literature reported that the hydrotreated naphthenic rubber base oil is a new product obtained by hydrogenation and it intensely discolors during UV radiation (Han et al., 2005). The properties, composition, and UV absorbance of hydrotreated (HT) naphthenic rubber oil are shown in Table 5.10.

The literature reported that with an increase in UV absorbance, the polar content of UV radiated HT naphthenic oil significantly increases from 0.1 to

TABLE 5.10
Properties, Composition, and UV Absorbance of HT Naphthenic Base Oil

HT Naphthenic Base Oil	Properties	Test Method
Boiling range, °C	380–450	
Kinematic viscosity, mm^2/s		ASTM D 4052
At 40°C	130.6	
At 100°C	9.67	
Pour point, °C	−27	ASTM D 97
Silica gel separation		ASTM D 2140
Saturates, wt%	95.6	
Aromatics, wt%	4.29	
Polars, wt%	0.11	
TAN, mg KOH/g	0.01	ASTM D 974
UV absorbance at 260 nm	0.0684	ASTM D 2008

Source: Reproduced with permission from Han, S. et al., *Energy and Fuels*, 19, 625, 2005. Copyright American Chemical Society.

TABLE 5.11

Elemental Analysis of Polar Content of UV Radiated HT Naphthenic Base Oil

Elemental Analysis	HT Naphthenic Oil	Separated Polar Content
Carbon, wt%	86.6	81.3
Hydrogen, wt%	14.4	11.2
Sulfur, ppm	5	96
Nitrogen, ppm	4	142
Oxygen, wt% (calc.)	1.86	7.46
H/C	1.86	1.66

Source: Reproduced with permission from Han, S. et al., *Energy and Fuels,* 19, 625, 2005. Copyright American Chemical Society.

3.1 wt%. To identify the polar content of HT naphthenic base stock, the oil was passed over silica gel and petroleum ether and benzene were used to wash off saturates and aromatics. The polar material retained on silica was removed using ethanol and analyzed (Han et al., 2005). Most of heteroatom content was present in the polar fraction and a drastic increase in O content, after UV radiation, indicated photooxidation by oxygen in air. The elemental analysis of UV radiated HT naphthenic oil and its polar content is shown in Table 5.11.

The molecular weight distribution of the polar fraction was reported to be a complex mixture containing many different compounds having a molecular weight distribution of 200–500. FTIR indicated the presence of –OH and –NH groups at 3385 cm^{-1}, C=O group at 1709 cm^{-1}, and the presence of sulfoxides at 1030 and 1070 cm^{-1}. XPS analysis also indicated the presence of –COOH, –COOR, and –N=N– groups (Han et al., 2005). The photooxidized products were identified in the polar fraction of HT naphthenic rubber oil; however, no sediment formation was reported. The most common petroleum solvents are mineral spirits, xylene, toluene, naphthas, hexane, and heptane. The solvency properties of hydrocarbons decrease from aromatics to naphthenes and paraffins. The use of different feed and more severe hydroprocessing leads to XHVI base stocks, having a very low naphthenic content and practically no aromatic and heteroatom content. A trace presence of coronenes was found. Upon prolonged exposure to UV light, XHVI base stocks form a precipitate, similar to a "haze" formation, also known as daylight stability. The properties, composition, and daylight stability of different XHVI base stocks are shown in Table 5.12.

XHVI 6 base stock, having a higher kinematic viscosity of 5.77 cSt at 100°C and a higher VI of 147, was found to contain a higher aromatic/polar content of 2.2 wt%, higher sulfur of 48 ppm, and N content of 2 ppm. It was also found to have a darker color and a daylight stability of only 3 days. XHVI 5 base stock, having a viscosity of 5.13 cSt at 100°C and a VI of 144, was found to contain a lower aromatic/polar content of less than 1 wt% and a lower sulfur content of only 2 ppm. It was found to have a lighter color and an excellent daylight stability of over 70 days. XHVI base

TABLE 5.12
Composition and Daylight Stability of Different
XHVI Base Stocks

Composition and Stability	XHVI 6	XHVI 5
Kinematic viscosity, cSt		
At 40°C	28.82	24.44
At 100°C	5.77	5.13
VI	147	144
Pour point, °C	−15	−21
Aromatics, wt%	2.2	<1
Sulfur, ppm	48	3
Nitrogen, ppm	2	2
Coronenes, ppm	0.6	0.1
Daylight stability, days	3	>70
Volatility, % off	14.2	11.1

Source: From Pillon, L.Z., *Petroleum Science and Technology*,
19, 1263, 2001; Pillon, L.Z., *Petroleum Science and
Technology*, 20, 223, 2002.

stocks were found to contain a trace amount of polynuclear aromatics, similar to
coronenes. XHVI 6 base stock, having a daylight stability of only 3 days, was also
found to contain an increased content of 0.6 ppm of coronenes. The UV absorption
data of petroleum products are usually obtained by dissolving 1 g of oil in 25 mL of
chloroform. In case of very low aromatic content, a sample of oil is extracted with
dimethyl sulfoxide (DMSO) and the UV absorbance of the extract is determined. The
UV absorbance of DMSO extracts from XHVI base stocks, having a different
daylight stability, is shown in Table 5.13.

TABLE 5.13
UV Absorbance of DMSO Extracts from Different XHVI
Base Stocks

Base Stocks, Daylight Stability	XHVI 6, 3 Days	XHVI 5, >70 Days
UV of DMSO extract		
260–289 nm	>2	0.3
290–299 nm	>2	1.9
300–329 nm	>2	3.2
330–350 nm	>2	1.1

Source: From Pillon, L.Z., *Petroleum Science and Technology*, 19, 1263,
2001. Reproduced by permission of Taylor & Francis Group, LLC,
http://www.taylorandfrancis.com.

TABLE 5.14
UV Absorbance of DMSO Extract from XHVI 5 and Coronene

XHVI 5-DMSO, Max UV	XHVI 5-DMSO, UV Absorbance	Coronene-DMSO, Max UV
294 nm	0.24	293 nm
305 nm	0.53	305 nm
342 nm	0.14	341 nm

Source: From Pillon, L.Z., *Petroleum Science and Technology*, 19, 1263, 2001. Reproduced by permission of Taylor & Francis Group, LLC, http://www.taylorandfrancis.com.

The aromatic fraction of petroleum products, separated and eluted with polar solvents, such as DMSO, may contain aromatics, condensed naphthenic-aromatics, aromatic olefins, and polar compounds containing sulfur, nitrogen, and oxygen. The UV absorbance of DMSO extract from XHVI 6 base stock indicated a high absorbance of 3–3.2 units in the wide range of 260–350 nm.

The UV absorbance of DMSO extract from XHVI 5 base stock indicated a high absorbance of 3.2 units in the narrow range of 300–329 nm but significantly lower UV absorbance in the range of 260–299 nm and 330–350 nm. While the composition of XHVI base stocks was found to vary, the composition of slack wax feed and the processing conditions, including the stabilization or the finishing step, might also be different. The UV absorbance of DMSO extract from XHVI 5 base stock and commercial coronene is shown in Table 5.14.

The UV absorbance of DMSO extract from XHVI 5 base stock indicated the presence of max absorbance at 294, 305, and 342 nm. The highest absorbance was observed at 305 nm which is characteristic of polynuclear aromatic structures. XHVI 5 base stock containing polynuclear aromatics, such as coronenes, was found to have an excellent daylight stability. Commercial coronene, diluted in chloroform, has three distinctive UV absorbance peaks at 293, 305, and 341 nm. The highest absorbance of coronene is observed at 305 nm. The DMSO extracts of base stocks contain aromatics and any other polar molecules containing S, N, or O. An attempt to use DMSO extracts for further analysis was difficult due to the high boiling point of the DMSO solvent and difficulties to remove it. The use of double extraction where the DMSO extract is extracted with chloroform can be used to perform FTIR analysis. The FTIR analysis of DMSO/CHCl3 extracts from differ-ent XHVI base stocks, having a different daylight stability, is shown in Table 5.15.

XHVI 5 base stock, having a light color and trace of coronenes (0.1 wppm), was found to have an excellent daylight stability of >70 days. The FTIR analysis indicated the presence of peaks at 1312, 1132, 953, 841, and 824 cm^{-1} and con-firmed the coronene molecular structure. The FTIR analysis of the DMSO/CHCl3 extract also indicated the presence of aromatic compounds, ketones, acids, and esters. The FTIR analysis of the DMSO/CHCl3 extract from darker XHVI 6 and having the daylight stability of only 3 days also confirmed the presence of polynuclear

TABLE 5.15
FTIR Analysis of DMSO/CHCl₃ Extracts from Different
XHVI Base Stocks

Base Stocks, Daylight Stability	XHVI 6, 3 Days	XHVI 5, >70 Days
FTIR	Aromatics	Aromatics
DMSO/CHCl3 extract	Coronenes	Coronenes
	Ketones	Ketones
	Acids	Acids
	Esters	Esters

Source: From Pillon, L.Z., *Petroleum Science and Technology*, 19, 1263, 2001. Reproduced by permission of Taylor & Francis Group, LLC, http://www.taylorandfrancis.com.

aromatics, such as coronenes. The FTIR analysis also indicated the presence of aromatic compounds, ketones, acids, and esters. Under hydrocracking conditions, many reactions can occur and, other than the presence of coronenes, XHVI base stocks were also found to contain polar oxidized aromatic compounds. The hydroisomerized base stocks, known as extra high viscosity index (XHVI) or slack wax isomerate (SWI), have a high saturate content, poor solvency properties and their daylight stability was found to vary.

5.3 SLACK WAX HYDROISOMERIZATION PROCESS

The petroleum industry often employs two-stage processes in which the feedstock undergoes the hydrotreating to protect the hydrocracking catalyst. Slack wax hydroisomerization is conducted over a catalyst containing a hydrogenating metal component, preferably platinum on a halogenated refractory metal oxide support. The halogenated metal oxide support is typically an alumina containing chlorides and fluorides (Pillon and Asselin, 1993). The fluoride and the platinum contents of catalysts can vary. At lower hydroisomerization temperature, the use of a high content of F in a range of 6%–7% and a Pt content in a range of 0.9% is required to increase the catalyst activity. Hydroisomerization can be conducted at the temperature of 270°C–400°C, at pressures of 500–3000 psi H₂, at hydrogen gas rates of 1000–10,000 SCF/bbl, and at a space velocity of 0.1–10 v/v/h (Pillon and Asselin, 1993). The commercial processes for the manufacture of slack wax isomerate base stocks are based on the hydrocracking and the hydroisomerization of HT slack wax feed followed by the fractionation and solvent dewaxing (Sequeira, 1994). Slack waxes contain 5–50 wt% of oil and might require the deoiling step to lower their oil content. After the deoiling step and before the hydroisomerization step, the hydrotreating of slack wax is required to remove the sulfur and nitrogen (Phillips, 1999). The effect of hydroprocessing on composition of slack wax feed and SWI is shown in Table 5.16.

TABLE 5.16
Effect of Hydroprocessing on Composition of Slack
Wax Feed and Isomerate

Processing	Deoiled Slack Wax	HT Slack Wax	Solvent Dewaxed SWI
Saturates, wt%	90.3	93.4	99.1
Aromatics, wt%	9.7	6.4	0.9
Sulfur, ppm	—	4	<1
Nitrogen, ppm	—	1	<1

Source: From Pillon, L.Z., *Petroleum Science and Technology*, 20, 223, 2002. Reproduced by permission of Taylor & Francis Group, LLC, http://www.taylorandfrancis.com.

Slack waxes from solvent dewaxing of waxy raffinates can contain up to 50 wt% of oil. The oil content of slack waxes, used as a feed, can vary from 0 to 45 wt% usually from 5 to 30 wt% (Pillon and Asselin, 1993). After hydrotreatment, the aromatic content of slack wax feed decreased from 9.7 to 6.4 wt%, the S content decreased to 4 ppm, and the N content decreased to 1 ppm. The use of hydrogen processing decreases the sulfur and nitrogen contents of the slack wax feed which protects the hydroisomerization catalyst. After the hydro-isomerization step, fractionation and solvent dewaxing, the aromatic content further decreased to 0.9 wt% and practically no presence of S and N was observed.

The literature reported on the use of mass spectrometry (MS) to analyze the composition of base stocks (Huo et al., 2000). The determination of hydrocarbon types by MS requires a separation of the petroleum sample into saturate and aromatic fractions using column chromatography and polar solvents. The separation of saturate and aromatic/polar fractions from petroleum high boiling oils can be achieved by elution chromatography. A weighted amount of petroleum oil is charged to the top of a glass chromatographic column and *n*-pentane can be used to elute the saturates. When all saturate hydrocarbons are eluted, polar solvents (diethyl ether, chloroform, and ethyl alcohol) are used to elute the aromatic and polar fractions. The solvents are completely evaporated and the residues are weighted. The recovery of the fractions might affect the results and it is usually low for waxy samples. In the case of not complete separation, the saturate fraction might contain some residual aromatic compounds. The MS analysis of the saturate fractions of slack wax feed and slack wax isomerate base stock is shown in Table 5.17.

The saturate fraction, eluted with *n*-pentane, usually contains a mixture of paraffinic, iso-paraffinic, and naphthenic hydrocarbons. The paraffinic hydrocarbons were reported as one lump called the paraffinic/iso-paraffinic content. Slack wax, containing 20 wt% of oil, was found to contain 71.3 vol% of paraffin/iso-paraffin content and 19.1 vol% of 1–6 ring naphthenes. After hydrotreating,

TABLE 5.17

Effect of Hydroprocessing on Saturate Hydrocarbon Composition

Hydroprocessing, Saturate Fractions	Deoiled Slack Wax	HT Slack Wax	Solvent Dewaxed SWI
MS analysis, vol%			
Paraffins/iso-paraffins	71.3	71.5	86.6
1-Ring naphthenes	10.6	12.3	7.6
2-Ring naphthenes	3.4	3.8	2.7
3-Ring naphthenes	2.2	2.5	0.3
4-Ring naphthenes	1.7	1.9	0.3
5-Ring naphthenes	0.7	0.7	0.5
6-Ring naphthenes	0.5	0.5	0.5
Average ring			
Number (per mole)	0.4	0.4	0.2

Source: From Pillon, L.Z., *Petroleum Science and Technology*, 20, 223, 2002. Reproduced by permission of Taylor & Francis Group, LLC, http://www.taylorandfrancis.com.

with a decrease in the aromatic content from 6.9 to 3 wt%, the 1-ring naphthenic content increased from 10.6 to 12.3 vol%, indicating that the hydrogen saturation of aromatics was taking place. With a decrease in the aromatic content, the average ring number per mole of 0.4 did not change indicating the absence of ring-opening reactions and no significant increase in the paraffin/iso-paraffin content of slack wax feed was observed.

After the hydroisomerization, with a further decrease in the aromatic content and a decrease in naphthenic content, an increase in the paraffin/iso-paraffin content of slack wax isomerate from 71.5 to 86.6 vol% is observed. Since *n*-paraffins crack easily, the paraffinic/iso-paraffinic content of slack wax isomerate will basically consist of iso-paraffins. The MS analysis showed a decrease in the average ring number per mole from 0.4 to 0.2, indicating that the ring-opening reactions were taking place. The MS analysis of the aromatic fractions of slack wax feed and slack wax isomerate base stock is shown in Table 5.18.

The aromatic fraction of petroleum products, separated and eluted with polar solvents, may contain aromatics, condensed naphthenic-aromatics, aromatic olefins, and polar compounds containing sulfur, nitrogen, and oxygen. The olefins can be in the saturate fraction, if they are aliphatic and cyclic olefins, or in the aromatic fraction, if they are aromatic olefins.

The slack wax feed was found to contain mostly 1-ring aromatics, such as alkyl benzenes and naphthenic benzenes, and a small content of 2–3 ring aromatics consisting of naphthalenes, acenaphthenes, fluorenes, and phenanthrenes. The slack wax feed also contained some unidentified aromatics and S compounds in the form of dibenzothiophenes. After hydrotreating, the 1-ring aromatic content decreased from 5.7 to 2.5 vol%, and the 2–3 ring aromatic

TABLE 5.18
Effect of Hydroprocessing on Aromatic Hydrocarbon Composition

Hydroprocessing, Aromatic Fractions	Deoiled Slack Wax	HT Slack Wax	Solvent Dewaxed SWI
MS analysis, vol%			
1-Ring aromatics	8.5	5.9	0.8
2-Ring aromatics	0.6	0.3	0.1
3-Ring aromatics	0.1	0	0
4-Ring aromatics	0.1	0	0
Unidentified aromatics	0.4	0.2	0

Source: From Pillon, L.Z., *Petroleum Science and Technology*, 20, 223, 2002. Reproduced by permission of Taylor & Francis Group, LLC, http://www.taylorandfrancis.com.

content decreased from 0.6 to 0.3 vol%. The HT slack wax contained basically only 1–2 ring aromatics and no presence of dibenzothiophenes was observed. After the hydroisomerization step, the 1-ring aromatic content further decreased to 0.3 vol% and the 2-ring aromatic content further decreased to 0.1 vol% and no other aromatics were found. The effect of different slack wax feed on VI of different SWI base stocks is shown in Table 5.19.

After solvent dewaxing, a liquid SWI boils in the lube oil boiling range and has a viscosity, VI, and other properties suitable for a lubricating oil. The hydrocarbon composition of slack waxes, used as feed, can vary depending on the crude oil and their oil content, which can lead to variation in VI. The reaction scheme for isomerization of *n*-paraffins reported in the literature shows that hydroisomerization of *n*-paraffins can lead to monobranched, dibranched, and tribranched cracking products. The isomerization of pure n-C_{16}, n-C_{28}, and n-C_{36} alkanes was reported to lead to the formation of a mixture of C_{16}, C_{28}, and

TABLE 5.19
Effect of Different Slack Wax Feed on VI of SWI Base Stocks

SWI 6 Base Stock	Slack Wax Feed # 1	Slack Wax Feed # 2	Slack Wax Feed # 3
Kinematic viscosity, cSt			
At 40°C	29.46	28.84	27.98
At 100°C	5.84	5.71	5.66
VI	146	143	148
Pour point, °C	−21	−21	−21

Source: From Pillon, L.Z., *Petroleum Science and Technology*, 20, 223, 2002. Reproduced by permission of Taylor & Francis Group, LLC, http://www.taylorandfrancis.com.

C_{36} isomers and the cracking products with a shorter chain length (Calemma et al., 2000). The use of a $Pt/amorphous$ $SiO_2–Al_2O_3$ catalyst was reported to produce high yields of up to 60% of iso-C_{16}, up to 50% of iso-C_{28}, and up to 40% of iso-C_{36}, respectively. The GC/MS analysis of n-C_{16} hydroisomerization products indicated the presence of mainly monobranched C_{16} isomers containing methyl, ethyl, propyl, and butyl side chains in different positions of the molecule (Calemma et al., 2000). The effect of different slack wax feed on the saturate hydrocarbon composition of SWI base stocks, having a different VI, is shown in Table 5.20.

SWI 6 base stock, solvent dewaxed to $-21°C$ and having a VI of 146, was found to contain 87.2 wt% of iso-paraffins and have an average ring number per mole of 0.2. Another SWI 6 base stock, also solvent dewaxed to $-21°C$ and having a lower VI of 143, was found to contain only 85.5 wt% of iso-paraffins and have a higher average ring number per mole of 0.3. A different SWI 6 base stock, solvent dewaxed to $-21°C$ and having a higher VI of 148, was found to contain 89.8 wt% of iso-paraffins and have an average ring number per mole of 0.2. With an increase in ring-opening reactions and an increase in iso-paraffinic content, an increase in VI of SWI is observed.

During hydroconversion processes, both cracking and isomerization occur simultaneously. The term hydroisomerization indicates that the isomerization predominates over hydrocracking (Calemma et al., 1999). Under optimum conditions, the hydroisomerization of slack wax derived from solvent refined oils can yield the lube oil base stock in a range of 55%–60% and having a high VI (above 145) and the yield can be further increased to 70% by recycling the unconverted

---

TABLE 5.20
Effect of Different Slack Wax Feed on the Saturate Hydrocarbon Composition

Hydroprocessing, Saturate Fractions	Slack Wax Feed # 1, SWI (VI 146)	Slack Wax Feed # 2, SWI (VI 143)	Slack Wax Feed # 3, SWI (VI 148)
MS analysis, vol%			
Paraffins/iso-paraffins	87.2	85.5	89.8
1-Ring naphthenes	8.1	8.3	5.7
2-Ring naphthenes	2.9	3.9	2.9
3-Ring naphthenes	0.4	0.7	0.3
4-Ring naphthenes	0.4	0.5	0.3
5-Ring naphthenes	0.5	0.6	0.5
6-Ring naphthenes	0.5	0.5	0.5
Average ring			
Number (per mole)	0.2	0.3	0.2

Source: From Pillon, L.Z., *Petroleum Science and Technology*, 20, 223, 2002. Reproduced by permission of Taylor & Francis Group, LLC, http://www.taylorandfrancis.com.

TABLE 5.21
Effect of Hydroisomerization Temperature on Wax Content
and Daylight Stability

Hydroisomerization	Low Temperature SWI 6	High Temperature SWI 6
Kinematic viscosity, cSt		
At 40°C	28.96	29.47
At 100°C	5.77	5.72
VI	146	139
Pour point, °C	−21	−21
DSC wax content		
−100°C to −40°C, wt%	24.38	17.68
−100°C to 5°C, wt%	38.91	28.84
Daylight stability, days	7	4
Volatility, % off	6.5	8.9

Source: From Pillon, L.Z., *Petroleum Science and Technology*, 20, 101, 2002. Reproduced by permission of Taylor & Francis Group, LLC, http://www.taylorandfrancis.com.

waxes (Calemma et al., 1999). The effect of the reaction temperature on the yield was reported to be negligible while a significant effect of the hydrogen pressure on the isomerization of heavy alkanes, such as n-C_{28}, was observed (Calemma et al., 2000). The effect of hydroisomerization temperature on the wax content and the daylight stability of SWI 6 base stocks are shown in Table 5.21.

Solvent dewaxed SWI 6 base stock, hydroisomerized at low temperature, was found to have a wax content of 24.38 wt%, measured in the temperature range from −100°C to −40°C, and a total wax of 38.91 wt%, measured in the temperature range from −100°C to 5°C. The solvent dewaxed SWI base stocks were found to have a varying daylight stability when hydroisomerized at low temperature. SWI 6 base stock, hydroisomerized at high temperature, was found to have a lower DSC wax content of 17.68 wt%, measured in the temperature range from −100°C to −40°C, and a lower total wax of 28.84 wt%, measured in the temperature range from −100°C to 5°C. With an increase in the hydroisomerization temperature and cracking taking place, the daylight stability decreased from 7 to 4 days, and an increase in volatility from 6.5% to 8.9% off was reported. During the slack wax hydroisomerization process, both cracking and isomerization occur simultaneously. Under selected processing conditions, such as a high temperature, the cracking reactions can predominate over the isomerization reactions which leads to a decrease in the wax content and a decrease in daylight stability.

5.4 USE OF HYDROFINING

Light stability is not a problem in lubricating oils, containing additives, but it is a quality requirement for some process oils. According to the early literature, following the hydroisomerization step, the isomerate may undergo a hydrogenation step to stabilize the oil and remove residual aromatics (Sequeira, 1994). The more recent literature on wax isomerization technology claims to convert slack waxes into lube oil base stocks and the product is "haze free" (ExxonMobil, 2002). The commercial processes for the manufacture of SWI base stock are based on the hydroisomerization of HT slack wax followed by the fractionation and solvent dewaxing. The typical dewaxing solvent can be a mixture of methyl ethyl ketone (MEK) and methyl isobutyl ketone (MIBK) (Prince, 1997). After the hydroisomerization and solvent dewaxing, SWI 6 base stocks were found to contain practically no aromatics, S and N, however, their daylight stability was found to vary and no presence of coronenes was found. The solvent dewaxed SWI 6 base stocks, produced from the same slack wax feed and hydroisomerized under the same processing conditions, were found to have different daylight stability. Solvent dewaxed SWI 6 base stock, containing 0.7 wt% of aromatic/polar content, was found to have a daylight stability of 24 days. Another solvent dewaxed SWI 6 base stock, containing 0.9 wt% of aromatic/polar content, was found to have a daylight stability of only 5 days. The composition and daylight stability of solvent dewaxed SWI 6 base stocks, hydroisomerized at low temperature, is shown in Table 5.22.

Hydrofining is usually used, as the last processing step, to remove the molecules that affect the color and the stability of base stocks. The typical hydrofining process, used to finish lube oil base stocks, requires the reactor temperature of 260°C–315°C and the reactor pressure of 500–1000 psig. The commercial hydrogen finishing catalysts consist of cobalt–molybdenum on alumina, nickel–molybdenum on alumina, iron–cobalt–molybdenum on alumina and nickel–tungsten on alumina or

TABLE 5.22
Daylight Stability of SWI 6 Hydroisomerized at Low Temperature

Low Temperature Hydroisomerization, Properties	Solvent Dewaxed, SWI 6	Solvent Dewaxed, SWI 6
Kinematic viscosity, cSt		
At 40°C	27.98	28.94
At 100°C	5.66	5.72
VI	148	143
Dewaxed pour point, °C	−21	−21
Saturates, wt%	99.3	99.1
Aromatics, wt%	0.7	0.9
Coronenes	0	0
Sulfur, ppm	<1	<1
Nitrogen, ppm	<1	<1
Daylight stability, days	24	5

Source: From Pillon, L.Z., *Petroleum Science and Technology*, 19, 1263, 2001. Reproduced by permission of Taylor & Francis Group, LLC, http://www.taylorandfrancis.com.

TABLE 5.23
Effect of Hydrofining on Saturate Hydrocarbon Composition

Hydrofining, Saturate Fractions	Solvent Dewaxed, SWI (VI 146)	Hydrofined, SWI (VI 146)
MS analysis, vol%		
Paraffins/iso-paraffins	86.6	87.1
1-Ring naphthenes	7.6	8
2-Ring naphthenes	2.7	3
3-Ring naphthenes	0.3	0.4
4-Ring naphthenes	0.3	0.4
5-Ring naphthenes	0.5	0.5
6-Ring naphthenes	0.5	0.4
Average ring		
Number (per mole)	0.2	0.2

Source: From Pillon, L.Z., *Petroleum Science and Technology*, 20, 223, 2002. Reproduced by permission of Taylor & Francis Group, LLC, http://www.taylorandfrancis.com.

silica–alumina. Promoters such as fluorides or phosphorous are sometimes added to enhance the catalyst performance (Sequeira, 1994). The hydrogen finishing process does not saturate aromatics when used at low pressures and low temperatures (Sequeira, 1994). Most hydrofining operations are operated at a severity set by the color improvement needed. The effect of hydrofining on the saturate hydrocarbon composition of SWI base stock is shown in Table 5.23.

After the solvent dewaxing, the paraffin/iso-paraffin content of SWI contained a total of 86.6 vol% of paraffinic/iso-paraffinic content. Since *n*-paraffins crack easily, the paraffinic/iso-paraffinic content of SWI will basically consist of iso-paraffins. After the hydrofining step, only a small increase in paraffinic/iso-paraffinic and naphthenic content was observed. The average ring number per mole of 0.2 did not change indicating the absence of ring-opening reactions. The effect of hydrofining on the composition of the aromatic fraction of SWI base stock is shown in Table 5.24.

After the solvent dewaxing, SWI was found to contain a 1-ring aromatic content of 0.8 wt% and a 2-ring aromatic content of 0.1 vol%. After the hydrofining step, a small decrease in the 1-ring aromatic content from 0.3 to 0.2 vol% was observed indicating some aromatic saturation taking place. The hydrofined SWI base stock was found to contain a small 2-ring aromatic content consisting of 0.1 vol% of naphthalenes. The effect of hydrofining on daylight stability and UV absorbance of DMSO extracts from SWI 6 base stock hydroisomerized at low temperature of 290°C is shown in Table 5.25.

Solvent dewaxed SWI, having a high VI of 148 and a dewaxed pour point of −21°C, was found to have a daylight stability of 24 days. The UV absorbance of the DMSO extract in a region of 260–350 nm was found to be varying from

TABLE 5.24

Effect of Hydrofining on the Aromatic Hydrocarbon Composition

Hydrofining, Aromatic Fractions	Solvent Dewaxed, SWI (VI 146)	Hydrofined, SWI (VI 146)
MS analysis, vol%		
1-Ring aromatics	0.8	0.2
2-Ring aromatics	0.1	0.1
3-Ring aromatics	0	0
4-Ring aromatics	0	0
Unidentified aromatics	0	0

Source: From Pillon, L.Z., *Petroleum Science and Technology*, 20, 223, 2002. Reproduced by permission of Taylor & Francis Group, LLC, http://www.taylorandfrancis.com.

0.9 to 2.6. After hydrofining, an improvement in the daylight stability from 24 days to over 50 days is observed and the UV absorbance of the DMSO extract in a region of 260–350 nm decreased to 0.1. The effect of hydrofining on daylight stability and UV absorbance of DMSO extract from another SWI 6 base stock hydroisomerized at a low temperature of 290°C is shown in Table 5.26.

Solvent dewaxed SWI, having a high VI of 143 and a dewaxed pour point of −21°C, was found to have a daylight stability of only 5 days. The UV absorbance of the DMSO extract in a region of 260–350 nm was found to be varying from 0.8 to 2.4. After hydrofining, an improvement in the daylight stability from 5 to 25 days is observed and the UV absorbance of the DMSO extract in a region of

TABLE 5.25

Effect of Hydrofining on Daylight Stability and UV Absorbance of SWI 6 Base Stock

SWI 6 Base Stock	Solvent Dewaxed	After Hydrofining
VI	148	148
Pour point, °C	−21	−21
Daylight stability, days	24	>50
UV of DMSO extract		
260–285 nm	2.6	0.1
290–299 nm	2.2	0.1
300–329 nm	1.6	0.1
330–350 nm	0.9	0.1

Source: From Pillon, L.Z., *Petroleum Science and Technology*, 19, 1263, 2001. Reproduced by permission of Taylor & Francis Group, LLC, http://www.taylorandfrancis.com.

TABLE 5.26
Effect of Hydrofining on Daylight Stability and UV Absorbance of SWI 6 Base Stock

SWI 6 Base Stock	Solvent Dewaxed	After Hydrofining
VI	143	143
Pour point, °C	−21	−21
Daylight stability, days	5	25
UV of DMSO extract		
260–285 nm	2.4	0.3
290–299 nm	1.8	0.2
300–329 nm	1.4	0.4
330–350 nm	0.8	0.2

Source: From Pillon, L.Z., *Petroleum Science and Technology*, 19, 1263, 2001. Reproduced by permission of Taylor & Francis Group, LLC, http://www.taylorandfrancis.com.

260–350 nm decreased to 0.2–0.4. The effect of hydrofining on daylight stability and UV absorbance of DMSO extract from another SWI 6 base stock hydro-isomerized at a low temperature of 290°C is shown in Table 5.27.

Solvent dewaxed SWI, having a high VI of 146 and a dewaxed pour point of −21°C, was found to have a daylight stability of only 9 days. The UV absorbance of the DMSO extract in a region of 260–350 nm was found to be varying from 1.2 to 2.7. After hydrofining, an improvement in the daylight stability from 9 to 38 days is observed and the UV absorbance of the DMSO extract in a region of 260–350 nm decreased to 0.2–0.3. The term hydroisomerization indicates that the

TABLE 5.27
Effect of Hydrofining on Daylight Stability and UV Absorbance of SWI 6 Base Stock

SWI 6 Base Stock	Solvent Dewaxed	After Hydrofining
VI	146	146
Pour point, °C	−21	−21
Daylight stability, days	9	38
UV of DMSO extract		
260–285 nm	2.7	0.3
290–299 nm	2.3	0.2
300–329 nm	1.8	0.3
330–350 nm	1.2	0.2

Source: From Pillon, L.Z., *Petroleum Science and Technology*, 19, 1263, 2001. Reproduced by permission of Taylor & Francis Group, LLC, http://www.taylorandfrancis.com.

TABLE 5.28
Effect of Hydroisomerization Temperature and Hydrofining
on Daylight Stability

Hydroisomerization, SWI 6 Base Stock	Low Temperature, Hydrofined	Low Temperature, Hydrofined	High Temperature, Hydrofined
Coronenes	0	0	0
Daylight stability, days	9	>44	3
UV of DMSO extract			
260–285 nm	1.3	0.4	>2
FTIR of DMSO/CHCl3	1–2 Ring aromatics	1–2 Ring aromatics	1–2 Ring aromatics
	Ketones	Ketones	Ketones
	Acids	Acids	Acids
	Esters	Esters	Esters

Source: From Pillon, L.Z., *Petroleum Science and Technology*, 19, 1263, 2001. Reproduced by permission of Taylor & Francis Group, LLC, http://www.taylorandfrancis.com.

isomerization process predominates over hydrocracking, however with an increase in temperature, an increase in cracking will take place. The effects of hydrogen finishing temperature and pressure are highly dependent on the quality of feedstock. The effect of hydrofining on daylight stability and UV absorbance of DMSO extract from SWI 6 base stocks hydroisomerized at different temperatures is shown in Table 5.28.

After hydrofining, SWI 6 base stock hydroisomerized at low temperature was found to have a daylight stability of only 9 days. The UV absorbance of DMSO extract, in the region of 260–289 nm, was found to be 1.3 indicating a relatively high presence of some aromatic/polar molecules. The FTIR analysis of the DMSO/CHCl3 extract indicated the presence of 1–2 ring aromatic compounds, ketones, acids, and esters. After hydrofining, another SWI 6 base stock, also hydroisomerized at low temperature, was found to have an excellent daylight stability of over 44 days. The UV absorbance of DMSO extract, in the region of 260–289 nm, was found to be only 0.4 indicating a significant decrease in the presence of some aromatic/polar molecules. The FTIR analysis of the DMSO/CHCl3 extract indicated the presence of 1–2 ring aromatic compounds, ketones, acids, and esters. After hydrofining, SWI 6 base stock hydroisomerized at high temperature was found to have a daylight stability of only 3 days. The UV absorbance of DMSO extract, in the region of 260–289 nm, was found to be over 2 indicating a high content of some aromatic/polar molecules. FTIR analysis of the DMSO/CHCl3 extract also indicated the presence of 1–2 ring aromatic compounds, ketones, acids, and esters. The hydrofining process was found effective in improving the color and, in most cases, was also found effective in improving the daylight stability of solvent dewaxed SWI 6 base stocks, if hydroisomerized at low temperatures.

5.5 USE OF HYDROREFINING

The selective hydrogenation process, known as hydrorefining, is a high pressure hydrofinishing process. The typical hydrorefining process requires the reactor temperature of 260°C–315°C and the reactor pressure of 1500–3000 psig. At low pressures and low temperatures, the hydrogen finishing process does not saturate aromatics nor break the carbon–carbon bonds. The use of higher pressure might saturate some aromatics while the use of higher temperature might lead to some cracking (Sequeira, 1994). The literature reported that an increase in the hydrogen pressure or temperature of HF process will usually improve neutralization, desulfurization, denitrification, color, and stability of base stocks. An increase in the temperature above a certain maximum might degrade the color, oxidation stability, and other properties (Sequeira, 1994). The typical hydrofinishing and hydrorefining processing conditions are shown in Table 5.29.

The early literature reported on hydroisomerization of paraffins and hydrorefined slack wax containing 10.5% of oil and 0.02 wt% of sulfur. Some oils were reported to contain aromatic molecules which have a negative VI (Karzhev et al., 1967). At the present time, hydrorefining is used to produce white oils from solvent neutral mineral oils or high quality naphthenic distillates. The single stage hydrorefining process is used to produce the technical grade white oils. In the hydrorefining process, the sulfur and the nitrogen contents of the technical white oils are removed and the aromatic content is reduced to a very low level (Sequeira, 1994). The effect of single stage hydrorefining on properties and composition of different viscosity technical white oils is shown in Table 5.30.

The use of a single hydrorefining step decreased the viscosity of a low viscosity feed from 21.6 to 19.45 cSt at 40°C and a small increase in VI from 106 to 107, with a small increase in pour point from −15°C to −12°C was observed. The S content drastically decreased from 7400 ppm to less than 1 ppm and some decrease in flash point from 96°C to 94°C was observed. The

---

TABLE 5.29
Typical Hydrofinishing and Hydrorefining Processing Conditions

Process Variables	Hydrofinishing	Hydrorefining
Pressure, psig	500–1000	1500–3000
Temperature, °C	260–315	260–315
Space velocity, Vo/Vc/Hr	1.0–1.5	0.5–1.0
Lube oil yield, vol%	98+	95–98
Catalyst life, years	1–2	1–2

Source: From Sequeira, A. Jr., *Lubricant Base Oil and Wax Processing*, Marcel Dekker, Inc., New York, 1994. Reproduced by permission of Routledge/Taylor & Francis Group, LLC.

TABLE 5.30
Effect of Hydrorefining on Properties and Sulfur Content of Technical White Oils

Hydrorefining, White Oils	Light Feed	Technical Light Oil	Heavy Feed	Technical Heavy Oil
Kinematic viscosity, cSt				
At 40°C	21.6	19.45	148	90.6
At 100°C	4.18	3.94	13.56	10.39
VI	106	107	94	106
Pour point, °C	−15	−12	−9	−3
Sulfur, ppm	7400	<1	10,100	7
Flash point, °C	96	94	134	122

Source: From Billon, A. et al., *Proceedings Refining Department, API*, 59, 168, 1980. Reproduced with permission from American Petroleum Institute.

use of a single hydrorefining step significantly decreased the viscosity of a high viscosity feed from 148 to 90.6 cSt at 40°C and a significant increase in VI from 97 to 114, with also a significant increase in pour point from −9°C to −3°C was reported. The S content also drastically decreased from 10,100 to 7 ppm, and a significant decrease in flash point from 134°C to 122°C was observed. The effect of hydrorefining on properties and the aromatic hydrocarbon composition of medium viscosity technical white oil are shown in Table 5.31.

TABLE 5.31
Effect of Hydrorefining on Aromatic Hydrocarbons of Technical White Oil

Hydrorefining, White Oil	Medium Viscosity Feed	Technical Medium Viscosity Oil
Kinematic viscosity, cSt		
At 40°C	82.98	45.96
At 100°C	9.35	6.83
VI	97	114
Pour point, °C	−12	−6
1-Ring aromatics, wt%	20.9	2.39
2-Ring aromatics, wt%	4.37	0.029
Polyaromatics, wt%	2.39	0.022
Sulfur, ppm	9500	5
Flash point, °C	124	100

Source: From Billon, A. et al., *Proceedings Refining Department, API*, 59, 168, 1980. Reproduced with permission from American Petroleum Institute.

The use of a single hydrorefining step significantly decreased the viscosity of a medium viscosity feed from 82.98 to 45.96 cSt at 40°C and a significant increase in VI from 97 to 114, with a significant increase in pour point from −12°C to −6°C was reported. With an increase in VI, the 1-ring aromatic content decreased from 20.9 to 2.39 wt%, the 2-ring aromatic content decreased from 4.37 to 0.029 wt%, and the polyaromatic content decreased from 2.39 to 0.022 wt%. A technical white oil is a highly refined mineral base stock which contains practically 100% saturate content. The use of a one-stage hydrorefining process decreased the total aromatic content from 27.7 to 2.4 wt%, and a drastic decrease in the sulfur content from 0.95 wt% to 5 ppm is observed. Different hydroprocessing technologies can be used to produce base stocks, having a high saturate content and practically no aromatic and S contents, similar to technical white oils.

A synthetic polyalfaolefin (PAO) base stock contains 100 wt% saturate content and no presence of heteroatom molecules. PAO base stocks can be produced by polymerizing a C_{10} monomer to form a mixture of three components such as C_{10} trimer (C_{30}), C_{10} tetramer (C_{40}), and C_{10} pentamer (C_{50}) (Pillon et al., 1995). They are classified according to their approximate kinematic viscosity at 100°C, similarly to other nonconventional base stocks, such as SWI. The distillation boiling points (BP), properties, and composition of different highly saturated base stocks, having a similar viscosity, are shown in Table 5.32.

A white oil is produced by high pressure hydrogenation used to saturate aromatics and remove essentially any sulfur and nitrogen from conventional

TABLE 5.32
Properties and Composition of Different Highly Saturated Base Stocks

Nonconventional Base Stocks	Hydrorefined White Oil	Hydrocracked Oil	Hydroisomerized SWI 6	Synthetic PAO 6
Distillation, °C				
Initial BP	340	323	341	408
Mid BP	433	426	465	481
Final BP	533	538	570	596
Kinematic viscosity, cSt				
At 40°C	32.7	35.4	29.4	30.4
At 100°C	5.6	—	5.8	5.8
VI	106	97	143	134
Saturates, wt%	99.7	96.9	>99.5	100
Aromatics, wt%	0.3	3.1	<0.5	0
Sulfur, ppm	<1	—	<1	0
Nitrogen, ppm	<1	—	<1	0

Source: From Pillon, L.Z., Asselin, A.E., and MacAlpine, G.A., *Lubricating Oil Having an Average Ring Number of Less Than 1.5 Per Mole Containing Succinic Anhydride Amine Rust Inhibitor*, US Patent 5,225,094, 1993; Pillon, L.Z., Reid, L.E., and Asselin, A.E., *Lubricating Oil for Inhibiting Rust Formation*, US Patent 5,397,487, 1995.

mineral base stocks. Different viscosity solvent refined mineral base stocks, containing 20–30 wt% of aromatics, have a typical VI of 85–95. The technical white oil, having an increased VI of 106, was found to contain 99.7 wt% of saturate content and practically no presence of S and N contents. The hydrocracked base stock is produced by hydrocracking rather than solvent extracting the aromatic and heteroatom molecules. The hydrocracked base stock, having a lower VI of 97, was found to contain a lower saturate content of 96.9 wt%. SWI is the lube oil fraction produced by hydroisomerization of slack wax. SWI 6 base stock has a significantly higher VI of 143 and was found to contain practically no aromatics, S and N. The synthetic PAO 6 base stock, having a lower VI of 134, contains 100 wt% of saturate content and different chemistry hydrocarbons. The MS analysis of the saturate fractions of different highly saturated base stocks is shown in Table 5.33.

The technical white oil, having a VI of 106, was found to contain 30.5 wt% of paraffinic/iso-paraffinic content, 68.8 wt% of naphthenic content, and 0.7 wt% of residual aromatics. The MS analysis indicated the average ring number per mole of 1.6. The hydrocracked base stock, having a lower VI of 97, was found to contain a lower paraffinic/iso-paraffinic content of 19.9 wt%, a higher naphthenic content of 77 wt%, and 3.1 wt% of residual aromatics. Under cracking conditions, aromatics can be saturated but not all rings will open. The MS analysis indicated a lower average ring number per mole of 1.5. The SWI base stock, having a high VI of 143, was found to have a high paraffinic/iso-paraffinic content of 89.9 wt% and a low naphthenic content of 10.1 wt%. Under cracking conditions, n-paraffins crack easily while cracking of iso-paraffins is more

---

TABLE 5.33

Saturate Hydrocarbon Composition of Different Nonconventional Base Stocks

Base Stocks, MS Analysis, vol%	Hydrorefined White Oil	Hydrocracked Oil	Hydroisomerized SWI 6	Synthetic PAO 6
VI	106	97	143	134
Paraffins/iso-paraffins	30.5	19.9	89.9	94.3
1-Ring naphthenes	23.1	27.8	8.8	2.5
2-Ring naphthenes	18.7	21.3	3.9	1.6
3-Ring naphthenes	11.6	14	0.9	0.2
4-Ring naphthenes	10.4	8.4	0.6	0.4
5-Ring naphthenes	3.8	4	0.5	0.4
6-Ring naphthenes	1.3	1.3	0.5	0.6
Residual aromatics	0.7	3.1	0	0
Average ring Number (per mole)	1.6	1.5	0.3	0.1

Source: From Pillon, L.Z., Asselin, A.E., and MacAlpine, G.A., *Lubricating Oil Having an Average Ring Number of Less Than 1.5 Per Mole Containing Succinic Anhydride Amine Rust Inhibitor,* US Patent 5,225,094, 1993.

difficult. The paraffin/iso-paraffin content of SWI base stocks is basically composed of iso-paraffins. The MS analysis indicated a low average ring number per mole of 0.3. The synthetic PAO 6 base stock, having a lower VI of 134, contains a higher 94.3 wt% of saturate hydrocarbons and some 5.7 wt% of ring containing molecules. A low average ring number per mole of 0.1 confirms the presence of other hydrocarbon molecules. The MS determination of the saturate hydrocarbon composition of mineral base stocks requires a preliminary separation of the petroleum sample into saturate and aromatic fractions. In the case of incomplete separation, the saturate fraction might contain some residual aromatic compounds. The viscosity and MS analysis of the saturate fractions, separated from different mineral base stocks, are shown in Table 5.34.

With an increase in the viscosity of hydrofined phenol extracted base stocks from 600N to 1400N, their total paraffinic/iso-paraffinic content was found to decrease from 22 to 13.3 vol%, and their 1–5 ring naphthenic content was found to increase from 72.6 to 81.5 vol%. Despite a significant difference in viscosity, some differences in composition and the presence of residual aromatics, the MS analysis indicated the same average ring number per mole of 1.6. The saturate fraction separated from 600N NMP extracted base stock indicated an increase in the paraffinic/iso-paraffinic content from 22 to 18.7 vol%, and an increase in the 1-ring naphthenic content from 32.3 to 37 vol%. Despite the use of different extraction solvents, some differences in composition and the presence of residual

TABLE 5.34
Composition of Saturate Fractions Separated from Different Mineral Base Stocks

Mineral Base Stocks, Extraction Solvent	600N, Phenol	600N, NMP	1400N, Phenol
Kinematic viscosity at 40°C, cSt	105.9	111.4	301.7
Saturate fractions			
Kinematic viscosity at 40°C, cSt	75.4	76.4	155.7
MS analysis, vol%			
Paraffins/iso-paraffins	22	18.7	13.3
1-Ring naphthenes	32.3	37	39.1
2-Ring naphthenes	18.7	18.4	20.4
3-Ring naphthenes	12.5	11.3	13.8
4-Ring naphthenes	8.2	7.9	7.3
5-Ring naphthenes	1.9	1.9	0.9
6-Ring naphthenes	0	0	0
Residual aromatics	4.4	4.7	5.2
Average ring			
Number (per mole)	1.6	1.6	1.6

Source: From Pillon, L.Z., Asselin, A.E., and MacAlpine, G.A., *Lubricating Oil Having an Average Ring Number of Less Than 1.5 Per Mole Containing Succinic Anhydride Amine Rust Inhibitor*, US Patent 5,225,094, 1993.

aromatics, the MS analysis indicated the same average ring number per mole of 1.6. While the viscosity and composition of different mineral base stocks might vary affecting their VI, their average ring number per mole of 1.6 is the same as found in technical white oils. The hydrorefining process leads to white oils having a higher VI but the same average ring number of 1.6 thus confirming that no significant ring-opening reaction is taking place and the saturation of aromatics leads to an increase in the VI. The hydrocracking followed by solvent dewaxing and hydrorefining was reported to stabilize the hydrocracked base stocks (Sequeira, 1994). The effect of hydrorefining on the UV and sunlight stability of 500N hydrocracked base stock is shown in Table 5.35.

The 500N hydrocracked base stock was reported to have a UV light stability of only 3 days and sunlight stability of 12 days. After hydrorefining, a small increase in VI from 95 to 97, with a decrease in the S content from 0.09 to 0.008 wt%, and a decrease in TAN and in the carbon residue from 0.12 to 0.07 wt% were reported. After hydrorefining, the UV light stability increased to 14 days and sunlight stability increased to over 30 days. The hydrocracked base stocks require stabilization, if used in some specialty products. Hydrorefining can be used to stabilize the SWI followed by fractionation to remove light ends and the solvent dewaxing (Sequeira, 1994). Other methods used to stabilize the hydrocracked base stock, after dewaxing, consist of clay treating and solvent refining. Some refiners use furfural extraction to stabilize the hydrocracked oils against discoloration and sludging. There are also some reports on alkylation of hydrocracked base stocks with olefins over acidic catalysts to increase their stability and the use of additives (Sequeira, 1994). The UV absorbance of DMSO extract from technical white oil is shown in Table 5.36.

TABLE 5.35
Effect of Hydrorefining on Daylight Stability of 500N Hydrocracked Base Stock

Hydrocracked 500N Oil	Before Hydrorefining	After Hydrorefining
API gravity	27.4	29.4
Viscosity SUS at 100°F	484	479
VI	95	97
Pour point, °C	−18	−18
Sulfur, wt%	0.09	0.008
TAN, mg KOH/g	0.03	<0.03
Carbon residue, wt%	0.12	0.07
UV light stability, days	3	14
Sunlight stability, days	12	>30

Source: From Sequeira, A. Jr., *Lubricant Base Oil and Wax Processing*, Marcel Dekker, Inc., New York, 1994. Reproduced by permission of Routledge/Taylor & Francis Group, LLC.

TABLE 5.36

UV Absorbance of DMSO Extract from White Oils

White Oils, DMSO Extracts	Technical, UV Absorbance	Technical, UV Specifications	Medicinal, UV Specifications
280–289 nm	0.36	Max 4	
290–299 nm	0.38	Max 3.3	
300–329 nm	0.34	Max 2.3	
330–350 nm	0.2	Max 0.8	
260–420			Max 0.11

Source: From Billon, A. et al., *Proceedings Refining Department, API*, 59, 168, 1980; Sequeira, A. Jr., *Lubricant Base Oil and Wax Processing*, Marcel Dekker, Inc., New York, 1994.

The level of severity of refining is tested following the UV absorbance in the range from 260 to 420 nm for different grade white oils. The use of a single stage hydrorefining process is sufficient to meet the UV absorbance requirements of DMSO extracts for technical white oils. The technical white oils are used in cosmetics, textile lubrication, insecticide vehicles, and paper impregnation (Speight, 1999). The technical white oils are also used in such products as waxes, agricultural spray oils, and mineral seal oils. Medicinal white oils need to be free of polycyclic aromatic hydrocarbons to meet the US Pharmacopoeia (USP) requirements for use in medicinal and cosmetic formulations. The level of severity of refining is also tested following the UV absorbance and the medicinal white oils require a two-stage hydrorefining process. The second stage hydrorefining treatment is used to saturate the last traces of aromatic compounds to produce the pharmaceutical and food grade white oils (Sequeira, 1994). The medicinal white oils are highly viscous, colorless, odorless, and tasteless. Hydrorefining is used to produce food and medicinal grade wax. The US FDA UV absorbance requirements for waxes and petrolatum are shown in Table 5.37.

The UV absorbance of DMSO extracts of wax is also used to determine their suitability for use in food, drug, and cosmetic applications. The level of severity of refining is tested following the UV absorbance in the range from 280 to 400 nm for petroleum wax and petrolatum. The literature reported that a treatment of molten wax with activated clay improves color, reduces odor and taste of the finished wax. The clay is regenerated before reuse by passing it through a multiple hearth furnace to remove the adsorbed color bodies (Sequeira, 1994). The decoloring operation, known as percolation, is a batch process. It also removes traces of possibly harmful polycyclic aromatic hydrocarbons which are considered potential carcinogens. At the present time, hydrorefining has replaced clay and percolation processes as the process of choice for the manufacture of wax which

TABLE 5.37

US FDA UV Absorbance Requirements for Waxes and Petrolatum

UV Absorbance	Petroleum Wax	Petrolatum
280–289 nm	Max 0.15	Max 0.25
290–299 nm	Max 0.12	Max 0.20
300–359 nm	Max 0.08	Max 0.14
360–400 nm	Max 0.02	Max 0.04

Source: From Sequeira, A. Jr., *Lubricant Base Oil and Wax Processing*, Marcel Dekker, Inc., New York, 1994. Reproduced by permission of Routledge/Taylor & Francis Group, LLC.

must meet the government purity specifications. The use of hydroprocessing is effective in removing aromatics, S and N molecules from any petroleum product ranging in viscosity from light oils to waxes.

REFERENCES

Aldrich Advancing Science, *Handbook of Chemicals and Laboratory Equipment*, 2007.
Bardasz, E.A. and Lamb, G.D., Additives for crankcase lubricant applications, in *Lubricant Additives Chemistry and Applications*, Rudnick, L.R., Ed., Marcel Dekker, Inc., New York, 2003.
Bijwaard, H.M.J. and Morcus, A., *Lubricating Base Oil Compositions*, CA Patent 1,185,962, 1985.
Billon, A. et al., *Proceedings Refining Department*, API, 59, 168, 1980.
Calemma, V. et al., *Preprints, Division of Petroleum Chemistry ACS*, 44(3), 241, 1999.
Calemma, V., Peratello, S., and Perego, C., *Applied Catalysis A: General*, 190, 207, 2000.
Cody, I.A. et al., *Raffinate Hydroconversion Process*, CA Patent 2,429,500, 2002.
Cooney, J.V. and Hazlett, R.N., *Heterocycles*, 22, 1513, 1984.
ExxonMobil, http://www.prod.exxonmobil.com (accessed October 2002).
Fetzer, J.C., Correlations between the spatial configuration and behaviour of large poly-nuclear aromatic hydrocarbons, in *Polynuclear Aromatic Compounds*, Ebert, L.B., Ed., American Chemical Society, Washington, DC, 1988.
Gary, J.H. and Handwerk, G.E., *Petroleum Refining, Technology and Economics*, 4th ed., Marcel Dekker, Inc., New York, 2001.
Han, S. et al., *Energy and Fuels*, 19, 625, 2005.
Huo, K.-F. et al., *Petroleum Science and Technology*, 18(7&8), 815, 2000.
Karzhev, V.I. et al., *Khim. Tekhnol. Topl. Masel*, 12(7), 10, 1967.
Naidu, S.K., Klaus, E.E., and Duda, J.L., *I&EC Product Research and Development*, 23, 613, 1984.
Pavia, D.L., Lampman, G.M., and Kriz, G.S., *Introduction to Spectroscopy*, Harcourt Brace College Publishers, Orlando, 1996.

Phillips, R.A., Highly refined mineral oils, in *Synthetic Lubricants and High-Performance Functional Fluids*, 2nd ed., Rudnick, L.R. and Shubkin, R.L., Eds., Marcel Dekker, Inc., New York, 1999.

Pillon, L.Z., *Petroleum Science and Technology*, 19(9&10), 1263, 2001.

Pillon, L.Z., *Petroleum Science and Technology*, 20(1&2), 101, 2002.

Pillon, L.Z., *Petroleum Science and Technology*, 20(1&2), 223, 2002.

Pillon, L.Z. and Asselin, A.E., *Wax Isomerate Having a Reduced Pour Point*, US Patent 5,229,029, 1993.

Pillon, L.Z., Asselin A.E., and MacAlpine, G.A., *Lubricating Oil Having an Average Ring Number of Less Than 1.5 Per Mole Containing Succinic Anhydride Amine Rust Inhibitor*, US Patent 5,225,094, 1993.

Pillon, L.Z., Reid, L.E., and Asselin, A.E., *Lubricating Oil for Inhibiting Rust Formation*, US Patent 5,397,487, 1995.

Pirro, D.M. and Wessol, A.A., *Lubrication Fundamentals*, 2nd ed., Marcel Dekker, Inc., New York, 2001.

Prince, R.J., Base oils from petroleum, in *Chemistry and Technology of Lubricants*, 2nd ed., Mortier, R.M. and Orszulik, S.T., Eds., Blackie Academic & Professional, London, 1997.

Sequeira, A. Jr., *Lubricant Base Oil and Wax Processing*, Marcel Dekker, Inc., New York, 1994.

Speight, J.G., *The Chemistry and Technology of Petroleum*, 3rd ed., Marcel Dekker, Inc., New York, 1999.

Speight, J.G., *The Chemistry and Technology of Petroleum*, 4th ed., CRC Press, Boca Raton, 2006.

6 Clay Treatment

6.1 FUEL STABILITY

The use of different feedstocks and processing conditions to produce petroleum fractions affects the composition and stability of petroleum products. According to the literature, about 500 original patents are published every year on various fuel additives with the majority related to detergent additives for automotive gasolines, diesel fuel wax depressants, antioxidants, metal deactivators, stabilizers, ignition, and combustion modifiers (Danilov, 2001). The patent literature shows an increased interest in antifoaming agents which are highly surface active additives. There is an increased interest to use antifoaming agents, such as polysiloxanes and quaternary ammonium salts, as suppressants of evaporation of gasoline (Danilov, 2001). There are many other additives used, such as biocides, deicers, and dyes. Some additives are known to cause instability problems and are recommended to be used at low treat rates. Some fuels use more than one additive which can lead to compatibility and stability problems. Diesel fuel was reported to contain wax and smoke depressants. Automotive gasoline was reported to contain antioxidant, detergent, and antiknock compounds. The literature reported on some incompatibility issues among some additives. The thickening effect and the fuel phase separation were reported when mixing Ca alkylphenolates with polybutene amine detergents. The filterability of diesel fuel was affected by mixing Ba alkylphenolates with polyvinylacetate wax depressant (Danilov, 2001). The type and the chemistry of some fuel additives are shown in Table 6.1.

The literature reported on different processing options used to produce different petroleum fractions (Speight, 1999). The determination of saturates, aromatics, and olefins is important in characterizing the quality of petroleum fractions as gasoline blending components and as feeds to catalytic reforming process. The olefin content is important in characterizing petroleum products from catalytic reforming, catalytic and thermal cracking as blending components of motor and aviation fuels. The aromatic hydrocarbon content and the naphthalene content of aviation turbine fuels affect their combustion characteristics and smoke-forming tendencies. The literature reported on the use of ultraviolet (UV) spectroscopy to determine the total aromatic content in kerosene (Harfoush and Shleiwit, 1999). The ASTM D 5186 is used to determine the aromatic content and polynuclear aromatic content of diesel fuels and aviation turbine fuels by

TABLE 6.1

Type and Chemistry of Different Fuel Additives

Fuel Additive Type	Fuel Additive Chemistry
Wax depressants	Polyacrylates
	Viny-acetate copolymers
Smoke suppressants	Ba, Mg, and Ca compounds
Ignition promoters	Catalytic gas oils
Combustion catalysts	Mg and Fe compounds
	Polybutene amines
Antiknock compounds	Alkyl–Pb compounds
	Aromatic amines
	Ferrocene derivatives
	Mg carbonyl compounds
Oxygenates	Alcohols, ethers, MTBE
Detergents	Polyether amines
Antiwear additives	Phosphorus compounds
	Alkali metal compounds
	Ester and amide derivatives
	Alcohol and phenol derivatives
Antifoaming agents	Siloxane derivatives
	Quaternary ammonium salts
Other	Biocides, deicers, and dyes

Source: From Danilov, A.M., *Chemistry and Technology of Fuels and Oils*, 37, 444, 2001.

supercritical fluid chromatogaphy (SFC). Monoaromatics include benzenes, alkyl benzenes, indanes, alkyl indanes, tetralins, and alkyl tetralins. Polynuclear aromatics include two or more aromatic rings and can be fused, such as naphthalene and phenantrene, or separate, as biphenyl. The variation in feedstocks and processing options used to produce kerosene are shown in Table 6.2.

Kerosene is used as a solvent, a fuel, or a jet fuel component. Many refineries use a straight-run kerosene fraction from the atmospheric crude unit. The kerosene fraction is a mixture of hydrocarbons containing C_{10}–C_{16} carbons per molecule which boils in a range of 140°C–320°C. Low contents of aromatics and unsaturated hydrocarbons are required to obtain low levels of smoke during burning (Speight, 1999). At the present time, kerosene is mostly produced by cracking the less volatile fraction of crude oil at atmospheric pressure and elevated temperature. The kerosene fraction is chemically processed to reduce the level of mercaptan sulfur, thiols, and reduce the acidic content to meet the specifications of jet fuels (Speight, 1999). Caustic washing was reported used when refining gasoline and diesel fuel (Wang et al., 2001). A study on removing mercaptans from gasoline with ammonia washing was reported. While caustic washing is more effective than ammonia washing, it produces alkaline residue and pollutes the environment (Yan, 2003). Jet fuel is blended from low sulfur or

TABLE 6.2
Different Feedstocks and Processing Options Used to Produce Kerosene

Feed Source	Conversion	Finishing
Atmospheric distillation		Hydrofining
Atmospheric distillation	Hydrocracking	
Atmospheric distillation	Catalytic cracking	
Atmospheric residue	Hydrocracking	
Vacuum distillation	Catalytic cracking	
Vacuum distillation	Catalytic cracking	Hydrotreating
Vacuum residue	Coking	Hydrotreating

Source: From Speight, J.G., *The Chemistry and Technology of Petroleum*, 3rd ed., Marcel Dekker, Inc., New York, 1999.

desulfurized kerosene, hydrocracked blending stocks, and hydrotreated light coker oil (Gary and Handwerk, 2001). With an increased environmental demand for cleaner burning fuels, some refineries use severe hydrotreating (HT) to reduce the aromatic and sulfur contents of fuels. The effect of different processing on the composition of jet fuels is shown in Table 6.3.

The aromatic and heteroatom content of jet fuels can vary depending on the crude type, the processing option, and its severity. Many refiners use the Merox process, which is proceeded by a caustic wash step and followed by a water wash step to remove carryover caustic and sodium carboxylate salts. The Merox treated–caustic washed jet fuel was reported to contain 24.2 wt% of aromatics, 0.23 wt% of S, 15 ppm of N, and also some olefins (Pillon, 2001b). The literature reported on commercial jet fuel containing 20.5 vol% of aromatics, 364 ppm of S, 6 ppm of N, and 1.5 vol% of olefins (Hernandez-Maldonado et al., 2004). Commercial jet fuels must be clean burning which limits the total aromatics and

TABLE 6.3
Effect of Different Processing on Composition of Jet Fuels

Jet Fuel Composition	Merox–Caustic Treatment	Hydroprocessing
Aromatics, wt%	24.3	20.5
Sulfur, wt%	0.23	364 ppm
Nitrogen, ppm	15	6
Olefins, vol%	>1.5	1.5

Source: From Pillon, L.Z., *Petroleum Science and Technology*, 19, 961, 2001b; Hernandez-Maldonado, A.J., Yang, R.T., and Cannella, W., *Industrial and Engineering Chemistry Research*, 43, 6142, 2004.

double ring compounds. The specifications limit total aromatic content to max 20 vol% and the naphthalene content to max 3 vol% (Gary and Handwerk, 2001). The smoke point and the aromatic content specifications limit the amount of cracked stocks which can be used to blend jet fuels. Hydrocracking saturates many double ring aromatics in cracked products and raises the smoke point. Kerosene from the hydrocracking unit can meet the jet fuel specifications (Gary and Handwerk, 2001).

The two basic types of jet fuels, also known as turbine fuels, are naphtha and kerosene. Naphtha jet fuel is more volatile and is produced mostly for the military. Safety requirements limit commercial jet fuels to less volatile and thus the narrower boiling range product. Commercial jet fuels are sold as Jet A, Jet A-1, JP-5, or JP-50 and the main differences are the freezing points (Gary and Handwerk, 2001). The operation of aircraft at high altitudes can subject the jet fuel to extremely low temperatures and lead to an increase in viscosity and the partial solidification of fuel. To reduce the freeze point, processing can be used to change the distillation range which can decrease the content of higher molecular weight alkanes or use additives such as pour point depressants. The literature reported on the use of ethylene vinyl-acetate (EVA) copolymers (Zabarnick et al., 2002). Jet fuels need to meet many specifications, including the freeze point, the distillation, the aromatic content, the sulfur content, the smoke point, and the flash point. The product specifications of different jet fuels are shown in Table 6.4.

Fuel instability is characterized by phase separation and sludge formation. The term stability is usually used when the formation of insoluble material during storage is observed. The early literature reported the analysis of fuel sediments which indicated the presence of different heteroatoms, such as S, N, and O. It was reported that the heteroatom containing molecules tend to be more thermally unstable at high temperatures than hydrocarbons which can lead to deposit formation. Elemental sulfur and disulfides were reported to be active promoters

TABLE 6.4
Product Specifications of Different Jet Fuels

Jet Fuels	Jet A	JP-5	JP-8
Distillation, D-86			
10% recovered, °C	Max 205	Max 205	Max 205
50% recovered, °C	Report	Report	Report
FBP, °C	Max 300	Max 290	Max 300
Freeze point, °C	Max −40	Max −46	Max −47
Aromatics, vol%	Max 20	Max 25	Max 22
Sulfur, wt%	Max 0.3	Max 0.4	Max 0.3
Flash point, °C	Min 38	Min 60	Min 38

Source: From Gary, J.H. and Handwerk, G.E., *Petroleum Refining, Technology and Economics*, 4th ed., Marcel Dekker, Inc., New York, 2001. Reproduced by permission of Routledge/Taylor & Francis Group, LLC.

of fuel instability (Thompson et al., 1949). The early literature also reported that the presence of alkyl mercaptans, thiophenes, sulfides, and disulfides accelerated the formation of deposits in cracked stocks (Schwartz et al., 1964). The presence of sulfonic acid was reported to be the most deleterious organo-sulfur compound affecting the storage stability of fuels (Hazlett et al., 1991).

The stability of jet fuels is tested at high temperatures. Certain N containing molecules present in jet fuels were reported responsible for their instability and color body formation. The presence of nitrogen heterocycles was reported to cause the formation of insoluble sediments and gums (Frankenfeld et al., 1983). Pyridine compounds have been reported present in practically all middle distillate fuels. The total nitrogen (TN) content of middle distillate fuel, produced from Gulf coast crude oil, was reported to contain 66% of pyridines in the form of 59 pyridine isomers (Mushrush et al., 2000). The literature reported on the synthesis of reactive pyridines and their use as model dopants to study the stability of fuels. The experimental data suggest that the interactive process between the basic and nonbasic nitrogen compounds might have a major effect on fuel stability (Mushrush et al., 2000). The more recent literature reported on a group JP-5 fuels which had passed all specification but were causing catastrophic engine failures (Morris et al., 2004a). The effect of sulfides and disulfides on the sediment formation in JP-5 fuel, under the conditions of the ASTM D 5304 test, is shown in Table 6.5.

The ASTM D 5304 test is used for determination of accelerated fuel stability at 90°C and in the presence of oxygen. The JP-5 fuel sample, producing 0.1 mg of solids per 100 mL of fuel, is considered to be stable fuel. The addition of hexyl disulfide was reported to lead to a higher sediment content of the fuel (Mushrush et al., 2001). The addition of 0.3 wt% of hexyl sulfide was required to increase the sediment formation of JP-5 fuel. The addition of only 0.1 wt% of hexyl disulfide was found to significantly increase the sediment formation. A low concentration of sulfonic acid was also reported (Mushrush et al., 2001). There are other procedures for testing the stability of fuels which require a higher temperature. The literature reported on the use of different chemistry additives which were found effective in improving the stability of jet fuel at high temperatures

TABLE 6.5
Effect of Sulfides and Disulfides on JP-5 Fuel Stability

JP-5 Fuel Sulfur Added, wt%	Hexyl Sulfide Solids, mg/100 mL	Hexyl Disulfide Solids, mg/100 mL
0	0.1	0.1
0.1	0.1	0.4
0.3	0.2	0.7

Source: From Mushrush, G.W. et al., *Petroleum Science and Technology*, 19, 561, 2001. Reproduced by permission of Taylor & Francis Group, LLC, http://www.taylorandfrancis.com.

(Silin et al., 2000). The literature also reported that middle distillate fuels have instability problems which lead to sediment initiated by a hydroperoxide induced oxidation process (Mushrush et al., 2002). Autoxidation of fuels has been extensively studied and oxygen dissolved in fuel was reported to react and form alkyl peroxides leading to ketones and aldehydes. These products were reported to form gums and varnishes which can constrict the fuel flow and lead to engine shutdown (Roan and Boehman, 2004). The peroxide content of different petroleum fractions is shown in Table 6.6.

Saturated light cycle oil (LCO), containing no aromatics, was reported to contain a total of 6.4 wt% of heteroatom and olefin content and 6 ppm of peroxide. Hydrotreated (HT) LCO, containing 25.2 wt% of aromatics, was reported to contain a total of 0.5 wt% of heteroatom and olefin content and 0.6 ppm of peroxide. Many products are blends and the peroxide content of HT product-blend was reported to increase from 46 to 631 ppm after being stored in opened barrel (Roan and Boehman, 2004). JP-8 jet fuel, containing 25.2 wt% of aromatics, was reported to contain a total of 1.2 wt% of heteroatom and olefin content and no presence of peroxide was detected. The use of nitrogen sparging was reported effective in decreasing the dissolved oxygen and the use of antioxidants, metal deactivators, detergents, and dispersants was effective in improving the thermal–oxidation stability (Roan and Boehman, 2004).

The dispersing properties of amino acid esters and their derivatives in inhibiting the gum and sediment formation in fuels were reported. The phenyl alanine dodecanoate and N-tert-dodecyl lysine dodecanoate were found to have good dispersing properties when tested in an accelerated fuel stability test (Juyal and Anand, 2002). The literature also reported on the use of dicyclohexylphenylphosphine to improve the thermal–oxidation stability of future jet fuels, known as JP-900, which are required to be stable at a high temperature of 480°C (Beaver et al., 2002). The literature also reported on the treating of stainless steel tubes to reduce the deposit formation. The chemical vapor deposition of a silica-based layer decreased the rate of deposit formation (Ervin et al., 2003). The use of additives and treated surfaces can only delay the oxidation process and decrease the rate of surface deposit formation.

TABLE 6.6
Peroxide Content of Different Petroleum Fractions

Petroleum Fraction Composition	Saturated Light Cycle Oil	Hydrotreated Light Cycle Oil	JP-8 Jet Fuel
Aromatics, wt%	0	25.2	21.3
Heteroatoms/alkenes, wt%	6.4	0.5	1.2
Peroxide, ppm	6	0.6	0

Source: Reprinted with permission from Roan, M.A. and Boehman, A.L., *Energy and Fuels*, 18, 835, 2004. Copyright American Chemical Society.

6.2 THERMAL–OXIDATION STABILITY OF JET FUEL

The use of the Merox process–caustic wash process is followed by the clay treatment. A salt drier is used to lower the free water content prior to the clay treating. Early literature reported that solids have the property of holding molecules at their surfaces and this property increases with an increase in their porosity and a decrease in the particle size. Various forces are involved including the physical and the chemical forces (Daniels et al., 1970). The natural clays are mined and commercially available, such as Attapulgite, ball clay, bentonite, calcium bentonite, common clay, fire clay, hectorite, kaolin, meerschaum, refractory clay, saponite sepiolite, shale, and sodium bentonite (USGS, 2007). The naturally occurring clays are layered crystalline materials which contain a large amount of water. Heating the clays above 100°C can drive out some of the water (Dolbear, 1998). According to the early literature, the natural Attapulgite clay can retain about 12 wt% of water even after prolonged drying at about 100°C. On dehydration at an optimum drying temperature, which is about 230°C, the clay structure remains intact and the Attapulgite clay contains 1 wt% of free water (McCarter et al., 1950). Seven types of clays are mined in the United States and their application is shown in Table 6.7.

Many different types of clays are available and used in various applications. Crude oils can be separated into saturates, aromatics, and resins using a chromatography column packed with Attapulgite clay. According to the literature, these fractions can be eluted and collected using different solvents. The saturates and aromatics were eluted with *n*-heptane and toluene solvent mixtures while more polar resins were eluted using a highly polar mixture of methanol, acetone, and chloroform (Leon et al., 2001). World production of Attapulgite clay is limited to only a few producers, with the United States accounting for 50% of the output. Next comes Senegal (35%), Spain (18%), and a small production in Australia and South Africa. A recent discovery of Attapulgite clay in Greece is a new market development (O'Driscoll, 2004).

TABLE 6.7

Use of Clays which Are Mined in USA

Clays Mined in USA	Use
Ball clays	Sanitary ware, ceramic tiles
Bentonite	Absorbents, drilling mud, foundry sand bonding agent, other
Common clay	Brick, lightweight aggregate, Portland cement clinker
Shale	Brick, lightweight aggregate, Portland cement clinker
Fire clay	Refractories
Kaolin	Paper and refractory markets
Attapulgite	Absorbents

Source: From USGS, *Clays Statistics and Information*, 2007.

TABLE 6.8
Variation in Composition of Attapulgite Clay

Components, wt%	Typical	Actual Sample
SiO_2	70.85	68.01
Al_2O_3	14.06	13.51
MgO	5.71	5.91
Fe_2O_3	5.34	5.31
CaO	1.62	2.71
K_2O	1.31	1.26
P_2O_3	0.84	1.22
Na_2O	0.25	0.18
SO_3	0.03	0.03

Source: From Pillon, L.Z., *Petroleum Science and Technology*, 19, 875, 2001a. Reproduced by permission of Taylor & Francis Group, LLC, http://www.taylorandfrancis.com.

Naturally occurring Attapulgite clay is a complex hydrated magnesium aluminum silicate and known as polygorskite in Europe or fullers' earth in the United States. Attapulgite clay is described as white or grey and forms microscopic needle-shaped crystals. Processing of granular clay was reported to include air drying, calcination, crushing, milling, sizing, packaging, storage, and shipping. Attapulgite clay is used as a sorbent or gellant. Its use as a sorbent also includes pet litter, industrial spill adsorbents, agrochemical, bleaching earths, and jet fuel treatment. Processing of powder/gellants was reported to include crushing, extrusion, milling, micronisation, air classification, packaging, storage, and shipping. Its use as a gellant includes catalyst binders, pharmaceuticals, cosmetics, fresh and salt water drilling muds (O'Driscoll, 2005). The typical composition and actual sample of Attapulgite clay, analyzed using the energy-dispersive x-ray fluorescence (XRF) spectrometry, is shown in Table 6.8.

Attapulgite clay was reported very effective in neutralizing traces of inorganic acids present in petroleum oils and, due to relatively large pores, it was also reported to be effective in removal of high molecular weight sulfonates, resins, and asphaltenes (Sequeira, 1994). The literature reported that variation in the nature, composition, and properties of the clays might lead to differences in their effectiveness to purify the jet fuel (Speight, 1999). The composition and the surface area of natural clays can vary and some are more acidic. The pH of clays can be tested by stirring 10 wt% of each clay in deionized water and testing the pH of water. Some clays are acidified and have a very acidic pH. The composition, surface, and pH of some natural and modified clays are shown in Table 6.9.

The natural Attapulgite clay, containing 71 wt% of silica oxide and 14 wt% of aluminum oxide, was reported to have a surface area of 110–120 m^2/g and was found to have a pH of 6.3. The calcined Attapulgite clay, having a similar

TABLE 6.9
Composition, Surface Area, and pH of Different Clays

Different Clays	SiO_2, wt%	Al_2O_3, wt%	Surface Area, m^2/g	Water, pH
Natural Attapulgite	71	14	120	6.3
Calcined Attapulgite	71	15	100	6.4
Acidic Attapulgite	77	12	150	4.8
Acidified Attapulgite	71	14	120	3.1
Acid activated	73	21	400	3.1
Natural zeolite	12	70	40	6.5

Source: From Pillon, L.Z., *Petroleum Science and Technology*, 19, 961, 2001b. Reproduced by permission of Taylor & Francis Group, LLC, http://www.taylorandfrancis.com.

composition, was reported to have a lower surface area of 100 m^2/g and was found to have a pH of 6.4. The natural acidic Attapulgite clay, containing 77 wt% of silica oxide and 12 wt% of aluminum oxide, was reported to have a higher surface area of 150 m^2/g and was found to have a more acidic pH of 4.8. The clay can be acidified to have a pH of 3.1 without a significant change in composition and the surface area. Another acid activated clay, containing 73 wt% of silica oxide and 21 wt% of aluminum oxide, was reported to have a high surface area of 400 m^2/g and was found to have a strong acidic pH of 3.1. The natural zeolite, containing only 12 wt% of silica and 70 wt% of alumina, was reported to have a low surface area of only 40 m^2/g and was found to have a pH of 6.5. The Merox treated–caustic washed jet fuel, containing 24.23 wt% of aromatics, 0.23 wt% of sulfur, and 15 ppm of N, was treated with 2.4 g of different clays and heated at 38°C for 1 h. The effect of different clays and basic type alumina adsorbents on the composition of jet fuel is shown in Table 6.10.

The ratio of basic nitrogen (BN) to total nitrogen (TN) of crudes is approximately constant, in the range of 0.3 +/− 0.05, irrespective of the source of crude oil (Speight, 1999). The ASTM D 2896 test method for base number of petroleum products by potentiometric perchloric acid titration can be used to determine the BN content of petroleum oils by titration with perchloric acid. The Merox treated–caustic washed jet fuel was found to have a high BN content of 12 ppm and a BN/TN ratio of 0.8 indicating a significant increase in polar basic type contaminants. According to the ASTM D 2896 test method, basic constituents of petroleum include BN compounds but also other compounds, such as salts of weak acids (soaps), basic salts of polyacidic bases, and salts of heavy metals. The total acid number (TAN) value of a petroleum product is the weight of potassium hydroxide (KOH) required to neutralize 1 g of oil. The Merox treated–caustic washed jet fuel was found to have a TAN below 0.01 mg KOH/g.

The ASTM D 3241 procedure for thermal–oxidation stability of aviation turbine fuels (JFTOT) is used to rate the tendencies of gas turbine fuels to deposit decomposition products within the fuel system, usually at 275°C. The fuel is

TABLE 6.10
Effect of Different Clays and Basic Alumina on Composition of Jet Fuel

Batch Treatment, 1 L Jet Fuel/2.4 g Clay/38°C/1 h	Aromatics, wt%	S, wt%	N, ppm	BN, ppm	TAN, mg KOH/g
Jet fuel feed	24.23	0.23	15	12	<0.01
Natural Attapulgite	24.35	0.22	12	8	<0.01
Calcined Attapulgite	24.33	0.23	13	12	<0.01
Acidic Attapulgite	24.28	0.22	11	9	<0.01
Acidified Attapulgite	24.32	0.23	9	7	<0.01
Acid activated clay	24.55	0.22	12	8	<0.01
Natural zeolite	24.20	0.23	15	12	<0.01
Basic alumina # 1	24.39	0.24	14	12	<0.01
Basic alumina # 2	24.35	0.23	13	9	<0.01

Source: From Pillon, L.Z., *Petroleum Science and Technology*, 19, 961, 2001b. Reproduced by permission of Taylor & Francis Group, LLC, http://www.taylorandfrancis.com.

pumped at a fixed volumetric flow rate through a heater after which it enters a precision stainless steel filter where fuel degradation products may become trapped. After 59 min at 275°C, the Merox treated–caustic washed jet fuel formed deposit, having a tube rating of 2, and failed the JFTOT test. After the batch treatment, the jet fuel was tested for its thermal–oxidation stability and none of the clays and typical laboratory adsorbents, such as basic activated alumina, were effective in improving the thermal–oxidation stability of the jet fuel. The effect of different clays and basic alumina, under column adsorption conditions, on the composition and the thermal–oxidation stability (JFTOT) of jet fuel is shown in Table 6.11.

TABLE 6.11
Effect of Different Clays and Alumina on Composition and JFTOT of Jet Fuel

Column Treatment, 1 L Jet Fuel/10 g Clay/RT	S, wt%	N, ppm	BN, ppm	JFTOT
Jet fuel feed	0.23	15	12	Fail
Natural Attapulgite	0.23	3	2	Pass
Calcined Attapulgite	0.23	5	5	Pass
Acidic Attapulgite	0.23	2	2	Pass
Natural zeolite	0.23	10	11	Fail
Basic alumina	0.23	9	9	Pass

Source: From Pillon, L.Z., *Petroleum Science and Technology*, 19, 961, 2001b. Reproduced by permission of Taylor & Francis Group, LLC, http://www.taylorandfrancis.com.

Under the column adsorption conditions, using 1 L of Merox treated–caustic washed jet fuel per 10 g of natural, calcined, or acidic Attapulgite clays, there is no change in the S content but a decrease in the N content to 2–5 ppm and a decrease in the BN content to 2–5 ppm are observed. Attapulgite clay treated jet fuels were found to have improved thermal–oxidation stability and pass the JFTOT test. Under the same column adsorption conditions, using 1 L of the same jet fuel per 10 g of natural zeolite, no significant decrease in N and BN contents and no improvement in thermal–oxidation stability of jet fuel were observed. Natural zeolite, having a higher alumina content and a lower surface area, was found not effective in purifying jet fuel. Natural zeolites were reported to be used as soil amendment and feed additive (ZEO Inc., 2002). Under the same column adsorption conditions, using 1 L of jet fuel per 10 g of basic alumina, no change in the S content with very small decrease in N content to 9 ppm and a decrease in the BN content to 9 ppm is observed. Basic alumina, having a very high surface area of 380 m^2/g, was found effective in improving the thermal–oxidation stability of jet fuel. More cost effective Attapulgite clay is used by refineries to finish Merox treated–caustic washed jet fuel and other products. The effect of a lower jet fuel ratio on N and BN contents of clay treated jet fuel is shown in Table 6.12.

Under the column adsorption conditions, using 1 L of untreated jet fuel per 15 g of natural Attapulgite clay, there is no change in S content but a decrease in the N content to 1 ppm and a decrease in the BN content to 1 ppm are observed. Under the same column adsorption conditions, using 1 L of untreated jet fuel per 15 g of calcined Attapulgite clay, there is no change in the S content but a decrease in the N content to 1 ppm and a decrease in the BN content to 1 ppm is also observed. While the composition of natural and calcined Attapulgite clays was reported the same, some decrease in the surface area of calcined Attapulgite clay was observed. The effect of calcined Attapulgite clay on the composition and the thermal–oxidation stability of the effluent jet fuel is shown in Table 6.13.

TABLE 6.12
Effect of Lower Jet Fuel Ratio on N and BN Contents of Clay Treated Jet Fuel

Column Treatment, 1 L Jet Fuel/15 g Clay/RT	S, wt%	N, ppm	BN, ppm
Jet fuel feed	0.23	15	12
Natural Attapulgite	0.23	1	1
Calcined Attapulgite	0.23	1	1

Source: From Pillon, L.Z., *Petroleum Science and Technology*, 19, 961, 2001b. Reproduced by permission of Taylor & Francis Group, LLC, http://www.taylorandfrancis.com.

TABLE 6.13
Effect of Calcined Attapulgite Clay on Thermal–Oxidation Stability of the Effluent

15 g Clay Effluent, L	Sulfur, wt%	Total N, ppm	Basic N, ppm	Olefins, vol%	JFTOT Test
1	0.23	1	1	>2	Pass
2	—	—	8		Pass
3	0.23	9	8	<2	Pass
4	—	—	9		Pass
5	0.23	10	10	1.5	Fail

Source: From Pillon, L.Z., *Petroleum Science and Technology*, 19, 961, 2001b. Reproduced by permission of Taylor & Francis Group, LLC, http://www.taylorandfrancis.com.

The initial effluent contains practically no N, has a low BN content of only 1 ppm and passed the JFTOT test. With an increase in the volume of jet fuel, from 1 to 5 L, no change in the S content was observed but the N content was found to gradually increase, from 1 to 10 ppm, and a gradual increase in BN content, from 1 to 10 ppm, is also observed. With an increase in the effluent volume, a gradual decrease in the olefin content indicates their high surface activity and retention on the clay. After passing 5 L of jet fuel through the adsorption column and an increase in N and BN contents, the effluent jet fuel failed the JFTOT test. In the case of Attapulgite clay, no significant benefit of calcination in BN content retention was observed. The effect of different clays on the BN content of the effluent jet fuel is shown in Table 6.14.

The untreated jet fuel was passed through the adsorption column containing 15 g of different clays, having a different surface area and pH. Initially, all fresh

TABLE 6.14
Effect of Different Clays on the BN Content of Effluent Jet Fuel

15 g Clay Effluent, L	Natural Attapulgite, BN, ppm	Acidic Attapulgite, BN, ppm	Acid Activated, BN, ppm
1	2	1	3
2	8	4	8
3	11	8	9
4	12	9	9
5	11	9	9
6	15	10	10

Source: From Pillon, L.Z., *Petroleum Science and Technology*, 19, 961, 2001b. Reproduced by permission of Taylor & Francis Group, LLC, http://www.taylorandfrancis.com.

clays were effective in decreasing the BN content of jet fuel from 12 to 1–3 ppm. With an increase in the volume of jet fuel, a gradual increase in the BN content of effluent was observed. The capacity of different clays to retain polar BN molecules was found to gradually decrease with an increase in the effluent volume. After passing 5–6 L of jet fuel through the column containing Attapulgite clay, the BN content of effluent was found to gradually increase from 2 to 12–15 ppm thus reaching the level of BN content present in untreated jet fuel. After passing 6 L of untreated jet fuel through the column containing acidic Attapulgite clay, the BN content of effluent was found to gradually increase from 1 to 10 ppm, thus indicating an increased retention of BN molecules. After passing 6 L of untreated jet fuel through the column containing acid activated clay, the BN content of effluent was also gradually increased from 3 to 10 ppm, thus confirming an increased retention of BN molecules. In the case of Attapulgite clay, no significant benefit of an increased acidity on BN content retention was observed. The effect of jet fuel/clay and basic alumina ratio on the sulfur and nitrogen contents of jet fuel is shown in Table 6.15.

Under the column adsorption conditions, using 1 L of jet fuel per 30 g of different Attapulgite clays, there is no significant change in the S content but a decrease in the N content, to below 1 ppm, and a decrease in the BN content, to below 1 ppm, is observed. Under the same column adsorption conditions, using 1 L of jet fuel per 30 g of natural zeolite, there is no significant change in the S content but a decrease in the N content to 2 ppm and a decrease in the BN content to 1 ppm are observed. Under the same column adsorption conditions, using 1 L of jet fuel per 30 g of basic alumina # 2, a decrease in the S content, from 0.23 to 0.15 wt%, with some decrease in the aromatic content is observed. While different clays and adsorbents have varying composition and surface area, these properties are critical when using a high jet fuel/adsorbent ratio. With a decrease in the jet fuel/adsorbent ratio, these properties are less significant and practically any clay

---

TABLE 6.15
Effect of Jet Fuel/Clay and Basic Alumina Ratio on Sulfur Content of Jet Fuel

Column Treatment, 1 L Jet Fuel/30 g Clay/RT	Aromatics, wt%	Sulfur, wt%	Total N, ppm	Basic N, ppm
Jet fuel feed	24.23	0.23	15	12
Natural Attapulgite	23.74	0.20	<1	<1
Calcined Attapulgite	19.98	0.22	<1	<1
Acidic Attapulgite	23.19	0.20	<1	<1
Natural zeolite	23.81	0.23	2	1
Basic alumina # 1	23.01	0.21	<1	<1
Basic alumina # 2	21.43	0.15	<1	<1

Source: From Pillon, L.Z., *Petroleum Science and Technology*, 19, 961, 2001b. Reproduced by permission of Taylor & Francis Group, LLC, http://www.taylorandfrancis.com.

and adsorbent are effective. Under the column adsorption conditions and a low jet fuel/adsorbent ratio, the surface activity of aluminas and Attapulgite clay, having different composition, surface area, and pH, becomes similar in adsorbing N and BN molecules. Attapulgite clay is used as a cost effective adsorbent, catalyst binder, and as a finishing step of jet fuel and other products.

6.3 CLAY LIFETIME

6.3.1 JFTOT FAILURE

Many refineries use a Merox process followed by clay treatment to remove "undesirable" molecules and meet the quality requirements of jet fuels. The literature reported that the efficiency of the Attapulgite clay declines with each successive adsorption–regeneration cycle (Sequeira, 1994). The point at which the effluent from the adsorption column no longer meets the quality requirements, dictating replacement or regeneration of the adsorbent, is called the "breakpoint" (Sleyko, 1985). The breakpoint for jet fuels would be defined by the effluent inability to meet the thermal–oxidation stability requirements, such as passing the JFTOT. Each refinery uses their own guidelines as to the quality requirements and the need for the clay replacement. For the same refinery, with an increase in nonconventional refining of crude oils and cracking of petroleum feedstocks, the lifetime of Attapulgite clay was reported to significantly decrease. Some refineries replace the Attapulgite clay only 1–2 times per year, others replace 5–6 times per year. Depending on the crude oils and the processing conditions, some refineries produce Merox treated–caustic washed jet fuel, having a good thermal–oxidation stability. The composition of Merox treated–caustic washed jet fuel, having a good thermal–oxidation stability, is shown in Table 6.16.

TABLE 6.16
Composition of Jet Fuel Having Good Thermal–Oxidation Stability

Merox Treated–Caustic Washed Jet Fuel # 1	Before Clay Treatment	After Clay Treatment
Density, g/cc at 15°C	0.8063	0.8061
Saturates, wt%	77.1	77.8
Aromatics, wt%	22.9	22.2
Sulfur, wt%	0.04	0.04
Total N, ppm	3	3
Basic N, ppm	3	2
Copper, ppb	<10	<10
JFTOT Stability (ASTM D 3241)	Pass	Pass

Source: From Pillon, L.Z., *Petroleum Science and Technology*, 19, 1109, 2001c. Reproduced by permission of Taylor & Francis Group, LLC, http://www.taylorandfrancis.com.

The Merox treated–caustic washed jet fuel # 1 was found to contain 22.9 wt% of aromatics, 0.04 wt% of sulfur, and 3 ppm of nitrogen content. The metal content, such as Cu, was found to be below 10 ppb which is below the detection limits. The 3 ppm of BN content indicated no significant presence of polar basic type contaminants. The jet fuel # 1 was found to have a good thermal stability and passed the JFTOT test before the clay treatment. After the clay treatment, some decrease in aromatic and BN contents but no significant change in the S and N contents was observed. The clay treated jet fuel # 1 was also found to have a good thermal–oxidation stability and passed the JFTOT test. In most cases, the Merox treated–caustic washed jet fuels have poor thermal–oxidation stability and require the clay treatment. The composition of the Merox treated–caustic washed jet fuel, having poor thermal–oxidation stability, is shown in Table 6.17.

The Merox treated jet fuel # 2 was found to contain 21.9 wt% of aromatics, 0.05 wt% of sulfur, and 5 ppm of nitrogen content. The metal content, such as Cu, was found to be below 10 ppb which is below the detection limits. The TAN of 0.01 mg KOH/g indicated no increase in acidic type contaminants but the presence of 5 ppm of BN content indicates an increase in the polar basic type contaminants. The jet fuel # 2 was found to have a poor thermal stability and failed the JFTOT test indicating a need for clay treatment. After the clay treatment, no significant change in composition and the polar content was observed; however, the thermal stability was found to improve and the clay treated jet fuel # 2 passed the JFTOT test. When the clay treatment does not meet the quality requirements of jet fuel, such as passing the JFTOT test, it needs to be replaced. The spent clay samples, collected from the top, middle, and the bottom of the clay bed, were washed with n-heptane solvent to remove the excess of hydrocarbons and analyzed for the presence of organic

TABLE 6.17
Composition of Jet Fuel Having Poor Thermal–Oxidation Stability

Merox Treated–Caustic Washed Jet Fuel # 2	Before Clay Treatment	After Clay Treatment
Density, g/cc at 15°C	0.8087	0.8087
Saturates, wt%	78.1	78.1
Aromatics, wt%	21.9	21.9
Sulfur, wt%	0.05	0.05
Total N, ppm	5	5
Basic N, ppm	5	5
Copper, ppb	<10	<10
JFTOT Stability (ASTM D 3241)	Fail	Pass

Source: From Pillon, L.Z., *Petroleum Science and Technology*, 19, 1109, 2001c. Reproduced by permission of Taylor & Francis Group, LLC, http://www.taylorandfrancis.com.

TABLE 6.18
Organic Content of Spent Attapulgite Clay Replaced Due to JFTOT Failure

Attapulgite Clay, JFTOT Failure	Carbon, wt%	Hydrogen, wt%	Sulfur, wt%	Nitrogen, wt%
Fresh clay	0.11	0.95	0.10	0.02
Spent clay				
Clay bed (top)	3.28	1.39	0.27	0.24
Clay bed (middle)	3.33	1.41	0.27	0.25
Clay bed (bottom)	3.22	1.28	0.28	0.23

Source: From Pillon, L.Z., *Petroleum Science and Technology*, 19, 875, 2001a. Reproduced by permission of Taylor & Francis Group, LLC, http://www.taylorandfrancis.com.

contaminants. The organic content of spent Attapulgite clay, replaced due to jet fuel JFTOT failure, is shown in Table 6.18.

The analysis of spent clay samples, collected from the top, middle, and the bottom of the clay bed, indicated an increase in the presence of organic contaminants which were found evenly spread from the top to the bottom of the clay treater. The C content increased from 0.11 wt% found in fresh clay to 3.22–3.33 wt%, and an increase in the H content from 0.95 wt% found in fresh clay to 1.28–1.41 wt% was observed. The S content increased from 0.1 wt% found in fresh clay to 0.27–0.28 wt%, indicating the adsorption of S containing molecules. The use of the same jet fuel containing only a ppm N content increases the N content of spent clays, from 0.02 wt% found in fresh clay to 0.23–0.25 wt%, indicating a drastic increase in surface activity of N containing molecules. The GC/MS analysis of the spent Attapulgite clay was carried out in a He purge gas in three steps. In the first step, the clay sample was heated from 50°C to 200°C and analyzed for any desorption of organic molecules. In the second step, the clay sample was heated from 50°C to 400°C and, in the third step, from 50°C to 600°C. The GC/MS analysis of spent clay, replaced due to JFTOT failure of jet fuel, is shown in Table 6.19.

At the temperature range of 50°C–600°C, the GC/MS analysis of fresh clay shows only desorption of water. The use of the GC/MS technique can only detect the presence of molecules having a high volatility. At a lower temperature of 50°C–200°C, the GC/MS analysis of spent clay shows desorption of C_9–C_{16} jet fuel hydrocarbons. At a higher temperature of 200°C–400°C, the GC/MS analysis shows desorption of pyridines, quinolines, other N molecules, and NO compounds. At a high temperature of 400°C–600°C, the GC/MS analysis shows desorption of naphthalene, alkylbenzenes, pyridines, quinolines, other N molecules, NO compounds, phenols, different acids, and acid esters. Only at the bottom of the clay bed, the presence of O containing molecules in the form of acids and acid esters was found. The presence of acids in some feedstocks might be related to the composition of crude oils or indicate the presence of oxidation by-products. While the presence of 0.27–0.28 wt% of S compounds on spent clay

TABLE 6.19
GC/MS Analysis of Spent Attapulgite Clay Replaced Due to JFTOT Failure

Attapulgite Clay, JFTOT Failure	GC/MS, 50°C–200°C	GC/MS, 200°C–400°C	GC/MS, 400°C–600°C
Fresh clay	Water	Water	Water
Spent Clay			
Clay bed (top)	C₉–C₁₆ hydrocarbons	Anilines	Naphthalenes
		Pyridines	Other hydrocarbons
		Quinolines	Quinolines
		Other N molecules	Other N molecules
		NO compounds	
Clay bed (middle)	C₉–C₁₆ hydrocarbons	Pyridines	Alkylbenzenes
		Quinolines	Naphthalenes
		Other N molecules	Quinolines
		NO compounds	Other N molecules
Clay bed (bottom)	C₉–C₁₆ hydrocarbons	Pyridines	Alkylbenzenes
		Quinolines	Pyridines
		Other N molecules	Quinolines
		NO compounds	Other N molecules
			Phenols
			Different acids
			Acid esters

Source: From Pillon, L.Z., *Petroleum Science and Technology,* 19, 875, 2001a. Reproduced by permission of Taylor & Francis Group, LLC, http://www.taylorandfrancis.com.

was found, no desorption of S compounds is observed indicating their low volatility. Despite a very low content of N compounds in jet fuel, their similar content on spent clay indicates their very high surface activity and volatility.

6.3.2 COPPER CONTAMINATION

While Attapulgite clay can be used to decolorize and neutralize any petroleum oil, it was reported only moderately effective in removing odorous compounds and trace metals (Sequeira, 1994). All fuels were reported to contain trace quantities of metals which can be naturally occurring or introduced during fuel handling. The presence of catalytic amounts of nickel in fuels was reported to produce compounds having a higher molecular weight and an increased polarity (Wynne et al., 2003). Trace amounts of copper can be introduced during the copper sweetening process or from contact with copper-bearing piping, brass fittings, and other copper-bearing alloys. In gas-drive fuel coker tests, traces of copper, iron, zinc, or lead were reported to have deleterious effect on jet fuel thermal stability (Collins et al., 2002). Metals, such as copper and iron, are known as pro-oxidants and the literature reported on the effect of copper leading to jet fuel

TABLE 6.20
Composition of Jet Fuel Contaminated with Copper

Merox Treated–Caustic Washed Jet Fuel Containing Copper	Before Clay Treatment	After Clay Treatment
Density, g/cc at 15°C	0.8181	0.8167
Saturates, wt%	78.7	78.6
Aromatics, wt%	21.3	21.4
Sulfur, wt%	0.21	0.19
Total N, ppm	7	6
Basic N, ppm	7	6
Copper, ppb	150	40
JFTOT Stability (ASTM D 3241)	Fail	Fail

Source: From Pillon, L.Z., *Petroleum Science and Technology*, 19, 1109, 2001c. Reproduced by permission of Taylor & Francis Group, LLC, http://www.taylorandfrancis.com.

instability. The literature reported on the development of filters to remove copper from jet fuel (Morris and Chang, 2000). Despite the fact that the copper sweetening process has been replaced by Merox treating and hydrotreating, the jet fuel copper contamination can occur from contact with copper containing alloys in fuel handling processes (Morris et al., 2004b). Some jet fuels were found to be contaminated with a trace of copper metal. The composition of Merox treated–caustic washed jet fuel, contaminated with copper, is shown in Table 6.20.

Before the clay treatment, the jet fuel was found to contain 21.3 wt% of aromatics, 0.21 wt% of sulfur, and 7 ppm of nitrogen content. The jet fuel was also found to contain 150 ppb of Cu indicating a metal contamination. According to the literature, the Cu content, above 10 ppb, degrades the thermal–oxidation stability of jet fuel. The jet fuel was found to have a poor thermal–oxidation stability and failed the JFTOT test. After the clay treatment, a decrease in the S content of jet fuel, from 0.21 to 0.19 wt%, and some decrease in the N content, from 7 to 6 ppm, were observed. Usually the aromatic and S contents of jet fuel are not affected by the clay treatment. After the clay treatment, the copper content of jet fuel was also found to decrease from 150 to 40 ppb, indicating the efficiency of the clay treater to remove some metal contamination. Despite a clay treatment and the reduction of the Cu content to 40 ppb, the jet fuel was found to have a poor thermal–oxidation stability and failed the JFTOT test. The spent Attapulgite clay, contaminated with copper, sampled from the top, middle, and the bottom of the clay treater was washed with *n*-heptane and analyzed for the presence of organic contaminants. The organic content of spent Attapulgite clay, contaminated with copper, is shown in Table 6.21.

The C content of the spent clays, sampled from the top of the clay bed, increased from 0.11 to 4.81 wt%, indicating a drastic increase in organic type contaminants. While the H content only increased from 0.95 to 1.24 wt%, the

TABLE 6.21
Organic Content of Spent Attapulgite Clay Contaminated with Copper

Attapulgite Clay, Cu Contamination	Carbon, wt%	Hydrogen, wt%	Sulfur, wt%	Nitrogen, wt%
Fresh clay	0.11	0.95	0.10	0.02
Spent Clay				
Clay bed (top)	4.81	1.24	0.42	0.36
Clay bed (middle)	4.06	1.23	0.23	0.23
Clay bed (bottom)	3.87	1.17	0.24	0.22

Source: From Pillon, L.Z., *Petroleum Science and Technology*, 19, 875, 2001a. Reproduced by permission of Taylor & Francis Group, LLC, http://www.taylorandfrancis.com.

S content increased from 0.1 to 0.42 wt%, and the N content increased from 0.02 to 0.36 wt%, indicating a significant increase in adsorption of heterocompounds. The C content of the spent clays, sampled from the middle and the bottom of the clay bed, only increased to 3.87–4.06 wt%, indicating some decrease in organic type contaminants. While the H content was found similar in a range of 1.17–1.23 wt%, the S content decreased to 0.23–0.24 wt%, and the N content also decreased to 0.22–0.23 wt%, indicating a lower content of heterocompounds in the middle and the bottom of the clay treater. The GC/MS analysis of spent Attapulgite clay, contaminated with copper, is shown in Table 6.22.

At a lower temperature of 50°C–200°C, the GC/MS analysis of spent clay shows only desorption of C_9–C_{16} jet fuel hydrocarbons. At a higher temperature

TABLE 6.22
GC/MS of Spent Attapulgite Clay Contaminated with Copper

Attapulgite Clay, Cu Contamination	GC/MS, 50°C–200°C	GC/MS, 200°C–400°C	GC/MS, 400°C–600°C
Fresh clay	Water	Water	Water
Spent Clay			
Clay bed (top)	C_9–C_{16} hydrocarbons	Sulfides	Alkylbenzenes
		Thiophenes	Pyridines
		Cyclohexamines	Quinolines
		Benzenamines	Other N molecules
		Pyridines	Phenols
		Other N molecules	
		NO compounds	

Source: From Pillon, L.Z., *Petroleum Science and Technology*, 19, 875, 2001a. Reproduced by permission of Taylor & Francis Group, LLC, http://www.taylorandfrancis.com.

of 200°C–400°C, the GC/MS analysis shows the desorption of water, thiophenes, sulfides, thiols, amines, pyridines, piperidines, quinolines, other N molecules, NO compounds, and alkylbenzenes. At a high temperature of 400°C–600°C, the GC/MS analysis shows desorption of water, alkylbenzenes, pyridines, quinolines, other N molecules, NO compounds, and phenols. With an increase in the S content of spent clay, collected from the top of the clay treater, the presence of volatile thiols, sulfides, and thiophenes is observed. The O containing molecules, such as phenols, were found to be present on the spent clays and desorbing at a higher temperature of 400°C–600°C. The literature reported that the pyridine compounds are readily oxidized to pyridine N-oxides at high temperatures, such as that of most nozzles and injectors (Mushrush et al., 2000). The presence of copper and other metals on spent clay was identified using the XRF technique. The inorganic content of spent Attapulgite clay, contaminated with copper, is shown in Table 6.23.

TABLE 6.23
Inorganic Content of Spent Attapulgite Clay Contaminated with Copper

XRF Analysis, wt%	Fresh Clay	Spent Clay, Top	Spent Clay, Bottom
SiO_2	68.0	39.1	65.8
Al_2O_3	13.5	8.7	13.9
MgO	5.9	3.4	5.1
Fe_2O_3	5.3	4.8	5.5
CaO	2.7	4.9	1.47
K_2O	1.26	0.94	1.26
P_2O_3	1.22	0.23	0.54
Na_2O	0.18	13.3	0.14
SO_3	0.03	1.5	0.18
TiO_2	0.58	0.48	0.61
MnO	0.05	0.04	0.05
V_2O_5	0.03	0.03	0.03
S	0.02	0.5	0.21
Cr_2O_3	0.02	0.02	0.02
ZrO_2	0.02	0.02	0.02
Cl	0.01	15.7	0.01
Co_3O_4	0.01	0.01	0.01
NiO	0.01	0.01	0.01
ZnO	0.01	0.02	0.02
Rb_2O	0.01	0.01	0.01
SrO	0.01	0.01	0.01
Y_2O_3	0	0.01	0.01
CuO	0	0.01	0

Source: From Pillon, L.Z., *Petroleum Science and Technology*, 19, 875, 2001a. Reproduced by permission of Taylor & Francis Group, LLC, http://www.taylorandfrancis.com.

The XRF analysis indicated the presence of S containing compounds concentrated mainly on the top of the clay treater; however, an increased S content was also found at the bottom of the clay treater thus indicating that the clay bed is saturated with inorganic and organic type contaminants. The inorganic content of spent clays was found to vary when sampled from the top versus the bottom of the clay treater. A decrease in silica oxide, from 68 wt% found in fresh Attapulgite clay to 39.1 wt% found on the top of the clay treater and 65.8 wt% found at the bottom of the clay treater, indicates that most of the inorganic contaminants concentrate on the top of the clay bed. The CaO content increased from 2.7 wt% found in fresh Attapulgite clay to 4.9 wt% found on the top of the clay treater. The Na_2O content drastically increased from 0.18 wt% found in fresh Attapulgite clay to 13.3 wt% found on the top of the clay treater. The Cl content also drastically increased from 0.01 wt% found in fresh Attapulgite clay to 15.71 wt% found on the top of the clay treater. The efficient performance of salt dryers is important in preventing contamination of the clay bed by inorganic molecules. The spent Attapulgite clay, sampled from the top of the clay treater, indicated the presence of 0.01 wt% of CuO, not found in fresh Attapulgite clay. No presence of Cu contamination was found at the bottom of the clay treater. The spent clay was found to have the most inorganic contamination on the top of the clay indicating significant changes in its composition and showing changes in its physical structure indicating clay deactivation.

6.4 CLAY DEACTIVATION

6.4.1 TOTAL ACID NUMBER

Attapulgite clay is very effective in adsorbing polar contaminants and neutralizing traces of inorganic acids. The usual components of TAN are organic soaps, soaps of heavy metals, organic nitrates, other compounds used as additives, and oxidation by-products. Naturally occurring acids and esters can be reactive at high temperatures leading to surface deposit formation causing poor stability of jet fuel (Taylor, 1996). Organic acids may form as a result of oxidation and the heavy metal soaps are the results of the reaction between acids and metals (Pirro and Wessol, 2001). The literature reported that the strong mineral acids react with clays leading to hydrolysis of aluminum–oxygen bond. The acidic protons also exchange with sodium and other alkali cations affecting the properties of clays (Dolbear, 1998). The presence of acidic Cl^- ion can lead to destruction of the clay's physical structure and the premature clay deactivation. ASTM D 4929 test method for organic chloride content of crude oils is used to determine the organically bound chlorine in crude oils. Organic chlorides are not known to be naturally present in crude oils and usually result from cleaning operations at producing sites, pipelines, and tanks. Hydrochloric acid can be produced in hydrotreating or reforming reactors and the acid can accumulate in condensing regions of the refinery. The typical TAN of jet fuel is below 0.01 mg KOH/g but, in some cases, an increase in TAN of jet fuel, before the clay treatment, was observed. The composition of jet fuel, having an increased TAN, is shown in Table 6.24.

TABLE 6.24

Composition of Jet Fuel Having an Increased TAN

Merox Treated–Caustic Washed Jet Fuel, Increased TAN	Before Clay Treatment	After Clay Treatment
Density, g/cc at 15°C	0.8195	0.8195
Saturates, wt%	77.1	77.2
Aromatics, wt%	22.9	22.8
Sulfur, wt%	0.17	0.16
Total N, ppm	11	10
Basic N, ppm	10	9
Copper, ppb	<10	<10
TAN, mg KOH/g	>0.01	>0.01

Source: From Pillon, L.Z., *Petroleum Science and Technology*, 19, 1109, 2001c. Reproduced by permission of Taylor & Francis Group, LLC, http://www. taylorandfrancis.com.

Before the clay treatment, the jet fuel was found to contain 22.9 wt% of aromatics, 0.17 wt% of sulfur, and 11 ppm of nitrogen content. No presence of metal contamination, such as Cu and Fe, was found. The BN content of 10 ppm indicated the presence of polar basic type contaminants. The TAN of >0.01 mg KOH/g indicated a significant increase in acidic type contaminants. After the clay treatment, while some decrease in TAN of jet fuel was observed, it continued to be >0.01 indicating the need for clay replacement. Based on the low volume of jet fuel passed through the clay treater and a significant pressure drop across the filter, a case of premature clay deactivation was observed. The clay needed to be replaced due to a high pressure drop across the clay filter. The analysis of the premature deactivated clay, sampled from the top of the clay treater, indicated a drastic increase in Cl content and a significant increase in CaO and Na_2O. The organic content of deactivated Attapulgite clay, replaced due to a high pressure drop across the clay filter, is shown in Table 6.25.

The analysis of samples, collected from the top, middle, and the bottom of the clay bed, indicated an increase in the presence of organic contaminants which were not evenly spread from the top to the bottom of the clay treater. The C content of spent clay was found to increase from 0.11 wt% for fresh clay to 2.69–5.14 wt%, and the highest concentration was observed on the top of the clay treater. The H content of spent clay was found to increase from 0.95 wt% for fresh clay to 0.86–1.52 wt%, and the highest concentration was also observed on the top of the clay treater. The S content of spent clay was found to increase from 0.1 wt% for fresh clay to 0.22–0.39 wt%, and the highest concentration was also observed on the top of the clay treater. The N content of spent clay was found to increase from only 0.02 wt% for fresh clay to 0.07–0.22 wt%, and the lowest concentration was observed on the top of the clay treater. An increase in adsorption of S compounds with a decrease in retention of N compounds is observed.

TABLE 6.25
Organic Content of Deactivated Attapulgite Clay Replaced
Due to a Pressure Drop

Attapulgite Clay, Pressure Drop	Carbon, wt%	Hydrogen, wt%	Sulfur, wt%	Nitrogen, wt%
Fresh clay	0.11	0.95	0.10	0.02
Spent Clay				
Clay bed (top)	5.14	1.52	0.39	0.07
Clay bed (middle)	3.53	1.16	0.22	0.22
Clay bed (bottom)	2.69	0.86	0.24	0.15

Source: From Pillon, L.Z., *Petroleum Science and Technology*, 19, 875, 2001a. Reproduced by permission of Taylor & Francis Group, LLC, http://www.taylorandfrancis.com.

In the case of spent clay, replaced due to JFTOT failure, the clay bed was evenly saturated with C, H, S, and N surface active contaminants, from the top to the bottom of the clay bed. In the case of spent clay, replaced due to a high pressure drop across the filter, all inorganic contaminants are concentrated on the top of the clay bed. The GC/MS analysis of deactivated Attapulgite clay, replaced due to a high pressure drop across the filter, is shown in Table 6.26.

At a lower temperature of 50°C–200°C, the GC/MS analysis shows desorption of C_9–C_{16} jet fuel hydrocarbons. At a higher temperature of 200°C–400°C, the GC/MS analysis shows the desorption of alkylbenzenes, thiophenes, pyridines, quinolines, other N molecules, and NO compounds. At a high temperature of 400°C–600°C, the GC/MS analysis shows desorption of alkylbenzenes, naphthalenes, indenes, pyridines, quinolines, and other N molecules. In the case of the S content, the spent clay was found to contain thiophenes concentrating on the top of the clay treater. With a decrease in the S content, no presence of volatile S compounds in the middle and the bottom of the clay treater was observed. While the organic content analysis indicated a significant presence of S compounds, the GC/MS of deactivated wet clay indicated a desorption of only hydrocarbons and N molecules, indicating a low volatility of S molecules. The surface active content of spent clays, replaced due to a high pressure drop across the filter, varied depending on the sample location. The absence of O containing molecules might be related to their low volatility or their presence as NO compounds. The GC/MS analysis of deactivated Attapulgite clay, contaminated with oxidation by-products, is shown in Table 6.27.

At a lower temperature of 50°C–200°C, the GC/MS analysis of premature deactivated Attapulgite clay was found to show desorption of water and C_9–C_{16} jet fuel hydrocarbons. At a higher temperature range of 200°C–400°C, the GC/MS analysis indicated the presence of water, alkylbenzenes, NO compounds, phenols, and different ketones. At a high temperature of 400°C–600°C, the GC/MS analysis shows desorption of water, alkylbenzenes, quinolines, other

TABLE 6.26
GC/MS Analysis of Deactivated Clay Replaced Due to a Pressure Drop

Attapulgite Clay, Pressure Drop	GC/MS, 50°C–200°C	GC/MS, 200°C–400°C	GC/MS, 400°C–600°C
Fresh clay	Water	Water	Water
Deactivated Clay			
Clay bed (top)	C_9–C_{16} hydrocarbons	Alkylbenzenes Naphthalenes Thiophenes Quinolines NO compounds	Alkylbenzenes Naphthalenes Indenes
Clay bed (middle)	C_9–C_{16} hydrocarbons	Pyridines Quinolines Other N molecules NO compounds	Alkylbenzenes Pyridines Quinolines Other N molecules
Clay bed (bottom)	C_9–C_{16} hydrocarbons	Quinolines Other N molecules	Alkylbenzenes Naphthalenes Indenes Quinolines Other N molecules

Source: From Pillon, L.Z., *Petroleum Science and Technology*, 19, 875, 2001a. Reproduced by permission of Taylor & Francis Group, LLC, http://www.taylorandfrancis.com.

TABLE 6.27
GC/MS of Deactivated Attapulgite Clay Contaminated with Oxidation By-Products

Attapulgite Clay, Oxidation By-Products	GC/MS, 50°C–200°C	GC/MS, 200°C–400°C	GC/MS, 400°C–600°C
Fresh clay	Water	Water	Water
Deactivated Clay			
Clay bed (top)	Water C_9–C_{16} hydrocarbons	Water Alkylbenzenes NO compounds Phenols Different ketones	Water Alkylbenzenes Quinolines Other N compounds Phenols Ketones

Source: From Pillon, L.Z., *Petroleum Science and Technology*, 19, 875, 2001a. Reproduced by permission of Taylor & Francis Group, LLC, http://www.taylorandfrancis.com.

N molecules, phenols, and ketones. Desorption of water during the GC/MS analysis of deactivated clay indicated that the water molecules were held inside the crystalline structure of the Attapulgite clay. The presence of ketones, such as hexanone and heptanone, indicates the presence of low temperature oxidation by-products. Many applications involving the use of lubricating oils are carried out at varying temperatures. Under low temperature oxidation conditions, peroxides, alcohols, aldehydes, ketones, and water are formed. Under high temperature conditions, acids are formed (Bardasz and Lamb, 2003).

6.4.2 BROMINE NUMBER

Clay treating is usually applied at increased temperature and pressure to thermally cracked naphthas to improve their color and stability. The literature reported that the stability of cracked naphtha is increased by clay adsorption of reactive diolefins (Gary and Handwerk, 2001). Olefins are formed during the catalytic and thermal cracking, and some presence of olefins during hydroprocessing was reported. The presence of olefins can affect the product stability and lead to sediment formation at high temperatures. Some clay treated jet fuels were found to be highly reactive and have a high bromine number. The Merox process, involving the catalytic oxidation of mercaptans with air, is proceeded by a caustic wash step to reduce the carboxylic acid content and followed by a water wash step to remove carryover caustic and sodium carboxylate salts from jet fuel. The Na content of clay treated jet fuels was found to be below 10 ppb and the Ca content was found to be below 50 ppb which is below detection limits. The ASTM D 1159 test method for bromine numbers of petroleum distillates and commercial aliphatic olefins by electrometric titration is used as an indication of the quantity of bromine-reactive constituents, including the presence of olefins. The bromine numbers of hydrocarbons, including different chemistry olefins, are shown in Table 6.28.

TABLE 6.28
Bromine Numbers of Hydrocarbons and Different Chemistry Olefins

Hydrocarbons	Purity, %	Bromine Number
Paraffins	>99	0–0.1
Straight chain olefins	>99	63–235
Branched chain olefins	>99	58–235
Cyclic olefins	>99	134–237
Diolefins	>99	185–352
Monocyclic aromatics	>99	0–0.7
Polycyclic aromatics	>99	0–12

Source: Reproduced with permission from ASTM D 1159-01 Standard Test Method for Bromine Numbers of Petroleum Distillates and Commercial Aliphatic Olefins by Electrometric Titration. Copyright ASTM International.

TABLE 6.29

Properties and Bromine Numbers of Some Fuel Products

Properties	Diesel Oil # 1	Diesel Oil # 2	FCC Cycle Oil	Diesel Blend
Density at 15°C, g/cc	0.8552	0.8588	0.8874	0.8751
Kin. Viscosity, cSt				
At 40°C	4.872	5.149	1.949	4.838
Pour point, °C	3	−3	−33	−3
Aromatics, wt%	26.7	27.6	—	35.5
Sulfur, wt%	1.47	1.65	2.35	2.06
Bromine number	4.5	3.7	35	10.5

Source: From Bhaskar, M. et al., *Petroleum Science and Technology*, 18, 851, 2000. Reproduced by permission of Taylor & Francis Group, LLC, http://www.taylorandfrancis.com.

Bromine number of petroleum products is an indication of the quantity of bromine-reactive constituents and is useful as a measure of aliphatic unsaturation. Thermal cracking, during distillation, can also lead to olefins. The properties and bromine numbers of some petroleum products, produced from Indian and Persian Gulf crude oils, indicated a significant decrease in pour point and an increase in bromine number of FCC cycle oil (Bhaskar et al., 2000). The properties and bromine numbers of some petroleum products, produced from Indian and Persian Gulf crude oils, are shown in Table 6.29.

Diesel oil # 1 was reported to have a pour point of 3°C, an S content of 1.47 wt%, and a bromine number of 4.5. Diesel oil # 2 was reported to have a pour point of −3°C, an S content of 1.65 wt%, and a bromine number of 3.7. The FCC cycle oil, having a low viscosity, was reported to have a very low pour point of −33°C, an S content of 2.35 wt%, and a higher bromine number of 35 indicating a significant increase in bromine-reactive molecules, such as olefins. The diesel blend was reported to have a pour point of −3°C, an S content of 2.06 wt%, and a higher bromine number of 10.5. The use of hydroprocessing decreases the bromine numbers of petroleum products. The composition and bromine numbers of conventionally refined spindle oil and hydroisomerized and solvent dewaxed SWI 6 base stocks are shown in Table 6.30.

A light vacuum gas oil (VGO) was reported to have a pour point of 0°C, an S content of 2.82 wt%, and a bromine number of 6. Light lube oil distillate, known as spindle oil, was reported to have a higher pour point of 12°C, an S content of 2.8 wt%, and a lower bromine number of 3. Hydroisomerization of slack wax followed by solvent dewaxing produces SWI 6 base stocks, containing less than 1 wt% of aromatics and practically no presence of S and N compounds. The bromine number of SWI 6 base stocks is below 0.01. The ASTM D 2710 method for bromine index of petroleum hydrocarbons by electrometric titration is

TABLE 6.30
Effect of Hydroisomerization on Bromine Number of SWI 6 Base Stocks

Properties	Light VGO	Spindle Oil	SWI 6 # 1	SWI 6 # 2
Kin. Viscosity, cSt				
At 40°C	6.730	13.660	28.84	27.98
Pour point, °C	0	12	−21	−21
Aromatics, wt%	41.1	—	0.7	0.9
Sulfur, wt%	2.82	2.80	0	0
Bromine number	6	3	<0.01	<0.01

Source: From Bhaskar, M. et al., *Petroleum Science and Technology*, 18, 851, 2000; Pillon, L.Z., *Petroleum Science and Technology*, 19, 1263, 2001d.

used to measure the bromine-reactive material, such as olefins, in petroleum oils having a very low bromine number. The bromine index of hydroisomerized and solvent dewaxed base stocks was found to be in a range of 4–6 which, after hydrofining, was found to further decrease to 1–2 indicating practically no

TABLE 6.31
Bromine Numbers of Some Hetero-Containing Molecules

Heterocompounds	Purity, %	Bromine Number
Sulfur Compounds		
Amyl mercaptan	99.92	83
Ethyl mercaptan	99.95	209
Ethyl sulfide	99.94	184
Diethyldisulfide	99.9	0.4
Thiophene	99.99	0.4
Oxygen Compounds		
Acetone	—	0
Methylethylketone	—	0
Nitrogen Compounds		
2-Methylpyridine	99.9	0.9
2,4,6-Trimethylpyridine	>99	2.7
Pyrrole	99.99	873
2,4-Dimethylpyrrole	>98	484
2,4-Dimethyl-3-ethylpyrrole	>98	248

Source: Reproduced with permission from ASTM D 1159-01 Standard Test Method for Bromine Numbers of Petroleum Distillates and Commercial Aliphatic Olefins by Electrometric Titration. Copyright ASTM International.

presence of any unsaturated or reactive molecules. Some Merox treated–caustic washed jet fuels were found to have poor water separation properties indicating the presence of surface active molecules adsorbing at the oil/water interface and have very high bromine numbers, above 500. The bromine numbers of some hetero-containing molecules are shown in Table 6.31.

The typical basic type N molecules are pyridines and quinolines which have a tendency to exist in the higher boiling fractions of crude oils and residue. The typical nonbasic N compounds found in crude oils are the pyrrole, indole, and the carbazole types (Speight, 1999). The literature reported on the presence of alkylated pyrroles in the gasoline fraction, along with alkylated pyridines, quinolines, tetrahydroquinolines, indoles, pyrroles, and carbazoles in the heavier fuels (Mushrush et al., 2000). The GC/MS analysis of different middle distillate fuels confirmed the presence of pyridines, pyrroles, carbazoles, quinolines, tetrahydroquinolines, indoles, and short chain alkyl substituted indoles (Bauserman et al., 2004). The literature also reported on many other N compounds found in heavy Russian crude oils. The N compounds, having strong basic properties, were identified to be quinolines, benzoquinolines, and NS compounds represented by thiophenoquinolines. The NO compounds, having weak basic properties, were represented by cyclic amides of the pyridone type, their hydrogenated analogs, such as lactams, quinoline carboxylic acids, and esters of quinoline carboxylic acids (Kovalenko et al., 2001). Many more heterocompounds present in petroleum products need to be tested for their bromine numbers to identify molecules responsible for poor stability and interfacial properties of some Merox treated–caustic washed jet fuels.

REFERENCES

ASTM D 1159-01 Standard Test Method for Bromine Numbers of Petroleum Distillates and Commercial Aliphatic Olefins by Electrometric Titration, *Annual Book of ASTM Standards*.

Bardasz, E.A. and Lamb, G.D., Additives for crankcase lubricant applications, in *Lubricant Additives Chemistry and Applications*, Rudnick, L.R., Ed., Marcel Dekker, Inc., New York, 2003.

Bauserman, J.W., Nguyen, K.M., and Mushrush, G.W., *Petroleum Science and Technology*, 22(11&12), 1491, 2004.

Beaver, B.D. et al., *Energy and Fuels*, 16, 1134, 2002.

Bhaskar, M. et al., *Petroleum Science and Technology*, 18(7&8), 851, 2000.

Collins, G.E. et al., *Energy and Fuels*, 16, 1054, 2002.

Daniels, F. et al., *Experimental Physical Chemistry*, 7th ed., McGraw-Hill Book Company, 1970.

Danilov, A.M., *Chemistry and Technology of Fuels and Oils*, 37(6), 444, 2001.

Dolbear, G.E., Hydrocracking: reactions, catalysts, and processes, in *Petroleum Chemistry and Refining*, Speight, J.G., Ed., Taylor and Francis, Washington, 1998.

Ervin, J.S. et al., *Energy and Fuels*, 17, 577, 2003.

Frankenfeld, J.W., Taylor, W.F., and Brinkman, D.W., *Industrial and Engineering Chemistry Product Research and Development*, 22, 615, 1983.

Gary, J.H. and Handwerk, G.E., *Petroleum Refining, Technology and Economics*, 4th ed., Marcel Dekker, Inc., New York, 2001.

Harfoush, A. and Shleiwit, H., *Petroleum Chemistry*, 39, 111, 1999.

Hazlett, R.N. et al., *Energy and Fuels*, 5, 269, 1991.

Hernandez-Maldonado, A.J., Yang, R.T., and Cannella, W., *Industrial and Engineering Chemistry Research*, 43, 6142, 2004.

Juyal, P. and Anand, O.N., *Petroleum Science and Technology*, 20(9&10), 1009, 2002.

Kovalenko, E.Yu. et al., *Chemistry and Technology of Fuels and Oils*, 37(4), 265, 2001.

Leon, O. et al., *Energy and Fuels*, 15, 1028, 2001.

McCarter, W.S.W., Krieger, K.A., and Heinemann, H., *Industrial and Engineering Chemistry*, 42, 529, 1950.

Morris, R.E. and Chang, E.L., *Petroleum Science and Technology*, 18(9&10), 1147, 2000.

Morris, R.E. et al., *Energy and Fuels*, 18, 485, 2004a.

Morris, R.E., Hughes, J.M., and Colbert, J.E., *Energy and Fuels*, 18, 490, 2004b.

Mushrush, G.W. et al., *Petroleum Science and Technology*, 18(7&8), 901, 2000.

Mushrush, G.W. et al., *Petroleum Science and Technology*, 19(5&6), 561, 2001.

Mushrush, G.W. et al., *Petroleum Science and Technology*, 20(5&6), 561, 2002.

O'Driscoll, M., *Industrial Minerals Magazine*, 6–7, April 2004.

O'Driscoll, M., *Industrial Minerals Magazine*, 56–58, January 2005.

Pillon, L.Z., *Petroleum Science and Technology*, 19(7&8), 875, 2001a.

Pillon, L.Z., *Petroleum Science and Technology*, 19(7&8), 961, 2001b.

Pillon, L.Z., *Petroleum Science and Technology*, 19(9&10), 1109, 2001c.

Pillon, L.Z., *Petroleum Science and Technology*, 19(9&10), 1263, 2001d.

Pirro, D.M. and Wessol, A.A., *Lubrication Fundamentals*, 2nd ed., Marcel Dekker, Inc., New York, 2001.

Roan, M.A. and Boehman, A.L., *Energy and Fuels*, 18, 835, 2004.

Schwartz, F.G., Whisman, M.L., and Ward, C.C., *US Bureau of Mines*, Bulletin, 626, 44, 1964.

Sequeira, A. Jr., *Lubricant Base Oil and Wax Processing*, Marcel Dekker, Inc., New York, 1994.

Silin, M.A. et al., *Petroleum Chemistry*, 40, 209, 2000.

Sleyko, F.L., *Adsorption Technology: A Step-by-Step Approach to Process Evaluation and Application*, Marcel Dekker, Inc., New York, 1985.

Speight, J.G., *The Chemistry and Technology of Petroleum*, 3rd ed., Marcel Dekker, Inc., New York, 1999.

Taylor, W.F., *Jet Fuel Chemistry and Formulation*, Workshop on Fuels with Improved Fire Safety, Washington, 1996.

Thompson, R.B., Druge, L.W., and Chenicek, J.A., *Industrial and Engineering Chemistry*, 41, 2715, 1949.

USGS, *Clays Statistics and Information*, http://minerals.er.usgs.gov (accessed February 2007).

Wang, Y. et al., *Petroleum Science and Technology*, 19(7&8), 923, 2001.

Wynne, J.H. et al., *Petroleum Science and Technology*, 21(7&8), 1327, 2003.

Yan, F., *Petroleum Science and Technology*, 21(11&12), 1879, 2003.

Zabarnick, S., Widmor, N., and Vangsness, M., *Energy and Fuels*, 16, 1565, 2002.

ZEO Inc., *Natural Zeolite Products and Technology*, www.zeoinc.com (accessed March 2002).

7 Foam Inhibition

7.1 SURFACE ACTIVITY OF ANTIFOAMING AGENTS

Crude oils foam when mixed with air and require the use of antifoaming agents. During crude oil separation and distillation, different petroleum fractions were reported to foam. The literature reported that the "foam flooding" observed in distillation towers was not related to any specific contaminants found in crude oils (Ross, 1987). The 60N and 150N lube oil distillates were reported to foam and their foaming tendency was reported to increase with an increase in their viscosity. With an increase in the temperature and a decrease in viscosity, their foaming tendency decreased (Butler and Henderson, 1990). Hydrofining, used as the last processing step, was found not effective in decreasing the foaming tendency of solvent refined mineral and hydroprocessed lube oil base stocks. The use of lube oil base stocks in formulated lubricating oils requires the use of antifoaming agents to meet the product specifications.

Foam inhibition of crude oils and petroleum products is a complex problem affected by many factors. Early literature reported that an effective antifoaming agent needs to enter the oil/air interface and spread on the oil surface (Bondi, 1951). The conditions required to enter (E) an interface can be defined in terms of the surface tension of the oil ST(oil), the surface tension of antifoaming agents ST (a/f), and the interfacial tension (IFT) between the oil and the antifoaming agent. For an antifoaming agent to enter the oil/air interface, the value of E needs to be greater than 0 and for an antifoaming agent to spread, the value of spreading coefficient (S) needs to be greater than 0.

$$E = \text{ST(oil)} - \text{ST(a/f)} + \text{IFT}$$
$$S = \text{ST(oil)} - \text{ST(a/f)} - \text{IFT}$$

The early literature also reported on the ST of different molecules and the combination of oil-insoluble glycerol, having a high ST, and a dispersing agent as an antifoaming agent. The role of dispersing agent was to keep the insoluble glycerol dispersed in the oil medium (Bondi, 1951). The ST of some molecules, including glycerol and polymethylsiloxane polymer, is shown in Table 7.1.

The term "silicone" covers the family of organosiloxane polymers containing Si–O–Si bonds of the same nature as in silicates but with organic radicals fixed on

TABLE 7.1

Surface Tension of Glycerol and Polymethylsiloxane Polymer

Molecule	ST, mN/m at 20°C
Triolein	28
Polyether oil	35
Ethylene glycol ($C_2H_6O_2$)	47.9
Glycerol ($C_3H_8O_3$)	63.4
Polymethylsiloxane	19.9
Perfluorocarbon oil	23

Source: From Bondi, A., *Physical Chemistry of Lubricating Oils*, Reinhold Publishing Corporation, New York, 1951; *CRC Handbook of Chemistry and Physics*, 86th ed., Lide, D.R., Editor-in-Chief, CRC Press, Boca Raton, 2005.

the silicone. The siloxane Si–O–Si bond is longer, more polar, and has a higher energy than a C–C bond (Rome, 2001). Silicones have been commercially available since 1943 and they are used as base fluids for a variety of products such as emulsions, solutions, greases, and compounds. The most common of the silicone polymers is polydimethylsiloxane used in petroleum applications as an antifoaming agent. The literature reported that the combination of flexible siloxane backbone and the presence of short-chain organic methyl groups make silicones surface active towards the oil/air surface (Pape, 1981). The degree of polymerization (n) of silicones, which can vary from 0 to 2500, is used to control their molecular weight (Pape, 1983).

$(CH_3)_3Si–O–[Si(CH_3)_2–O–]_nSi(CH_3)_3$
Silicone fluids (polydimethylsiloxane)

Dimethyl silicones have the lowest ST values and these are largely independent of viscosity (Quinn et al., 1999). The present laboratory methods for fluid ST measurements are Wilhelmy Plate, Du Nouy ring, Sessile Drop, Pendant Drop, Capillary Rise, Drop Weight, and the Maximum Bubble Pressure methods. According to the literature, all these methods should produce the same results; however, difficulties in the mathematical treatment of the phenomena can lead to discrepancies (SensaDyne, 2000).

Silicone antifoaming agents are used in applications which require a high surface activity and a great spreading power. Water has a high ST of 72.8 mN/m at 20°C and early literature reported on a high spreading coefficient of low viscosity silicone antifoaming agents on the surface of water (Bondi, 1951). Silicone antifoaming agents are nonpolar, nonionic, and insoluble in water. In fact, silicones are hydrophobic to the extent that they are used as water repellents (Pape, 1981). Different viscosity silicone fluids are available and they are

TABLE 7.2
Spreading Coefficient of Low Viscosity Silicones on Water Surface

Water/Silicone Interface at 20°C	5 cSt Silicone	35 cSt Silicone
Silicone ST, mNm	19.0	19.9
Water/silicone IFT, mN/m	42.2	43.1
Silicone spreading (S), mN/m	11.6	9.6

Source: From Bondi, A., *Physical Chemistry of Lubricating Oils*, Reinhold Publishing Corporation, New York, 1951.

identified by referring to their viscosity in cSt at RT. The spreading coefficient of different silicones in water, as reported in the literature, is shown in Table 7.2.

Despite a high water/silicone IFT of 42.2 mN/m, low viscosity 5 cSt silicone has a high spreading coefficient (S) of 11.6 mN/m due to a high ST of water. With an increase in silicone viscosity from 5 to 35 cSt and an increase in ST from 19.0 to 19.9 mN/m, a decrease in spreading from 11.6 to 9.6 mN/m is observed. The early literature reported that the presence of surface active contaminants might decrease the water/silicone IFT and increase its spreading coefficient (Bondi, 1951). The literature reported on the critical factors affecting the performance of an antifoaming agent which were listed as ST, droplet size, and solubility. Under the optimum conditions, the use of dimethyl silicone, having an ST of about 20 mN/m, can effectively defoam a water-based surfactant having an ST of 35 mN/m (ProQuest, 1996). The recent literature reported on the foaming tendency and foam stability of aqueous surfactant solutions (Rosen, 2004). The foaming tendency of aqueous surfactant solutions is shown in Table 7.3.

The literature indicated that an increase in the length of the hydrophobic group increases the efficiency of the surfactant as a foaming agent. The effect of surfactant chemistry and molecular weight on the foam stability of aqueous solutions is shown in Table 7.4.

TABLE 7.3
Foaming Tendency of Aqueous Surfactant Solutions

Surfactants	Temperature, °C	Concentration, M	Foam Height, mm
$C_{12}H_{25}SO_3{}^-Na^+$	46	5×10^{-3}	205
$C_{12}H_{25}SO_3{}^-Na^+$	60	11×10^{-3}	210
$C_{10}H_{21}C_6H_4SO_3{}^-Na^+$	60	5×10^{-4}	185
$C_{12}H_{25}C_6H_4SO_3{}^-Na^+$	60	4×10^{-4}	205
$C_5H_{11}CH(C_5H_{11})SO_4{}^-Na^+$	60	10×10^{-3}	130

Source: From Rosen, M.J., *Surfactants and Interfacial Phenomena*, 3rd ed., Wiley-Interscience, Hoboken, 2004. Reproduced by permission of John Wiley & Sons, Inc.

TABLE 7.4
Effect of Surfactant Chemistry on Foam Stability of Aqueous Solutions

Surfactant Chemistry	Concentration, %	Foam Height, Initial (mm)	Foam Height, After 5 min (mm)
$C_{10}H_{21}SO_3^-Na^+$	0.68	160	5
$C_{12}H_{25}SO_3^-Na^+$	0.32	190	125
$C_{15}H_{31}COO^-Na^+$	0.25	236	232
$C_{12}H_{25}OC_2H_4SO_4^-Na^+$	0.14	246	241
$C_{12}H_5(OC_2H_4)SO_4^-Na^+$	0.11	180	131

Source: From Rosen, M.J., *Surfactants and Interfacial Phenomena*, 3rd ed., Wiley-Interscience, Hoboken, 2004. Reproduced by permission of John Wiley & Sons, Inc.

Silicones are known to prevent foaming by spreading on the liquid surface and the literature reported on the spreading coefficient of polydimethylsiloxane on the surface of different aqueous surfactant solutions. The selected surfactants were anionic aerosol-OT (AOT), nonionic penta(ethylene glycol)-mono-n-decyl ether ($C_{10}E_5$), cationic alkyltrimethylammonium bromides (CnTAB), and zwitterionic fluorinated betaine (Zonyl FSK) (Bergeron et al., 1997). The surface activity of polydimethylsiloxane (PDMS), having an ST of 20.6 mN/m, in different aqueous surfactant solutions is shown in Table 7.5.

A high efficiency of silicone to spread on the surface of water of 13.1 mN/m was reported to gradually decrease in different aqueous surfactant solutions,

TABLE 7.5
Spreading Coefficient of Polydimethylsiloxane in Aqueous Surfactant Solutions

Interface at 20°C, Surfactant Chemistry	Surfactant, ST, mN/m	Surfactant/PDMS, IFT, mN/m	PDMS, S, mN/m
Water	72.8	39.1	13.1
AOT	28.0	4.7	2.7
$C_{10}E_5$	31.5	3.5	7.4
C_9TAB	41.0	10.7	9.7
C_{10}TAB	39.8	10.4	8.8
C_{12}TAB	38.8	9.8	8.4
C_{14}TAB	37.3	9.4	7.3
C_{16}TAB	37.7	9.8	7.3
FSK	16.0	10.1	−14

Source: Reprinted from Bergeron et al., *Colloids and Surfaces, A: Physicochemical and Engineering Aspects*, 122, 103, 1997. With permission from Elsevier.

having a lower ST ranging from 16.0 to 41.0 mN/m. While the surfactant/silicone IFT significantly decreased from 39.1 mN/m in water to 3.5–10.7 mN/m in aqueous surfactant solutions, silicone was reported to have a negative spreading coefficient of -14 mN/m in FSK aqueous surfactant solution, having a low ST of only 16 mN/m (Bergeron et al., 1997). For the same viscosity and chemistry silicone, its surface activity can vary depending on the ST of foaming liquid, the foaming liquid/silicone IFT, and the temperature. The effect of temperature on the density and the ST of water is shown in Table 7.6.

The literature reported that the methyl radicals in polydimethylsiloxane can be substituted by many other organic groups such as hydrogen, alkyl, allyl, trifluoropropyl, glycol ether, epoxy, alkoxy, carboxy, and amino (Rome, 2001). To produce lower surface energy materials than the dimethylsiloxanes, fluorosilicone fluids were commercialized. Fluorosilicone lubricants are used in applications in which the metal surface and the lubricant are exposed to hydrocarbon solvents (Quinn et al., 1999). The literature reported that replacing some of the methyl groups with fluorine-containing alkyl groups required a hydrocarbon bridge to be placed between the CF_3-group and the siloxane backbone to achieve adequate thermal stability (Owen, 1980).

$(CH_3)_3Si-O-[Si(CH_3)(CH_2CH_2CF_3)-O-]_nSi(CH_3)_3$
Fluorosilicone fluids (trifluoropropylmethyl siloxane)

The inclusion of a $-(CH_2)_2-$ group was reported to affect the properties of the molecule to the extent that the trifluoropropylmethyl siloxanes have a higher ST and

TABLE 7.6
Effect of Temperature on Density and Surface Tension
of Water

Interface	Temperature, °C	Water Density, g/cm^3	ST, mN/m
Water/air	0	0.99984	75.64
Water/air	10	0.99970	74.23
Water/air	20	0.99821	72.75
Water/air	30	0.99565	71.20
Water/air	40	0.99222	69.60
Water/air	50	0.98803	67.94
Water/air	60	0.98320	66.24
Water/air	70	0.97778	64.47
Water/air	80	0.97182	62.67
Water/air	90	0.96535	60.82
Water/air	100	0.95840	58.91

Source: From *CRC Handbook of Chemistry and Physics*, 86th ed., Lide, D.R., Editor-in-Chief, CRC Press, Boca Raton, 2005. Reproduced by permission of Routledge/Taylor & Francis Group, LLC.

TABLE 7.7

Surface Tension of Silicone and Fluorosilicone Polymers

Polymers	Viscosity	ST, mN/m
Polydimethylsiloxane	$10{,}000 \text{ m}^2 \text{ s}^{-1}$	20.4
Polytrifluoropropylmethylsiloxane	$3 \text{ m}^2 \text{ s}^{-1}$	23.5
Polydimethylsiloxane	$300 \text{ mm}^2 \text{ s}^{-1}$	20.4
Trifluoropropylmethylsiloxane	$300 \text{ mm}^2 \text{ s}^{-1}$	24.0

Source: From Owen, M.J., *Industrial and Engineering Chemistry Product Research and Development*, 19, 97, 1980; Rome, F.C., *Hydrocarbon Asia*, 42, May/June 2001.

different solubility (Pape, 1981). For the same viscosity polymer of $300 \text{ mm}^2 \text{ s}^{-1}$, an increase in ST from 20.4 to 24 mN/m, for trifluoropropylmethyl siloxane, was reported (Rome, 2001). Different viscosity silicone and fluorosilicone fluids are available and their ST was reported to vary as shown in Table 7.7.

With an increase in the viscosity of silicones from 350 to 12500 cSt, their ST was found to increase from 23.9 to 24 mN/m, thus indicating practically no increase when measured at 24°C. With an increase in the viscosity of fluorosilicones from 300 to 10000 cSt, their ST was found to increase from 26.2 to 28.1 mN/m, measured at 24°C. When compared to Si 350, similar viscosity FSi 300 has an increased ST by about 2 mN/m. When compared to Si 12500, similar viscosity FSi 10000 has an increased ST by about 4 mN/m. The inclusion of a $-(CH_2)_2-$ group was reported to increase a resistance to hydrocarbon solubility while maintaining a relatively low ST.

The literature reported that the chemistry of silicone-free antifoaming agents can vary from polyethylene glycols, polyethers to organic copolymers (Mang, 2001). The ST of polymers is fundamental to adhesion, coating, wetting, dewetting, foaming, and blending and for polymers that are solids at 20°C, the ST is an extrapolated value from studies of molten polymers at higher temperatures. The literature reported on the effect of the molecular weight on the ST of polypropylene at high temperatures (Moreira and Demarquette, 2001). The effect of high temperatures on the ST of polypropylene polymer is shown in Table 7.8.

With an increase in molecular weight, their ST increases. With an increase in the temperature, their ST decreases. The literature also reported that at very high temperatures, some degradation might take place which can affect their molecular weight and their ST. The literature reported on the ST extrapolated to 20°C from studies on melts for a variety of polymers, including their molecular weight (Owen, 1980). The ST extrapolated to 20°C from studies of melts for a variety of polymers is shown in Table 7.9.

Beyond the chain length of C_{14}, hydrocarbons become waxes or solids while silicones, at the same chain length, are still liquids. The paraffin wax was reported to have a critical ST of 23 mN/m measured at 20°C (Owen, 1980). The ST

TABLE 7.8
Effect of High Temperatures on Surface Tension of Polypropylene Polymer

Polypropylene, Molecular Weight	180°C, ST, mN/m	200°C, ST, mN/m	220°C, ST, mN/m
3,400	31.9	30.2	28.1
12,400	31.6	30.4	28.5
18,100	33.3	31.9	30.2
29,100	35.1	34.5	32.1
41,200	35.3	34.7	32.7
107,200	35.9	34.8	32.3
200,600	35.5	34.3	32.2

Source: From Moreira, J.C. and Demarquette, N.R., *Journal of Applied Polymer Science*, 82, 1907, 2001. Reproduced with permission from John Wiley & Sons, Inc.

extrapolated to 20°C for poly(methyl methacrylate), having a molecular weight of 3000, was reported to be 41.1 mN/m. Hydrocarbons are less flexible than silicones and require polar side chains to be surface active. The presence of ester side chains makes the polyacrylate polymer surface active towards the oil/air surface (Pape, 1981).

$[-CH(COOR)-CH_2-]_n$
Acrylate copolymers

The polyacrylate antifoaming agent is commercially available in the form of 40 wt% dispersion in a petroleum solvent, such as kerosene. The ST of 40 wt% polyacrylate dispersion was found to be 27.2 mN/m, measured at 24°C, which is

TABLE 7.9
Surface Tension Extrapolated to 20°C for Different Polymers

Polymers	Molecular Weight	Extrapolated ST, mN/m
Polypropylene	3000	28.3
Poly(propylene oxide)	4000	30.4
Polyisobutylene	2700	33.6
Poly(methyl methacrylate)	3000	41.1

Source: Reprinted with permission from Owen, M.J., *Industrial and Engineering Chemistry Product Research and Development*, 19, 97, 1980. Copyright American Chemical Society.

below the ST of mineral base stocks. Many polymer additives are highly viscous materials and are used as concentrated solutions in diluent oils. Polyacrylate dispersion is used as an antifoaming agent in some lubricating oils.

7.2   FOAMING OF HYDROFINED MINERAL BASE STOCKS

Crude oils are the source of the feedstocks used to manufacture lube oil base stocks. Light lube oil distillate is known as spindle oil, the medium distillate is known as a light machine oil, and the heavy distillate is known as a medium machine oil. Lube oil fractions, from most crude oils, contain low viscosity index (VI), high cloud point, and high pour point components which have undesirable characteristics for finished lubricating oils. The low VI components, which are mostly aromatics and undesirable as lube oil molecules, are removed during the solvent extraction step. The wax type molecules, undesirable as lube oil molecules due to their high pour points, are removed during the dewaxing step. The viscosity of solvent refined mineral base stocks varies from light spindle oils to heavy bright stocks, produced from residual oils. The solvent dewaxing process is suitable for dewaxing the entire range of different viscosity oils. With an increase in viscosity, a decrease in VI from 100–103 to 92–97, with an increase in dewaxed pour point from −12°C to −9°C, was reported. Lower viscosity base stocks are usually used in applications which require good low temperature properties and are dewaxed to lower pour points. The properties and the S content of different viscosity mineral base stocks, produced from Arabian crude oil, are shown in Table 7.10.

With an increase in viscosity, an increase in S content is observed. An increase in viscosity and in S content of mineral base stocks usually indicates

TABLE 7.10
Properties and S Content of Different Viscosity Paraffinic Base Stocks

Mineral Base Stocks	Spindle Oil	150SN	500SN	Bright Stock
Density at 20°C, g/L	0.85	0.87	0.89	0.93
Viscosity, cSt				
At 40°C	12.7	27.3	95.5	550
At 100°C	3.1	5.0	10.8	33
VI	100	103	97	92
Pour point, °C	−15	−12	−9	−9
S, wt%	0.4	0.9	1.1	1.5

Source: From Prince, R.J., *Chemistry and Technology of Lubricants*, 2nd ed., Mortier, R.M. and Orszulik, S.T., Eds., Blackie Academic & Professional, London, 1997. Reproduced with permission of Springer Science and Business Media.

an increase in the aromatic content. Solvent refined base stocks also contain some N content which increases with an increase in their S content. With an increase in the viscosity, an increase in the carbon residue with a decrease in volatility is also observed. Without a finishing step, lube oil fractions have a poor color, a high organic acidity, a poor thermal stability, sludge forming properties, and a high carbon residue (Gary and Handwerk, 2001). Hydrogen finishing, also known as hydrofining, are mild hydrogenation processes. Hydrofining is used to reduce odor, improve color, and improve the thermal and oxidative stability of solvent refined mineral base stocks.

The ASTM D 892 foam test is widely used by the industry to test foaming characteristics of lube oil base stocks and formulated lubricating oils. The ASTM D 892 foam test consists of a 1000 mL graduated cylinder fitted with a lead ring to overcome the buoyancy and an air-inlet tube to the bottom which is fastened to a gas diffuser. The sample is blown with air for 5 min and the foaming tendency is the volume of foam measured immediately after the cessation of airflow. The foam stability is the volume of foam measured 10 min after disconnecting the air supply. The ASTM D 892 foam test covers the determination of the foaming characteristics of lube oil base stocks and lubricating oils at 24°C (Seq. I) and 93.5°C (Seq. II). The effect of hydrofining on the properties and the foaming characteristics of solvent dewaxed 150N phenol extracted base stock is shown in Table 7.11.

After hydrofining, a small increase in VI from 89 to 90, with an increase in saturate content from 78 to 79.1 wt%, was observed. With an increase in saturate content, a decrease in the aromatic content from 22 to 20.9 wt% was also observed indicating that some hydrogen saturation of aromatics is taking place. After hydrofining, the sulfur content of 150N phenol extracted base stock was also found to decrease from 0.33 to 0.25 wt% and the nitrogen content was found to decrease from 72 to 57 ppm. Despite some changes in the composition of 150N base stock, after hydrofining, no change in the ST and no significant decrease in the foaming tendency were observed. The extensive testing of foaming characteristics of different hydrofined phenol extracted base stocks, using the ASTM D 892 foam test procedure, and the statistical analysis of the results indicated a good repeatability of the test when strictly controlling the temperature and the airflow rate during the test (Pillon, 1994). The surface activity of Si 12500, having a viscosity of 12500 cSt and an ST of 24 mN/m, on the oil surface of different 150N mineral base stocks is shown in Table 7.12.

At 24°C, Si 12500 was found to have a low IFT of 1.5 mN/m, a high efficiency to enter the oil/air interface of 150N mineral base stock of 11.2 mN/m, and a high efficiency to spread of 8.2 mN/m. The Si 12500 was also found to have a low IFT of 1.8 mN/m, a high efficiency to enter the oil/air interface of another 150N mineral base stock of 11.5 mN/m, and a high efficiency to spread of 7.9 mN/m. With an increase in the viscosity of silicones from 300 to 60,000 cSt, only a small increase in their ST from 23.9 to 24.1 mN/m, measured at 24°C, was observed. The patent literature reported that only 1–2 ppm of different viscosity silicones are required to prevent foaming in 150N mineral base stocks (Pillon and Asselin,

TABLE 7.11
Effect of Hydrofining on Foaming Characteristics of 150N Mineral Base Stock

150N Mineral Base Stock	Before Hydrofining	After Hydrofining
Kin. Viscosity, cSt		
At 40°C	29.64	29.6
At 100°C	4.99	4.99
VI	89	90
Saturates, wt%	78.0	79.1
Aromatics, wt%	22.0	20.9
Sulfur, wt%	0.33	0.25
Nitrogen, ppm	72	57
Surface tension, mN/m	33.5	33.5
ASTM D 892 Foam Test		
Seq. I at 24°C		
Foaming tendency, mL	310	305
Foam stability, mL	0	0
Seq. II at 93.5°C		
Foaming tendency, mL	35	30
Foam stability, mL	0	0

Source: From Pillon, L.Z., *Optimization of the ASTM D 892 Foam Test*, ASTM Committee Meeting on Petroleum Products and Lubricants, Boston, 1994.

TABLE 7.12
Surface Activity of Si 12500 on the Oil Surface of 150N Mineral Base Stocks

Oil/Silicone Interface at 24°C	150N Oil #1/Si 12500	150N Oil #2/Si 12500
150N mineral oil ST, mN/m	33.7	33.7
Si 12500 ST, mNm	24.0	24.0
Oil/Si 12500 IFT, mN/m	1.5	1.8
Si 12500 entering (E), mN/m	11.2	11.5
Si 12500 spreading (S), mN/m	8.2	7.9

Source: From Pillon, L.Z. and Asselin, A.E., *Antifoaming Agents for Lubricating Oils*, US Patent 5,766,513, 1998; Pillon, L.Z., Asselin, A.E., and Vernon, P.D.F., *Method for Reducing Foaming of Lubricating Oils*, US Patent 6,090,758, 2000.

TABLE 7.13
Effective Treat Rates of Silicones in 150N Mineral Base Stock

Mineral Oil, HF 150N	Treat Rate, ppm (Active)	Seq. I at 24°C, mL	Seq. II at 93.5°C, mL	Seq. III at 24°C, mL
None	0	280/0	20/0	240/0
Si 350	1	35/0	30/0	290/0
	2	5/0	25/0	10/0
	3	10/0	35/0	10/0
Si 12500	1	45/0	10/0	45/0
	2	0/0	10/0	5/0
	3	0/0	10/0	5/0
Si 60000	1	65/0	10/0	15/0
	2	5/0	10/0	15/0
	3	25/0	15/0	25/0

Source: From Pillon, L.Z. and Asselin, A.E., *Antifoaming Agents for Lubricating Oils*, US Patent 5,766,513, 1998.

1998). The ASTM D 892 foam test covers the determination of the foaming characteristics of lube oil base stocks and lubricating oils at 24°C (Seq. I) and 93.5°C (Seq. II), and then, after collapsing the foam, at 24°C (Seq. III). The use of Seq. III foam data allows confirmation of the Seq. I results. The effect of different viscosity nondiluted silicones on the foaming tendency of 150N mineral base stock is shown in Table 7.13.

The 150N mineral base stock was found to have a high Seq. I and a Seq. III foaming tendency of 240–280 mL, measured at 24°C, and a low Seq. II foaming tendency of 20 mL, measured at 93.5°C. No foam stability was observed. The use of 1 ppm of low viscosity silicone, Si 350, was found to decrease the Seq. I foaming tendency to 35 mL but an increase in the Seq. III foaming tendency from 240 to 290 mL was observed. The use of 2–3 ppm of Si 350 was found effective in decreasing the Seq. I foaming tendency to 5–10 mL and Seq. III foaming tendency to 10 mL. However, the use of 1–3 ppm of low viscosity Si 350 was found to increase the Seq. II foaming tendency from 20 to 30–35 mL. The use of 1 ppm of the higher viscosity Si 12500 was found effective in decreasing the Seq. I and Seq. III foaming tendency to 45 mL. However, the use of 2–3 ppm of Si 12500 was found effective in decreasing the Seq. I and Seq. III foaming tendency to below 5 mL and no increase in Seq. II foaming was observed. The use of 1 ppm of the high viscosity Si 60000 was found effective in decreasing the Seq. I and Seq. III foaming tendency to 15–65 mL. However, the use of 2–3 ppm of Si 60000 was found effective in decreasing the Seq. I and Seq. III foaming tendency to 5–25 mL and no increase in Seq. II foaming was observed. The effect of different viscosity silicones on the Seq. I foaming tendency of hydrofined 150N mineral base stock, at 24°C, is shown in Figure 7.1.

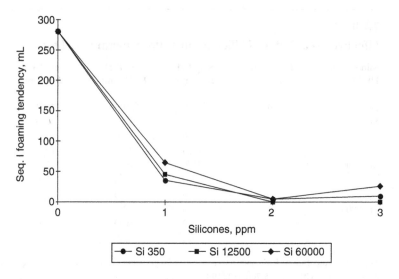

FIGURE 7.1 Effect of different viscosity silicones on the Seq. I foaming tendency of hydrofined 150N mineral base stock measured at 24°C.

The use of 2 ppm of different viscosity nondiluted silicones was found to be the most effective in preventing foaming of 150N mineral base stock. At the treat rate of 3 ppm, the use of Si 350 and Si 60000 was found to increase the Seq. I foaming tendency of 150N mineral base stock indicating overtreatment. The use of fluorosilicone, having a higher ST, was tested in different 150N mineral base stocks and a significant decrease in its surface activity was observed.

With an increase in the ST from 24 to 26.2 mN/m, the use of 3 ppm of FSi 300 was not effective in preventing foaming of 150N mineral base stocks. The performance of nondiluted FSi 300 in different 150N mineral base stocks is shown in Table 7.14.

TABLE 7.14

Performance of Fluorosilicone in Different 150N Mineral Base Stocks

Mineral Oil, HF 150N	Treat Rate, ppm (Active)	Seq. I at 24°C, mL	Seq. II at 93.5°C, mL	Seq. III at 24°C, mL
HF 150N (phenol)	0	345/0	30/0	—
FSi 300	3	340/0	25/0	—
HF 150N (phenol)	0	280/0	20/0	240/0
FSi 300	3	300/0	0/0	265/0

Source: From Pillon, L.Z. and Asselin, A.E., *Antifoaming Agents for Lubricating Oils*, US Patent 5,766,513, 1998.

TABLE 7.15
Surface Activity of PA Dispersions on the Oil Surface of 150N Mineral Base Stock

150N Mineral Oil/PA Interface at 24°C	150N Oil/PA, 40%	150N Oil/PA, 92%
150N mineral oil ST, mN/m	33.7	33.7
PA dispersion ST, mNm	27.2	32.5
Oil/PA IFT, mN/m	1.5	1.5
PA entering (E), mN/m	8	2.7
PA spreading (S), mN/m	5	−0.3

Source: From Pillon, L.Z. and Asselin, A.E., *Antifoaming Agents for Lubricating Oils*, US Patent 5,766,513, 1998.

The use of only 1−2 ppm of Si 12500 was found effective in preventing foaming of different 150N mineral base stocks. The use of 3 ppm of low viscosity FSi 300 was not found effective in preventing foaming and only a small decrease was observed. To decrease the Seq. I foaming tendency of 150N base stocks, below 50 mL, the use of 10 ppm of FSi 300 was required. However, the use of only 5 ppm of FSi 300 was needed to prevent foaming in higher viscosity 600N mineral base stock, having a higher ST. The surface activity of different poly-acrylate (PA) dispersions on the surface of 150N mineral base stock is shown in Table 7.15.

An increase in polyacrylate concentration from 40% to 92% leads to an increase in ST from 27.7 to 32.5 and a decrease in surface activity is observed. A significant variation in concentration of PA dispersion can decrease its effi-ciency to enter the oil interface of the same 150N mineral base stock from 8 to 2.7 mN/m, and decrease its efficiency to spread from 5 mN/m to a negative of −0.3 mN/m, which means no spreading will take place. PA (40%) was found effective in preventing foaming; however, high treat rates were required. The same dispersion, having a higher ST might be highly surface active in higher viscosity mineral base stocks, having a higher ST. The use of PA (40%) in higher viscosity 600N mineral base stocks required significantly lower treat rates. The effect of hydrofining on the properties and the foaming characteristics of solvent dewaxed 600N phenol extracted base stock is shown in Table 7.16.

After hydrofining, an increase in VI from 93 to 95, with an increase in saturate content from 77.7 to 78.9 wt%, was observed. With an increase in saturate content, a decrease in the aromatic content from 22.3 to 21.1 wt% was also observed. After hydrofining, the sulfur content of 600N phenol extracted base stock was also found to decrease from 0.3 to 0.18 wt%, and the nitrogen content was found to decrease from 41 to 36 ppm. With a decrease in aromatic content, an increase in VI is observed indicating that some hydrogen saturation of aromatics is taking place. Higher viscosity 600N mineral base stock was found to have a higher ST of

TABLE 7.16
Effect of Hydrofining on Foaming Characteristics of 600N Mineral Base Stock

600N Mineral Base Stock	Before Hydrofining	After Hydrofining
Kin. Viscosity, cSt		
At 40°C	112.33	108.36
At 100°C	11.82	11.68
VI	93	95
Saturates, wt%	77.7	78.9
Aromatics, wt%	22.3	21.1
Sulfur, wt%	0.3	0.18
Nitrogen, ppm	41	36
Surface tension, mN/m	34.3	34.3
ASTM D 892 Foam Test		
Seq. I at 24°C		
Foaming tendency, mL	560	560
Foam stability, mL	280	260
Seq. II at 93.5°C		
Foaming tendency, mL	60	65
Foam stability, mL	0	0

Source: From Pillon, L.Z., *Optimization of the ASTM D 892 Foam Test*, ASTM Committee Meeting on Petroleum Products and Lubricants, Boston, 1994.

34.3 mN/m and, despite some changes in the composition of 600N base stock, no change in the ST after hydrofining was observed. Before and after hydrofining, higher viscosity 600N mineral base stocks were found to form stable foams, lasting over 10 min, when measured at 24°C. The effect of viscosity and the aromatic content on the ST of mineral base stocks are shown in Table 7.17.

TABLE 7.17
Effect of Viscosity and Aromatic Content on Surface Tension of Mineral Base Stocks

Processing	Viscosity	Aromatics, wt%	ST, mN/m at 24°C
Solvent refined	100N	15.2	33.2
Solvent refined	150N	17.2	33.7
Solvent refined	150N	20.1	33.8
Solvent refined	600N	19.5	34.7
Solvent refined	600N	19.6	34.8
Bright stock	2500N	34–35	35–36

Lower viscosity 100N mineral base stock, containing an aromatic content of 15.2 wt%, was found to have an ST of 33.2 mN/m. Higher viscosity 150N base stocks, containing 17.2–20.1 wt% of aromatic content, were found to have a higher ST of 33.7–33.8 mN/m. With an increase in the viscosity of 600N mineral base stocks, having a similar aromatic content of 19.5–19.6 wt%, their ST was found to further increase to 34.7–34.8 mN/m, indicating the presence of higher molecular weight molecules which are known to have higher ST. Very high viscosity 2500N bright stock, containing a high aromatic content of 34–35 wt%, was found to have a significantly higher ST of 35–36 mN/m. Most hydrofining processes are operated at a severity set by the color improvement needed. While the color of hydrofined mineral base stocks was found to improve, solvent refined base stocks foam when mixed with air and higher viscosity oils form stable foams. In finished lubricants, containing additives, the color of base stocks has little significance but foaming is a serious problem affecting their quality and the equipment.

7.3 EFFECT OF HYDROPROCESSING

The literature reported on different processing options used to produce base stocks, having a higher VI, which include the use of vacuum gas oil (VGO) as feed followed by hydrotreatment (HT) upgrade, hydrodewaxing, and hydrofinishing (Pirro and Wessol, 2001). In some cases, the VGO hydrotreatment upgrade might involve more severe hydrocracking (HC) conditions and depending on the feed, no significant increase in VI will be observed. The literature reported that the conventional VI base stocks, obtained by hydrocracking processes, are similar to solvent refined base stocks with the exception of lower aromatic, sulfur, and nitrogen contents. Their color is lighter and their performance in formulated products, containing additives, is improved (Sequeira, 1994). Different processing options used to produce and finish nonconventional base stocks are shown in Table 7.18.

The feed and the severity of hydroprocessing affect the properties and hydrocarbon composition of nonconventional base stocks. The literature reported on the composition and VI of nonconventional and synthetic base stocks and their classification in four different API groups (Mang, 2001). The API classification

TABLE 7.18
Different Processing Options Used to Produce
Nonconventional Base Stocks

Conventional VGO-Feed	Nonconventional VGO-Feed	Nonconventional HT Slack Wax-Feed
Solvent extraction	Hydrotreatment or HC	Hydroisomerization
Solvent dewaxing	Solvent dewaxing	Solvent dewaxing
Hydrofining	Hydrofining	Hydrofining optional

TABLE 7.19
API Classification and VI of Different Base Stocks

API Group Properties	I, 100N Mineral	II, HC	III, HC	IV, PAO
Viscosity at 100°C, cSt	4	4	4	4
VI	100	105	130	125
Pour point, °C	−15	−15	−20	−65
n- and iso-paraffins, wt%	25	30	75	96
Monocycloparaffins, wt%	20	35	15	4
Polycycloparaffins, wt%	30	34	10	0
Aromatics, wt%	24	0.5	0.1	0
Thiophenes, wt%	0.5	0	0	0
Volatility, % off	23	18	13	12

Source: From Mang, T., *Lubricants and Lubrications*, Mang, T. and Dresel, W., Eds., Wiley-VCH, Weinheim, 2001. Reproduced by permission of Wiley-VCH Verlag GmbH & Co KG.

and VI of low viscosity solvent refined, nonconventional base stocks and synthetic polyalfaolefin (PAO), having the same kinematic viscosity at 100°C, are shown in Table 7.19.

The use of hydroprocessing decreases the aromatic, sulfur, and nitrogen contents of petroleum derived base stocks. A small increase in VI from 100 to 105 leads to an increase in paraffinic and iso-paraffinic contents from 25 to 30 wt%, a significant increase in total naphthenic content from 50 to 69 wt%, and a drastic decrease in the aromatic content from 24 to 0.5 wt%. With a further increase in VI to 130, a drastic increase in paraffinic and iso-paraffinic contents from 25 to 75 wt %, with a drastic decrease in total naphthenic content from 50 to 25 wt%, is observed. Synthetic base stocks, such as PAO, have a high VI of 125 and contain 100 wt% saturate content. For the same viscosity base stocks, with an increase in VI a decrease in volatility is observed. 100N hydrocracked base stock, having a VI of 103, contained only 7.4 wt% of aromatics, 82 ppm of sulfur, and 2 ppm of nitrogen. At 24°C, low viscosity 100N hydrocracked base stock was found to have a low foaming tendency of 30 mL and no foam stability. At a higher temperature of 93.5°C, with a further decrease in viscosity, 100N hydrocracked base stock was found not to foam. 100N solvent refined and hydrocracked base stocks foam but the foaming tendency of 100N hydrocracked base stock was lower. The properties and the foaming characteristics of hydrocracked and hydroisomerized base stocks are shown in Table 7.20.

SWI 6 base stock, having a VI of 143, was reported to have less than 0.5 wt% of aromatics and practically no heteroatom content. At 24°C, SWI 6 base stock was found to have a foaming tendency of 145 mL and no foam stability. At a higher temperature of 93.5°C, with a decrease in viscosity, the foaming tendency further decreased to 10 mL. Another SWI 6 base stock, having a lower VI of 138, was also reported to have less than 0.5 wt% of aromatics but it contained 17 ppm of S

TABLE 7.20
Foaming Characteristics of Hydrocracked and Hydroisomerized
Base Stocks

Hydroprocessing, Base Stocks	Hydrocracked, 100N	Hydroisomerized, SWI 6	Hydroisomerized, SWI 6
Kin. Viscosity, cSt			
At 40°C	18.29	29.38	31.35
At 100°C	3.87	5.77	5.96
VI	103	143	138
Saturates, wt%	92.6	>99.5	>99.5
Aromatics, wt%	7.7	<0.5	<0.5
Sulfur, ppm	82	<1	17
Nitrogen, ppm	2	<1	2
ASTM D 892 Foam Test			
Seq. I at 24°C			
Foaming tendency, mL	30	145	170
Foam stability, mL	0	0	0
Seq. II at 93.5°C			
Foaming tendency, mL	0	10	15
Foam stability, mL	0	0	0

Source: From Pillon, L.Z. and Asselin, A.E., *Antifoaming Agents for Lubricating Oils*, US Patent 5,766,513, 1998; Pillon, L.Z., Asselin, A.E., and Vernon, P.D.F., *Method for Reducing Foaming of Lubricating Oils*, US Patent 6,090,758, 2000.

and 2 ppm of N. At 24°C, SWI 6 base stock was found to have an increased foaming tendency of 170 mL and no foam stability. At a higher temperature of 93.5°C, an increased foaming tendency of 15 mL was also observed. Slack wax isomerate (SWI 6) base stocks, produced by slack wax hydroisomerization, were also found to foam but their foaming tendency was found to be lower when compared to similar viscosity 150N solvent refined base stocks. The effect of hydrofining on the properties and the foaming characteristics of SWI 6 base stock, hydroisomerized at low temperature, is shown in Table 7.21.

After hydrofining, no change in VI and dewaxed pour point was observed. Hydrofining was found effective in decreasing the aromatic/polar content of SWI 6 base stock from 0.9 to 0.7 wt%, indicating some saturation taking place and some improvement in daylight stability was observed. Similarly to hydrofining of solvent refined base stocks, no change in ST was observed. However, hydrofining was found to increase the foaming tendency of SWI 6 base stock from 105 to 130 mL, when measured at 24°C. Despite the elimination of aromatics and heteroatom molecules, the use of hydroprocessing does not prevent foaming and the use of antifoaming agents is required. The ST of different hydrocarbons, measured at 20°C, is shown in Table 7.22.

TABLE 7.21
Effect of Hydrofining on Foaming Characteristics of SWI 6 Base Stock

SWI 6 Base Stock	Before Hydrofining	After Hydrofining
Kin. Viscosity, cSt		
At 40°C	28.94	28.84
At 100°C	5.72	5.71
VI	143	143
Dewaxed pour point, °C	−21	−21
Saturates, wt%	99.1	99.3
Aromatics, wt%	0.9	0.7
Sulfur, ppm	<1	<1
Nitrogen, ppm	<1	<1
Daylight stability, days	5	25
Surface tension, mN/m	32.8	32.8
ASTM D 892 Foam Test		
Seq. I at 24°C		
Foaming tendency, mL	105	130
Foam stability, mL	0	0
Seq. II at 93.5°C		
Foaming tendency, mL	10	10
Foam stability, mL	0	0

Source: From Pillon, L.Z., Asselin, A.E., and Vernon, P.D.F., *Method for Reducing Foaming of Lubricating Oils*, US Patent 6,090,758, 2000; Pillon, L.Z., *Petroleum Science and Technology*, 19, 1263, 2001; Pillon, L.Z., *Petroleum Science and Technology*, 20, 223, 2002.

With an increase in molecular weight from *n*-pentane to *n*-octane, an increase in ST from 20.5 to 25.7 mN/m is observed. The presence of cyclic hydrocarbons will further increase the ST of organic liquids. The addition of alkyl side chains leads to an increase in the ST of benzene. The use of slack wax hydroisomerization technology leads to base stocks having a very high VI and practically no aromatic and heteroatom content. The effect of hydroprocessing on the composition and the ST of different base stocks, having a similar viscosity at 40°C, is shown in Table 7.23.

The use of hydrocracking and hydroisomerization refining technologies produces base stocks, having a low or practically no aromatic and heteroatom content which leads to a lower ST. With a significant decrease in the aromatic content to below 0.5 wt%, the ST of SWI 6 base stocks was found to decrease to 32.7–32.8 mN/m which is lower by about 1 mN/m when compared to 150N mineral base stocks. With a change in the saturate hydrocarbon chemistry, a further decrease in ST of synthetic PAO 6 base stock to 32.4 mN/m is observed.

TABLE 7.22
Surface Tension of Different Hydrocarbons at 20°C

Hydrocarbons	Chemistry	ST, mN/m at 20°C
Aromatic	Benzene	28.8
	Toluene	28.5
	Ethylbenzene	29.0
	Butylbenzene	29.2
Cyclic	1,2,3,4-Tetrahydronaphthalene (tetralin)	35.2
	Decahydronaphthalene (decalin)	29.9
	Cyclopentane	22.4
	Cyclohexane	25.0
Normal	Octane	25.7
	Heptane	24.4
	Hexane	22.6
	Pentane	20.5
Iso	Iso-pentane	13.7

Source: From Speight, J.G., *The Chemistry and Technology of Petroleum*, 3rd ed., Marcel Dekker, Inc., New York, 1999. Reproduced by permission of Routledge/Taylor & Francis Group, LLC.

The surface activity of different viscosity silicones on the oil surface of SWI 6 base stocks is shown in Table 7.24.

At 24°C, low viscosity Si 350 was found to have a very low IFT of only 0.8 mN/m leading to high efficiency to enter and spread on the oil surface. Higher viscosity Si 12500, which is usually used in lubricant products, was tested in different SWI 6 base stocks and showed some increase in IFT to 1.8–2.2 mN/m but still a high efficiency to enter and spread on the oil surface was observed. Even the use of high viscosity Si 60000 did not increase the IFT and a high efficiency to

TABLE 7.23
Effect of Hydroprocessing on Composition and Surface Tension of Base Stocks

Processing	Viscosity, cSt at 40°C	Aromatics, wt%	ST, mN/m at 24°C
150N	29.48	17.2	33.7
150N	29.60	20.1	33.8
SWI 6	29.38	<0.5	32.7
SWI 6	31.35	<0.5	32.8
PAO 6	30.43	0	32.4

TABLE 7.24

Surface Activity of Silicones on Oil Surface of SWI 6 Base Stock

SWI 6/Silicones at 24°C	Si 350	Si 12500 #1	Si 12500 #2	Si 60000
SWI 6 ST, mN/m	32.8	32.8	32.8	32.8
Silicone ST, mNm	23.9	24.0	24.0	24.1
Oil/silicone IFT, mN/m	0.8	1.8	2.2	2.2
Silicone (E), mN/m	9.7	10.6	11.0	10.9
Silicone (S), mN/m	8.1	7.0	6.6	6.5

Source: From Pillon, L.Z. and Asselin, A.E., *Antifoaming Agents for Lubricating Oils*, US Patent 5,766,513, 1998; Pillon, L.Z., Asselin, A.E., and Vernon, P.D.F., *Method for Reducing Foaming of Lubricating Oils*, US Patent 6,090,758, 2000.

enter and spread is observed. The effect of nondiluted Si 12500 on the foaming tendency of different batches of SWI 6 base stocks is shown in Table 7.25.

At the treat rate of 1 ppm, Si 12500 was found to decrease the Seq. I and Seq. III foaming tendency of SWI 6 base stock # 1 from 140–145 to 10–15 mL. The use of 3 ppm of Si 12500 was sufficient to develop a total resistance to foaming. At the treat rate of 1 ppm, Si 12500 was found to decrease the Seq. I and Seq. III foaming tendency of SWI 6 base stock # 2 from 170–175 to 5–15 mL. The use of 3 ppm of Si 12500 was also sufficient to develop a total resistance to foaming. While 2 ppm of silicones were sufficient to prevent foaming of 150N mineral base stock, the use of 3 ppm was required to totally prevent foaming of SWI 6 base stocks. The use of hydroprocessing and a decrease in ST of SWI 6 base stocks were found to have no significant effect on the surface activity of silicone. Despite a high saturate content and a decrease in the solvency properties of SWI 6 base

TABLE 7.25

Performance of Si 12500 in Different SWI 6 Base Stocks

Hydroisomerized Oil	Treat Rate, ppm (Active)	Seq. I at 24°C, mL	Seq. II at 93.5°C, mL	Seq. III at 24°C, mL
SWI 6 # 1	0	140/0	20/0	145/0
Si 12500	1	15/0	0/0	10/0
Si 12500	3	0/0	0/0	0/0
SWI 6 # 2	0	175/0	25/0	170/0
Si 12500	1	15/0	5/0	5/0
Si 12500	3	5/0	0/0	0/0

Source: From Pillon, L.Z. and Asselin, A.E., *Antifoaming Agents for Lubricating Oils*, US Patent 5,766,513, 1998.

TABLE 7.26
Surface Activity of FSi 300 on Oil Surface of 150N Mineral and SWI 6 Base Stocks

Oil/FSi 300 at 24°C	150N Mineral	SWI 6
Base stock ST, mN/m	33.7	32.8
FSi 300 ST, mNm	26.2	26.2
Oil/FSi 300 IFT, mN/m	5.8	6.8
Entering (E), mN/m	13.3	13.4
Spreading (S), mN/m	1.7	−0.2

Source: From Pillon, L.Z. and Asselin, A.E., *Antifoaming Agents for Lubricating Oils*, US Patent 5,766,513, 1998.

stock, nondiluted silicones were found surface active and effective in preventing foaming at low treat rates. The surface activity of FSi 300, having a higher ST, on the oil surface of 150N mineral and SWI 6 base stocks is shown in Table 7.26.

The use of different chemistry silicone, such as FSi 300 having a higher ST and different solubility properties, was found to significantly increase the oil/FSi 300 IFT to 5.8 mN/m in 150N mineral base stock and to 6.8 mN/m in SWI 6 base stock. Replacing some of the methyl groups with fluorine-containing alkyl groups increased the ST of fluorosilicones and also increased IFT leading to a significant decrease in spreading on the oil surface of base stocks. With a significant increase in IFT, a significant decrease in spreading to 1.7 mN/m in 150N mineral and to a negative of −0.2 mN/m in SWI 6 was observed. The performance of nondiluted FSi 300 in SWI 6 base stock, having a negative spreading (S), is shown in Table 7.27.

At the treat rate of 3 ppm, FSi 300 was found effective in decreasing the Seq. I foaming tendency of SWI 6 base stock from 175 to 35 mL. At a higher treat rate of 5 ppm, FSi 300 was found effective in decreasing the Seq. I foaming tendency

TABLE 7.27
Performance of FSi 300 in SWI 6 Base Stock

Hydroisomerized Oil	Treat Rate, ppm (Active)	Seq. I at 24°C, mL	Seq. II at 93.5°C, mL
SWI 6	0	175/0	15/0
FSi 300	3	35/0	0/0
FSi 300	5	20/0	0/0

Source: From Pillon, L.Z. and Asselin, A.E., *Antifoaming Agents for Lubricating Oils*, US Patent 5,766,513, 1998.

TABLE 7.28

Surface Activity of PA Dispersions on the Oil Surface of SWI 6 Base Stock

SWI 6/PA Interface at 24°C	SWI 6/PA, 40%	SWI 6/PA, 92%
SWI 6 oil ST, mN/m	32.8	32.8
PA dispersion ST, mN/m	27.2	32.5
Oil/PA IFT, mN/m	5.6	5.6
PA entering (E), mN/m	11.2	5.9
PA spreading (S), mN/m	0	−5.3

Source: From Pillon, L.Z. and Asselin, A.E., *Antifoaming Agents for Lubricating Oils*, US Patent 5,766,513, 1998.

of SWI 6 base stock to 20 mL and no foaming tendency at 93.5°C was observed. Despite a higher ST and a negative spreading coefficient of −0.2 mN/m, FSi 300 was found effective in preventing foaming of SWI 6 base stock. FSi 300 was found to have a high efficiency to enter the SWI 6/FSi 300 interface which might be related to the high saturate content and poor solvency properties of SWI 6 base stock leading to an increase in its adsorption at the oil/air interface without spreading. The surface activity of different polyacrylate (PA) dispersions on the surface of SWI 6 base stock is shown in Table 7.28.

The use of PA dispersion in SWI 6 leads to a significant increase in IFT from 1.5 mN/m in 150N mineral base stock to 5.6 mN/m in a highly saturated base stock. With a decrease in ST of SWI 6 and an increase in IFT, no spreading on the oil surface is observed. PA (40%) was found effective in preventing foaming; however, very high treat rates were required. The literature reported that if antifoaming agents cannot spread on the oil surface, they might still prevent foaming. If silicones cannot spread at the oil/air interface, they might adsorb and form a mixed layer which can destabilize the foam (Owen, 1980). Surface active antifoaming agents were reported to adhere to the surface of foam film and change its rheological properties (Sun et al., 1992). Slack wax isomerate base stock was developed to compete with expensive synthetic base stocks, such as PAO, which are 100% saturate and made by oligomerization of the decene molecule. The composition and foaming characteristics of different viscosity synthetic PAO base stocks are shown in Table 7.29.

Synthetic PAO 4, PAO 6, and higher viscosity PAO 8 base stocks were found not to foam when tested at 24°C and 93.5°C. The manufacturing of synthetic base stocks, such as PAO, leads to lube oil base stocks which are more expensive but have improved low temperature fluidity and interfacial properties, such as resistance to foaming. The majority of lubricating oils use mineral base stocks which are solvent refined, solvent dewaxed, and hydrofined. However, to produce

TABLE 7.29
Composition and Foaming Characteristics of Synthetic
PAO Base Stocks

Synthetic	PAO 4	PAO 6	PAO 8
Kin. Viscosity, cSt			
At 40°C	16.68	30.89	46.30
At 100°C	3.84	5.98	7.74
VI	124	143	136
Saturates, wt%	100	100	100
Aromatics, wt%	0	0	0
Sulfur, wt%	0	0	0
Nitrogen, ppm	0	0	0
ASTM D 892 Foam Test			
Seq. I at 24°C			
Foaming tendency, mL	0	0	0
Foam stability, mL	0	0	0
Seq. II at 93.5°C			
Foaming tendency, mL	0	0	0
Foam stability, mL	0	0	0

Source: From Rudnick, L.R. and Shubkin, R.L., *Synthetic Lubricants and High-Performance Functional Fluids*, 2nd ed., Rudnick, L.R. and Shubkin, R.L., Eds., Marcel Dekker, Inc., New York, 1999.

lubricant products having improved performance at the minimum cost, synthetic base stocks are often blended with mineral oils to improve their low temperature fluidity. Foaming of mineral oils and the use of PAO base stocks, having lower ST, might lead to a decrease in the surface activity of antifoaming agents and an increase in foaming of part synthetic lubricating oils.

7.4 HIGH TEMPERATURE APPLICATIONS

7.4.1 LUBRICATING OILS

Many crude oils and petroleum products are stored, processed, or used at different temperatures which will affect their viscosity. Early literature reported on the effect of viscosity on foaming of mineral base stocks (Tourret and White, 1952). The literature reported on the effect of temperature on foaming of paraffinic, naphthenic, and synthetic lubricating oils. With an increase in the temperature from −10°C to 150°C, the foaming was reported to decrease and no foaming was observed at high temperatures (Hubmann and Lanik, 1988). With a decrease in the temperature from 24°C to 10°C, the viscosity of 150N mineral base stock

TABLE 7.30
Effect of Temperature on ST of Low Molecular Weight Normal Hydrocarbons

Normal Hydrocarbons	ST, mN/m at −18°C	ST, mN/m at 20°C	ST, mN/m at 38°C	ST, mN/m at 93°C
Pentane	20.5	16.0	14.0	8.0
Hexane	22.6	18.4	16.5	10.9
Heptane	24.4	20.3	18.6	13.1
Octane	25.7	21.8	20.2	14.9

Source: From Speight, J.G., *The Chemistry and Technology of Petroleum*, 3rd ed., Marcel Dekker, Inc., New York, 1999. Reproduced by permission of Routledge/Taylor & Francis Group, LLC.

increases from 68.2 to 162.6 cSt, leading to a significant increase in foaming tendency from 295 to 470 mL. With an increase in viscosity, an increase in ST from 33.9 to 35.1 mN/m is observed. Despite an increase in foaming, an increase in ST will lead to an increase in surface activity of antifoaming agents. With an increase in the temperature from 24°C to 93.5°C, the viscosity of 150N mineral base stock decreases from 68.2 to 5.9 cSt, leading to a significant decrease in foaming tendency from 295 to 30 mL. Despite a decrease in ST from 33.9 to 28.8 mN/m, the surface activity of antifoaming agents at 93.5°C is not that important because the foaming tendency is low, below 50 mL. The effect of temperature on the ST of low molecular weight normal hydrocarbons is shown in Table 7.30.

With a decrease in the temperature from 20°C to −18°C, the ST of low molecular weight hydrocarbons, such as pentane, hexane, heptane, and octane, increases from 16.0–21.8 to 20.5–25.7 mN/m. A decrease in the temperature will increase the viscosity and ST leading to an increase in the surface activity of antifoaming agents. With an increase in the temperatures from 20°C to 93°C, their ST decreases to 14.0–20.2 mN/m at 38°C and a further decrease to 8.0–14.9 mN/m at 93°C is observed. An increase in the temperature will decrease the viscosity and ST leading to a decrease in the surface activity agents. The foaming characteristics of low and high viscosity oils are different. With an increase in the temperature from 24°C to 93.5°C, the viscosity of 600N mineral base stock decreases from 280.12 to 13.44 cSt, leading to a significant decrease in foaming tendency from 570 to 60 mL, and a decrease in ST from 34.8 to 29.6 mN/m is observed. Despite a significant decrease in foaming, the surface activity of antifoaming agents at 24°C and 93.5°C is important to reduce the foaming tendency of higher viscosity base stocks to below 50 mL. The early literature reported on the effect of increased temperatures on ST of high molecular weight n-paraffins (Bondi, 1951). The effect of increased temperatures on ST of high molecular weight n-paraffins is shown in Table 7.31.

TABLE 7.31
Effect of High Temperatures on ST of High Molecular Weight *n*-Paraffins

Temperature, °C	n-$C_{26}H_{54}$	n-$C_{32}H_{66}$	n-$C_{60}H_{122}$
75	26.1	27.0	26.5
100	24.1	25.2	23.1
150	20.0	21.6	22

Source: From Bondi, A., *Physical Chemistry of Lubricating Oils*, Reinhold Publishing Corporation, New York, 1951.

With an increase in the temperatures, the ST of higher molecular weight hydrocarbons is higher. While the ST of low molecular weight normal C_5–C_8 hydrocarbons was reported to be in a range of 8–14.9 mN/m at 93°C, the ST of n-C_{26}, n-C_{32}, and n-C_{60} hydrocarbons was in a significantly higher range of 23.1–25.2 mN/m at 100°C. The effect of the temperature on the foaming tendency of different crude oils and petroleum products will vary depending on their viscosity and composition and each product needs to be tested. The patent literature reported on the effect of temperature on the foaming tendency of different viscosity solvent refined mineral base stocks, ranging from low viscosity 100N oil to very high viscosity bright stock (Butler and Henderson, 1990). The effect of temperature on the foaming tendency of different viscosity mineral base stocks is shown in Table 7.32.

Lower viscosity mineral base stocks, such as 100N and 150N, have a high foaming tendency at 24°C, but a low foaming tendency, below 50 mL at 93.5°C, and require an effective antifoaming agent which is surface active at RT. Higher viscosity 600N mineral base stocks have a high foaming tendency at 24°C

TABLE 7.32
Effect of Temperature on Foaming Tendency of Different Viscosity Base Stocks

Foaming Tendency, Mineral Base Stocks	Seq. I at 24°C, mL	Seq. II at 93.5°C, mL	Seq. III at 24°C, mL
100N	385	25	370
150N	535	20	470
600N	460	100	600
Bright stock	30	220	120

Source: From Butler, K.D. and Henderson, H.E., *Adsorbent Processing to Reduce Base Stock Foaming*, US Patent 1,266,007, 1990.

TABLE 7.33

Effect of Temperature on Spreading Coefficient (S)
of Different Silicones in SWI 6

Interface, SWI 6/Silicones	Spreading (S), mN/m at 24°C	Spreading (S), mN/m at 93.5°C
Si 350	8.9	0.9
Si 12500	6.6	1.6
Si 60000	7.9	7.1
Si 100000	7.7	6.1
FSi 300	1.5	3.6

Source: From Pillon, L.Z., Asselin, A.E., and Vernon, P.D.F., *Method for Reducing Foaming of Lubricating Oils*, US Patent 6,090,758, 2000.

and 93.5°C, and require an antifoaming agent which is surface active at RT and 93.5°C. High viscosity bright stocks have a high foaming tendency at 93.5°C and a high Seq. III foaming, indicating the need for an antifoaming agent which is surface active at high temperatures and RT. The majority of lubricating oils, such as industrial oils, require the foam inhibition at 24°C and 93.5°C and require the measurement of ST of oil, ST of antifoaming agents, and the oil/antifoaming IFT at these temperatures. One of the instruments commercially available, such as Processor Tensiometer K12, allows measurements to be taken at different temperatures, ranging from −10°C to 100°C. The effect of temperature on spreading coefficient (S) of different viscosity silicones and FSi 300 on the oil surface of SWI 6 base stock is shown in Table 7.33.

With an increase in the temperature from 24°C to 93.5°C, there is a drastic decrease in the surface activity of low viscosity silicones. The spreading coefficient of Si 350 decreased from 8.9 to only 0.9 mN/m. The spreading coefficient of Si 12500 decreased from 6.6 to 1.6 mN/m. With an increase in the temperature from 24°C to 93.5°C, there is only a small decrease in the surface activity of high viscosity silicones. The spreading coefficient of Si 60000 decreased from 7.9 to only 7.1 mN/m. The spreading coefficient of Si 100000 decreased from 7.7 to only 6.1 mN/m. At higher temperature, higher viscosity silicones have a higher spreading coefficient and will be more effective in foam inhibition. In industrial applications, the lubricating oils are exposed to varying temperatures and engine oils require foam inhibition at low temperatures and very high temperatures, above 100°C. The effect of high temperatures on the ST of mineral lubricating oils is shown in Table 7.34.

At 24°C, different mineral lubricating oils were reported to have a typical ST of 31.0–32.3 mN/m which decreased to 25.1–26.3 mN/m at 100°C. With a further increase in the temperature to 200°C, the ST of lubricating oils further

TABLE 7.34
Effect of High Temperatures on Surface Tension of Mineral Lubricating Oils

Lubricating Oil Temperature, °C	Oil # 1, ST, mN/m	Oil # 2, ST, mN/m	Oil # 3, ST, mN/m
25	31.0	32.3	31.6
100	25.1	26.3	25.6
200	18.3	19.5	19.1

Source: From Bondi, A., *Physical Chemistry of Lubricating Oils*, Reinhold Publishing Corporation, New York, 1951.

decreased to 18.3–19.5 mN/m. Foam inhibition of engine oils at high temperatures is a difficult problem, even when using the most surface active silicone antifoaming agents. To prevent foaming, the silicone fluid needs to spread on the oil surface. When its ST is too low, silicone cannot spread and it might become a foam promoter. The typical viscosity grade used in lubricating oils, Si 12500, is not always effective at very high temperatures indicating the need for higher viscosity silicones. The conventional methods for the experimental ST measurements of polymer liquids are limited due to their high viscosity and many theoretical models are used. The literature reported on a good correlation between experimental and theoretical data, from density gradient theory, for some polymer liquids including polydimethylsiloxane (Dee and Sauer, 1997). The effect of high temperatures on the ST of different polymers, having different chemistry and molecular weight, is shown in Table 7.35.

TABLE 7.35
Effect of High Temperatures on Surface Tension of Different Polymers

ST, mN/m, Temp., °C	Poly(ethylene glycol), Mw 1,000	Polystyrene, Mw 21,000	Polydimethyl Siloxane, Mw 32,000
25	45	—	22
50	42	—	19
100	38	35	17
150	35	32	15
250	—	25	10
300	—	—	5

Source: From Dee, G.T. and Sauer, B.B., *TRIP*, 5, 230, 1997.

With an increase in the temperature from 25°C to 100°C, the ST of poly (ethylene glycol), having a low molecular weight of 1,000, was reported to decrease from 45 to 38 mN/m. With an increase in the temperature from 100°C to 250°C, the ST of polystyrene, having a molecular weight of 21,000, was reported to decrease from 35 to 25 mN/m. The polymer molecules are rarely pure in composition and structure, such as the presence of different long-chain branching, which can affect the surface properties of polymer liquids (Dee and Sauer, 1997). With an increase in the temperature from 25°C to 300°C, the ST of polydimethylsiloxane, having a higher molecular weight of 32,000, was reported to gradually decrease from 22 to only 5 mN/m. At low temperatures, while the ST of petroleum oils is significantly lower than water, the silicones are usually effective due to low oil/silicone IFT. The early literature reported on low IFT between different petroleum oils and silicone which varies in a range of 2.4–4.8 mN/m (Owen, 1980). At high temperatures, a drastic decrease in ST of petroleum products and decrease in ST of lower viscosity silicones might prevent them from spreading on the oil surface and foam inhibition will not occur.

7.4.2 COKING PROCESS

To prevent foaming of crude oils, petroleum feedstocks, and products, the foam inhibition process requires the measurement of the viscosity, the foaming tendency, and the surface activity of antifoaming agents at the temperature of application. A change in the temperature will change the viscosity and affect their tendency to foam and their ST. The literature reported that silicones prevent foam formation in petroleum oils by spreading on the oil/air surface (Mannheimer, 1991). At low temperatures, the ST of different viscosity Gulf of Mexico crude oils was reported to vary in a range of 26–29 mN/m (Poindexter et al., 2002). The literature reported on higher ST of Athabasca bitumen of 35.3 mN/m measured at 21.1°C and 34.7 mN/m measured at 23.3°C (Speight, 2006). The literature reported on the ST of petroleum products, such as gasoline (26 mN/m) and kerosene (30 mN/m) (Speight, 2006). The ST of mineral lube oil base stocks was also found to vary depending on their viscosity. The ST of mineral lube oils can increase from as low as 30 mN/m to as high as 36 mN/m, for high viscosity bright stocks. The ST of different crude oils and petroleum products is shown in Table 7.36.

During the coking process, silicones are used to prevent foaming and the measurement of ST at such high temperatures is very difficult. The vacuum reduced crude (VRC) is the main feedstock to a delayed coker and the temperature ranges from 343°C to 382°C without coke formation, but the temperature of the thermal cracking process can be as high as 500°C (Ellis and Paul, 2000). The literature reported on an upgrader coking unit. The mechanism of coking was reported to be based on three stages, such as partial vaporization and mild cracking of feed as it passes through a coking furnace, cracking of vapors as they pass through the coke drum, and a successive cracking and polymerization of heavy hydrocarbon liquid trapped in the coke drum (Lloydminster, 2004). The literature reported on the properties and composition of Athabasca

TABLE 7.36
Surface Tension of Crude Oils and Different Petroleum Products

Crude Oils and Petroleum Products	Surface Tension, mN/m
Gulf of Mexico crude oils	26–29
Athabasca bitumen	35
Gasoline	26
Kerosene	30
Lube oil base stocks	30–36

vacuum residue and the effect of high temperatures on the ST of Athabasca vacuum residue at coking process conditions (Aminu et al., 2004). The properties and ST of Athabasca vacuum residue at high coking temperatures are shown in Table 7.37.

While Athabasca bitumen was reported to have a high ST of 35 mN/m, at 150°C, Athabasca vacuum residue has a lower ST of 21 mN/m which further decreases to 5.8–14.7 mN/m under a high temperature coking condition of 300°C–500°C. The literature reported that the ST decreased with an increase in the temperature but stayed constant with an increase in the processing time. While the ST stayed the same with an increase in the processing time, the viscosity was reported to drastically increase indicating other effects of cracking and coking

TABLE 7.37
Surface Tension of Athabasca Vacuum Residue at Coking Temperatures

Athabasca Vacuum Residue (bp > 524°C)	Composition and Properties
Density, kg/m^3	1086.8
Sulfur, wt%	5.8
Nitrogen, wt%	0.7
n-Pentane insolubles, wt%	24.7
Toluene insolubles, wt%	1.8
Microcarbon residue (MCR), wt%	27.7
Ash, wt%	1.25
Surface Tension, mN/m	
At 150°C	21
At 312°C–400°C	14.0–14.7
At 466°C–530°C	5.8–6.4

Source: From Aminu, M.O. et al., *Industrial and Engineering Chemistry Research*, 43, 2929, 2004.

TABLE 7.38
Thermal Degradation of Different Viscosity Silicones
under Coking Conditions

PDMS Viscosity	Weight Loss, %	Mn before Heating	Mn after Heating
60,000 cSt	33.7	82,070	55,747
600,000 cSt	18.3	155,462	91,110

Source: From Kremer, L.N. and Hueston, T.G., *PTQ Summer*, 65–69, 2002.
Reproduced by permission of Crambeth Allen Publishing Ltd. and
Dow Corning Corporation, Midland, Michigan, USA.

taking place (Aminu et al., 2004). Silicone antifoaming agents, such as polydimethylsiloxane (PDMS), are usually added to the top of the coker drum to prevent
foaming. The literature reported on the benefits of using very high viscosity
polydimethylsiloxanes, such as Si 100000 and Si 600000, in delayed cokers
and their thermal degradation which starts at 315°C (Kremer and Hueston,
2002). The thermal degradation of different viscosity PDMS, heated for 2 h at
400°C (750°F) under a nitrogen blanket, is shown in Table 7.38.

Under the coke drum conditions, the thermal breakdown of lower viscosity
PDMS 60,000 cSt leads to a weight loss of 33.7% and a reduction in the
molecular weight. Under the same conditions, the thermal breakdown of higher
viscosity PDMS 600,000 cSt leads to a lower weight loss of 18.3% and a
reduction in molecular weight (Kremer and Hueston, 2002). The literature
reported that the temperature of the thermal cracking process in delayed coking
can go up to 450°C–500°C, while its full cycle time is about 24 h which can
further increase the degradation process of silicone polymers (Rome, 2001).
Yields and product quality vary due to different feedstocks charged to a delayed
coker and different processing conditions. The coker fractionator or combination
of distillation tower separates the coker overheads into gases, gasoline, diesel,
heavy coker gas oil, and recycle (Ellis and Paul, 2000). The effect of different
viscosity silicone antifoaming agents on the Si content of coking products is
shown in Table 7.39.

An increase in PDMS viscosity from 100,000 to 600,000 cSt and a decrease
in PDMS treat rate lead to a decrease in Si content of heavy coker naphtha (HCN)
by 72%, reduction in Si content of light coker gas oil (LCGO) by 75%, and a
decrease in Si content of heavy coker gas oil (HCGO) by 44% (Kremer and
Hueston, 2002). Coking is often used in preference to catalytic cracking because
of the presence of nitrogen components and metals in the feed that poison
catalysts. The literature reported that due to the economic loss from Si poisoning
of hydrotreater catalyst, there has been an effort to reduce or eliminate the use of
silicone antifoaming agents in the coking process. Field experience was reported

TABLE 7.39
Effect of PDMS Viscosity on Silicone Content
of Coking Products

Coking Process	PDMS 100,000 cSt	PDMS 600,000 cSt	Si Reduction
Products	Si, ppm	Si, ppm	%
HCN	32.3	9.2	72
LCGO	11.8	2.9	75
HCGO	4.0	2.2	44

Source: From Kremer, L.N. and Hueston, T.G., *PTQ Summer*, 65–69, 2002. Reproduced by permission of Crambeth Allen Publishing Ltd. and Baker Petrolite Corporation, Sugar Land, Texas, USA.

to confirm that less high viscosity silicone antifoaming agents are required to control the foam in the coke drum resulting in lower Si contamination of the coker products (Kremer and Hueston, 2002). Silicone antifoaming agents are used during different steps of crude oils refining; however, their presence in petroleum feedstocks and finished products is of concern due to their high surface activity.

REFERENCES

Aminu, M.O. et al., *Industrial and Engineering Chemistry Research*, 43, 2929, 2004.
Bergeron, V. et al., *Colloids and Surfaces, A: Physicochemical and Engineering Aspects*, 122(1–3), 103, 1997.
Bondi, A., *Physical Chemistry of Lubricating Oils*, Reinhold Publishing Corporation, New York, 1951.
Butler, K.D. and Henderson, H.E., *Adsorbent Processing to Reduce Base Stock Foaming*, US Patent 1,266,007, 1990.
CRC Handbook of Chemistry and Physics, 86th ed., Lide, D.R., Editor-in-Chief, CRC Press, Boca Raton, 2005.
Dee, G.T. and Sauer, B.B., *TRIP*, 5(7), 230, 1997.
Ellis, P.J. and Paul, C.A., *Delayed Coking Fundamentals*, AIChE 2000 Spring National Meeting, Atlanta, May 5–9, 2000.
Gary, J.H. and Handwerk, G.E., *Petroleum Refining, Technology and Economics*, 4th ed., Marcel Dekker, Inc., New York, 2001.
Hubmann, A. and Lanik, A., *Tribol. Schmierungstech.*, 35(3), 138, 1988.
Kremer, L.N. and Hueston, T.G., *PTQ Summer*, 65–69, 2002.
Lloydminster, http://www.heavyoil.com, *Upgrader Delayed Coking Unit* (accessed December 2004).
Mang, T., Base oils, in *Lubricants and Lubrications*, Mang, T. and Dresel, W., Eds., Wiley-VCH, Weinheim, 2001.
Mannheimer, R.J., *Factors that Influence the Coalescence of Air Bubbles in Oils that Contain Silicone Antifoams*, Interpec China '91, Beijing, 1991.

Moreira, J.C. and Demarquette, N.R., *Journal of Applied Polymer Science*, 82, 1907, 2001.

Owen, M.J., *Industrial and Engineering Chemistry Product Research and Development*, 19(1), 97, 1980.

Pape, P.G., *Silicones: Unique Chemicals for Petroleum Processing*, SPE 10089, 56th Annual Fall Technical Conference and Exhibition of SPE, San Antonio, 1981.

Pape, P.G., *Journal of Petroleum Technology*, 35, 1197, 1983.

Pillon, L.Z., *Optimization of the ASTM D 892 Foam Test*, ASTM Committee Meeting on Petroleum Products and Lubricants, Boston, 1994.

Pillon, L.Z., *Petroleum Science and Technology*, 19(9&10), 1263, 2001.

Pillon, L.Z., *Petroleum Science and Technology*, 20(1&2), 223, 2002.

Pillon, L.Z. and Asselin, A.E., *Antifoaming Agents for Lubricating Oils*, US Patent 5,766,513, 1998.

Pillon, L.Z., Asselin, A.E., and Vernon, P.D.F., *Method for Reducing Foaming of Lubricating Oils*, US Patent 6,090,758, 2000.

Pirro, D.M. and Wessol, A.A., *Lubrication Fundamentals*, 2nd ed., Marcel Dekker, Inc., New York, 2001.

Poindexter, M.K. et al., *Energy and Fuels*, 16, 700, 2002.

Prince, R.J., Base oils from petroleum, in *Chemistry and Technology of Lubricants*, 2nd ed., Mortier, R.M. and Orszulik, S.T., Eds., Blackie Academic & Professional, London, 1997.

ProQuest Document ID 10552585, *Chemical Engineering*, 103(12), 99, 1996.

Quinn, C., Traver, F., and Murthy, K., Silicones, in *Synthetic Lubricants and High-Performance Functional Fluids*, 2nd ed., Rudnick, L.R. and Shubkin, R.L., Eds., Marcel Dekker, Inc., New York, 1999.

Rome, F.C., *Hydrocarbon Asia*, 42, May/June 2001.

Rosen, M.J., *Surfactants and Interfacial Phenomena*, 3rd ed., Wiley-Interscience, Hoboken, 2004.

Ross, S., Foaminess and capilarity in apolar solutions, in *Interfacial Phenomena in Apolar Media*, Eicke, H.F. and Parfitt, G.D., Eds., Marcel Dekker, Inc., New York, 1987.

Rudnick, L.R. and Shubkin, R.L., Poly(alfa-olefins), in *Synthetic Lubricants and High-Performance Functional Fluids*, 2nd ed., Rudnick, L.R. and Shubkin, R.L., Eds., Marcel Dekker, Inc., New York, 1999.

SensaDyne Instrument Division, *Comparison of Surface Tension Measurements Methods*, 2000.

Sequeira, A. Jr., *Lubricant Base Oil and Wax Processing*, Marcel Dekker, Inc., New York, 1994.

Speight, J.G., *The Chemistry and Technology of Petroleum*, 3rd ed., Marcel Dekker, Inc., New York, 1999.

Speight, J.G., *The Chemistry and Technology of Petroleum*, 4th ed., CRC Press, Boca Raton, 2006.

Sun, H. et al., *Huadong Huagong Xueyuan Xuebao*, 18(5), 589, 1992.

Tourret, R. and White, N., *Aircraft Engineering*, 24, 122, 1952.

8 Air Entrainment

8.1 CONVENTIONAL REFINING

Foaming occurs when air mixes with oils. Air entrainment is a dispersion of small air bubbles in oil. The first comprehensive account of interfacial phenomena in liquid hydrocarbons covered various topics, including foaminess, foam inhibition, and air entrainment (Ross, 1987). The constant velocity with which a spherical particle falls in a liquid is expressed by a relatively simple law which is known as Stoke's equation. The literature reported that pure liquids allow entrained air to escape with no delay other than the rate of rise described by Stoke's equation. The rate of air rise is controlled by the diameter of the bubble and the viscosity of the bulk liquid (Ross, 1987). Stoke's theory was extended to gas bubbles and it is one of the methods used to measure the viscosity of liquids. According to the literature, pure liquids allow entrained air to escape with no delay. Certain solutes are able to form a thin layer, known as lamella, at the liquid/air interface and adversely affect the air separation properties of liquids. If these solutes are present, the escape of the entrained air bubbles is delayed and stable foam is produced (Ross, 1987). The review of early literature on foaming of petroleum oils reported that the presence of certain acids and phenols increased foaming of crude oils. It was also reported that an increase in foam stability of medicinal grade paraffin oils and engine oils was found to increase with an increase in their kinematic viscosity (Zaki et al., 2002). The effect of conventional refining on inorganic and organic polar content of mineral base stocks is shown in Table 8.1.

The conventional refining of crude oils to produce mineral base stocks is based on many different processing steps during which metals, asphaltenes, and many other undesirable molecules are removed. To produce high quality base stocks, the viscosity index (VI), the wax content, and the sulfur content of vacuum gas oil (VGO) are important. Solvent extraction of lube oil distillates, based on physical and chemical separation, produces refined oils known as solvent raffinates. Aromatic hydrocarbons present in crude oils have a low VI and the solvent extraction removes some aromatic molecules to increase their VI. The n-paraffins found in crude oils have a high VI but they are undesirable as lube oil molecules due to their high pour points. The processes, such as solvent dewaxing, are usually used to decrease the cloud and pour point of mineral base stocks. Color variation can result from differences in crude oils, viscosity, and degree of

TABLE 8.1
Effect of Conventional Refining on Polar Content of Mineral Base Stocks

Conventional Refining	Processing Effect
Desalting	Reduces salts and metals
Distillation	Asphaltenes left in vacuum residue
Deasphalting	Reduces carbon residue and sludge forming tendency of residual oils
Solvent extraction	Reduces aromatics and heteroatom content
Solvent dewaxing	Reduces waxy type molecules
Clay treatment	Reduces acidity
Hydrofinishing	Improves color and stability

Source: From Gary, J.H. and Handwerk, G.E., *Petroleum Refining: Technology and Economics,* 4th ed., Marcel Dekker, Inc., New York, 2001.

refining. Many older refineries use activated clays at elevated temperatures to improve the color and the stability of lube oil base stocks. Clay percolation is an adsorption process which might be a continuous process or most commonly is a static bed of clay. Polar compounds are adsorbed on the clay and removed by filtration (Sequeira, 1994). Hydrogen finishing processes are used in place of more costly acid and clay finishing processes for the purpose of improving color, odor, thermal, and oxidative stability of mineral base stocks.

The selection of crude oil and the processing severity will affect the properties and the trace polar content of mineral base stocks. Naphthenic base stocks are made in small quantities from selected crude oils which have low wax content and they generally do not require dewaxing. The literature reported that naphthene pale oils (NPO) are vacuum distilled and solvent extracted to remove aromatics and, in many cases, hydrofinishing is used as the last processing step (Sequeira, 1992). The effect of solvent extraction and hydrofinishing on the properties and composition of 100N naphthenic base stock is shown in Table 8.2.

The naphthenic distillate was reported to have a low VI of only 17, an aromatic content of 34 wt%, and an S content of 0.1 wt%. After solvent extraction, the VI increased to 61 and the pour point increased from $-37°C$ to $-29°C$. With a decrease in aromatic content from 34 to 24 wt%, a decrease in the S content from 0.1 to 0.03 wt% is also observed. The solvent extraction process increases VI and also improves the color of naphthenic raffinates. After hydrofinishing, there is no change in the VI, pour point, and the aromatic content; however, some decrease in the S content to 0.02 wt% with an improvement in color is observed. Very low viscosity naphthenic base stocks are used in lubricating oils which do not require high VI but require good low temperature properties.

Different batches of naphthenic base stocks, having a low viscosity of only 8–9 cSt at 40°C, were found to have a similar VI, in a range of 57–64, and a low natural pour point, varying from as low as $-57°C$ to $-39°C$. These low viscosity

TABLE 8.2
Effect of Hydrofinishing on Properties of 100N Naphthenic Base Stock

100N Naphthenic Oil	Distillate	Raffinate	Hydrofinished Oil
API gravity	23.0	28.8	28.8
Viscosity, cSt at 38°C	108	100	100
VI	17	61	61
Natural pour point, °C	−37	−29	−29
Aromatics, wt%	34	24	24
Sulfur, wt%	0.10	0.03	0.02
ASTM color	7	1.5	L0.5
COC flash, °C	179	179	179

Source: From Sequeira, A., *Pre-Prints Division of Petroleum Chemistry, ACS,* 37, 1286, 1992. Reproduced by permission of Chevron Corporation.

naphthenic base stocks, containing 20–22 wt% of aromatics, were found to have an S content of 0.1–0.2 wt% and contain a trace N content in the range of 2–7 ppm. Due to their low viscosity, their surface tension (ST) was low, about 30 mN/m measured at RT.

The ASTM D 892 foam test covers the determination of the foaming characteristics of lubricating oils at 24°C (Seq. I) and 93.5°C (Seq. II), and then, after collapsing the foam, at 24°C (Seq. III). The sample is blown with air for 5 min and the foaming tendency is the volume of foam measured immediately after the cessation of airflow. The foam stability is the volume of foam measured 10 min after disconnecting the air supply. When mixed with air, low viscosity naphthenic base stocks were found to foam. Due to their low viscosity, their foaming tendency was low. At 24°C, different naphthenic base stocks were found to have a Seq. I foaming tendency varying in a range of 30–60 mL, and a Seq. III foaming tendency varying in a range of 30–70 mL. At 93.5°C, with a decrease in the viscosity, their foaming tendency was found to decrease to 10–20 mL.

The ASTM D 3427 test method for air release properties of petroleum oils covers the ability of industrial oils, such as turbine, hydraulic, and gear oils, to separate the entrained air and this test might not be suitable for ranking oils in applications where residence times are short and gas contents are high. Following the ASTM D 3427 test method for air release properties of petroleum oils, compressed air is blown through the test oil which has been heated to 25°C, 50°C, or 75°C. After the airflow is stopped, the time required for the air entrained in the oil to reduce in volume to 0.2% is recorded as the air release time. The air release time of lube oil base stocks, following the ASTM D 3427 test method, is usually measured at 50°C. Despite some variation in composition and ST, the air release time of very low viscosity naphthenic base stocks was below 1 min indicating excellent air separation properties. The literature reported that paraffinic distillates are vacuum distilled, solvent extracted, solvent dewaxed, and hydrofinishing is used as the last processing step (Sequeira, 1992). The effect of

TABLE 8.3
Effect of Hydrofinishing on Properties of 100N Paraffinic Base Stock

100N Paraffinic Oil Properties	Distillate	Raffinate	Dewaxed and Hydrofinished Oil
API gravity	26.7	34.7	32.4
Viscosity, cSt at 38°C	103	85.1	100
VI	86	110	95
Dewaxed pour point, °C	24	29	−18
Aromatics, wt%	34	17	16
Sulfur, wt%	1.1	0.69	0.17
ASTM color	4.5	1.5	L0.5
COC flash, °C	193	193	193

Source: From Sequeira, A., *Pre-Prints Division of Petroleum Chemistry*, ACS, 37, 1286, 1992. Reproduced by permission of Chevron Corporation.

solvent refining and hydrofinishing on the properties of 100N paraffinic base stock is shown in Table 8.3.

The paraffinic distillate was reported to have a higher VI of 86 and a higher pour point of 24°C. After solvent extraction, the VI increased to 110 and the pour point increased to 29°C. With an increase in VI, a decrease in aromatic content from 34 to 17 wt% and a decrease in the S content from 1.1 to 0.39 wt% were reported. The solvent extraction process increases VI and also improves the color of paraffinic raffinates. After solvent dewaxing to −18°C dewaxed pour point and hydrofinishing, a decrease in VI to 95 is observed. With a small decrease in the aromatic content to 16 wt% and the S content to 0.17 wt%, an improvement in color is also observed. For the same viscosity, the flash point of paraffinic base stocks is higher.

Different extraction solvents are used to decrease the aromatic content and increase the VI of solvent raffinates. Commonly used extraction solvents are phenol and N-methylpyrrolidone (NMP) and the use of NMP extraction solvent is increasing (Prince, 1997). The use of NMP solvent was reported to increase the raffinate yield by 6–8 wt% and reduce the solvent/feedstock ratio by 15 wt%. The literature reported that in comparison to phenol, NMP solvent does not form azeotropic mixtures with water, is less toxic, and has a higher selectivity with higher dissolving power and a lower viscosity (Ivanov et al., 2000). The literature also reported that the use of NMP is increasing because of lower construction costs for a grass-root plant and the recovery of NMP solvent is better than that for the phenol solvent. A lower viscosity of the NMP solvent gives a greater throughput for a given size tower (Gary and Handwerk, 2001). The solvent refined and solvent dewaxed mineral oils are usually referred to as solvent neutrals (N). Hydrofining is usually used to improve the color of mineral base

TABLE 8.4
Effect of Extraction Solvent on Composition of Hydrofined Mineral Base Stocks

Base Stocks, Solvent Extraction	HF 150N Phenol	HF 600N Phenol	HF 150N NMP	HF 600N NMP
Kin. Viscosity, cSt				
At 40°C	29.5	105.9	29.6	111.4
At 100°C	5	11.3	4.99	11.6
VI	94	92	90	89
Saturates, wt%	82.8	80.5	79.9	80.4
Aromatics, wt%	17.2	19.5	20.1	19.6
Sulfur, wt%	0.09	0.12	0.12	0.11
Total nitrogen, ppm	8	30	66	100
Basic nitrogen, ppm	4	16	53	88
BN/TN ratio	0.50	0.53	0.80	0.88

Source: From Pillon, L.Z., Reid, L.E., and Asselin, A.E., *Lubricating Oil for Inhibiting Rust Formation*, US Patent 5,397,487, 1995.

stocks and the use of different extraction solvents affects their composition. The effect of the extraction solvent on composition and the polar content of hydrofined mineral base stocks is shown in Table 8.4.

With an increase in the viscosity of mineral base stocks, an increase in the aromatic and heteroatom content is observed. With an increase in viscosity of phenol extracted base stocks from 150N to 600N, their S content increased from 0.09 to 0.12 wt%, and their N content also increased from 8 to 30 ppm. With an increase in viscosity of NMP extracted from 150N to 600N, no significant increase in the aromatic and the S contents was observed; however, the N content increased from 66 to 100 ppm. The ASTM D 2896 test method is used to determine the basic nitrogen (BN) content of petroleum oils by titration with perchloric acid. Basic constituents of petroleum include not only BN compounds but also other compounds, such as salts of weak acids (soaps), basic salts of polyacidic bases, and salts of heavy metals. With an increase in viscosity of hydrofined phenol extracted base stocks from 150N to 600N, their polar BN content was found to increase from 4 to 16 ppm. With an increase in viscosity of hydrofined NMP extracted base stocks from 150N to 600N, their N content increased and the polar BN content was found to increase from 53 to 88 ppm. According to the literature, the ratio of basic type nitrogen compounds (BN) to total nitrogen (TN) of crude oils is approximately constant in the range of 0.3 +/− 0.05, irrespective of the source of crudes (Speight, 1999). The BN/TN ratio of 0.5 indicated some increase in polar basic type contaminants of hydrofined phenol extracted base stocks. The BN/TN ratio of 0.8–0.9 indicated a higher content of polar basic type molecules in hydrofined NMP extracted base

TABLE 8.5

Foaming Characteristics and Air Release Time of HF 150N Mineral Base Stocks

Hydrofined 150N Base Stocks	Phenol Extracted	NMP Extracted
Surface tension, mN/m	33.7	33.8
ASTM D 892 Foam Test		
Seq. I at 24°C		
Foaming tendency, mL	345	280
Foam stability, mL	0	0
Seq. II at 93.5°C		
Foaming tendency, mL	30	20
Foam stability, mL	0	0
ASTM D 3427, 50°C		
Air release time, min	1.6	2.1

Source: From Pillon, L.Z. and Asselin, A.E., *Antifoaming Agents for Lubricating Oils*, US Patent 5,766,513, 1998.

stocks. The foaming characteristics and the air release time of different hydrofined 150N mineral base stocks are shown in Table 8.5.

Hydrofined 150N phenol extracted base stock was found to have a high Seq. I foaming tendency of 345 mL, measured at 24°C, and a low Seq. II foaming tendency of 20 mL, measured at 93.5°C. The air release time, measured at 50°C, was found to be in a range of 1.6 min. Hydrofined 150N NMP extracted base stock, having a lower VI, was found to have a higher aromatic content of 20.1 wt%, a higher S content of 0.18 wt%, and a higher N content of 66 ppm. It was found to have a lower Seq. I foaming tendency of 280 mL, measured at 24°C, and a lower Seq. II foaming tendency, measured at 93.5°C. With an increase in the BN content, no increase in the foaming tendency, measured at 24°C, was observed, but the air release time, measured at 50°C, was found to increase to 2.1 min. Higher viscosity hydrofined 600N mineral base stocks have a higher tendency to foam and form stable foams at 24°C. The literature reported that NMP solvent is less thermally stable than phenol. As a result of the presence of air and oxidation taking place, the circulating NMP solvent was reported to contain 0.1–10 wt% of oxidation by-products, mainly *N*-methylsuccinimide. The total acid number, TAN, was reported to increase from 0.15 to 0.6 mg/KOH due to breakdown of amines and hydrolysis with water (Ivanov et al., 2000). The foaming characteristics and air release time of hydrofined 600N mineral base stocks are shown in Table 8.6.

Hydrofined 600N phenol extracted base stock was found to have a high Seq. I foaming tendency of 630 mL and the foam stability of 165 mL at 24°C and a low Seq. II foaming tendency of 40 mL at 93.5°C. The air release time, measured at 50°C, was found to be in a range of 6–7 min. Hydrofined 600N NMP extracted

TABLE 8.6
Foaming Characteristics and Air Release Time of HF 600N Mineral Base Stocks

Hydrofined 600N Base Stocks	Phenol Extracted	NMP Extracted
Surface tension, mN/m	34.7	34.8
ASTM D 892 Foam Test		
Seq. I at 24°C		
Foaming tendency, mL	630	620
Foam stability, mL	165	250
Seq. II at 93.5°C		
Foaming tendency, mL	40	50
Foam stability, mL	0	0
ASTM D 3427, 50°C		
Air release time, min	6–7	8–9

Source: From Pillon, L.Z. and Asselin, A.E., *Antifoaming Agents for Lubricating Oils*, US Patent 5,766,513, 1998; Pillon, L.Z., Asselin, A.E., and Vernon, P.D.F., *Method for Reducing Foaming of Lubricating Oils*, US Patent 6,090,758, 2000.

base stock was found to have a similar high Seq. I foaming tendency of 620 mL and an increased foam stability of 250 mL at 24°C and a low Seq. II foaming tendency of 50 mL at 93.5°C. With a significant increase in polar BN content to 88 ppm, a longer air release time of 8–9 min was measured at 50°C.

To adsorb at the oil/air interface, surface active contaminants, similar to antifoaming agents, need to have a low ST, below the ST of petroleum oil. With an increase in viscosity and the ST of base stocks from 33.7–33.8 mN/m for 150N base stocks to 34.7–34.8 mN/m for 600N, more polar molecules can become surface active and adsorb at the oil/air interface. For the same viscosity, hydrofined NMP extracted base stocks have a longer air release time due to a higher polar BN content. Surface active contaminants might be introduced during the refining process. The effect of temperature on the ST of different extraction solvents and ketone dewaxing solvent is shown in Table 8.7.

If residual extraction solvents are present, their ST at 50°C is higher than that of mineral base stocks. Before hydrofining, the solvent dewaxed base stocks are often contaminated with the dewaxing solvents, such as methyl ethyl ketone (MEK), methyl isobutyl ketone (MIBK), or their mixture. The ketone dewaxing solvents, such as MEK, have a low ST but do not have the surfactant structure which requires the hydrophobic group of a minimum C_{10}–C_{12} carbons. The BN content, separated from HF NMP extracted base stocks, was analyzed by FTIR which indicated the presence of amines, pyridines, and carboxylic acids. According-ing to the early literature, various alkylated pyridines are occurring in crude oils

TABLE 8.7
Effect of Temperature on ST of Some Refining Solvents

ST, mN/m	Formula	25°C	50°C	75°C	100°C
Extraction Solvents					
Furfural	$C_5H_4O_2$	43.09	39.78	36.46	33.14
Phenol	C_6H_6O	—	38.20	35.53	32.86
N-methyl-2-pyrrolidinone (NMP)	C_5H_9NO	40.21	37.33	34.45	31.57
Dewaxing Solvent					
Methyl ethyl ketone (MEK)	C_4H_8O	23.97	21.16	—	—

Source: From *CRC Handbook of Chemistry and Physics*, 86th ed., Lide, D.R., Editor-in-Chief, CRC Press, Boca Raton, 2005. Reproduced by permission of Routledge/Taylor & Francis Group, LLC.

(Klingsberg, 1969). The effect of temperature on the ST of different amines, pyridines, and acids is shown in Table 8.8.

All aliphatic amines, including tributylamine which contains C_{12} carbons, have a low ST and can potentially adsorb at the oil/air interface of mineral base stocks. Pyridines have a high ST at 25°C but alkylated pyridines have a significantly lower ST at 50°C. Similar to low molecular weight amines, low molecular

TABLE 8.8
Effect of Temperature on the ST of Some Pure Amines, Pyridines, and Acids

ST, mN/m	Formula	25°C	50°C	75°C
Dipropylamine	$C_6H_{15}N$	22.31	19.75	17.20
Dibutylamine	$C_8H_{19}N$	24.12	21.74	19.36
Diisobutylamine	$C_8H_{19}N$	21.72	19.44	17.16
Tributylamine	$C_{12}H_{27}N$	24.39	22.32	20.24
Cyclohexylamine	$C_6H_{13}N$	31.22	28.25	25.28
Aniline	C_6H_7N	42.12	39.41	36.69
Benzylamine	C_7H_9N	39.30	36.27	33.23
Pyridine	C_5H_5N	36.56	33.29	30.03
2-Methylpyridine	C_6H_7N	33.00	29.90	26.79
2,3-Dimethylpyridine	C_7H_9N	32.71	30.04	27.36
Propanoic acid	$C_3H_6O_2$	26.20	23.72	21.23
Butanoic acid	$C_4H_8O_2$	26.05	23.75	21.45
Heptanoic acid	$C_7H_{14}O_2$	27.76	25.64	—

Source: From *CRC Handbook of Chemistry and Physics*, 86th ed., Lide, D.R., Editor-in-Chief, CRC Press, Boca Raton, 2005. Reproduced by permission of Routledge/Taylor & Francis Group, LLC.

weight aliphatic acids having C_3–C_7 carbons have a low ST but also would require a bigger hydrophobic group to act as a surfactant and affect the properties of petroleum oils at the oil/air interface. The literature reported that amines and acids found in mineral base stocks form salts (Sequeira, 1994). The salt formation of low molecular weight basic aliphatic amines and pyridines, with low molecular weight acids, can increase their molecular weight and hydrophobic content and make them powerful surfactants.

8.2 NONCONVENTIONAL REFINING

The literature reported that air release values of petroleum base oils are adversely affected by minor constituents which adsorb on the dispersed air bubbles. The air release time of mineral base stocks was reported to decrease after separating a fraction containing aromatics, naphthenics, sulfur, and nitrogen compounds (Basu et al., 1985). Usually the profoaming solute is a legitimate component of the system and such a component might not be a surfactant but becomes surface active. Under certain conditions of temperature and concentration, the solubility of certain components in a medium might decrease leading to an increase in their surface activity (Ross, 1987). The use of processing might change the chemistry of molecules and their properties, including surface activity. The literature reported on the effect of hydrotreatment (HT) on the polar and the heteroatom content of VGO, containing a high polar content of 18.6 wt% (Ali, 2001). The presence of 18.6 wt% of polars and the total of S and N of less than 3 wt%, without the presence of asphaltenes, indicates a significant presence of O containing molecules. The effect of hydrotreating on the polar and heteroatom content of VGO is shown in Table 8.9.

The HT of VGO was reported to decrease the boiling point from the initial 0.5% (IBP) of 343°C to 121°C, the 50% (Mid BP) from 499°C to 393°C, and the 99.5% (FBP) from 641°C to 546°C indicating a lower viscosity product having an increased volatility. HT of VGO was reported to contain less than 0.1 wt% of S, 0.1 wt% of N, and no asphaltenes but still a high polar content of 7.7 wt% (Ali, 2001). After the HT of VGO, with a decrease in boiling point, an increase in the saturate content from 13.13 to 28.9 wt%, a reduction in the aromatic content from 68.1 to 63.4 wt%, and a reduction in the polar content from 18.6 to 7.7 wt% are observed. The actual S content decreased from 2.3 to <0.01 wt%, and the N content also decreased from 0.22 to 0.1 wt%. After the HT, the presence of 7.7 wt% of polars and the total of S and N of only 0.1 wt% indicate a significant content of O containing molecules.

The majority of lubricating oils use solvent refined mineral base stocks which have a typical VI, below 100, and a TAN below 0.01 mg KOH/g. The literature reported on properties of lube oil base stocks used to formulate machine oil cutting fluid for band saws. With an increase in the VI, above 100, an increase in the TAN from a typical value of 0.01 to 0.02 mg KOH/g was reported (Men et al., 2001). The effect of processing severity on VI and TAN of different 100SN and 150SN neutral base stocks is shown in Table 8.10.

TABLE 8.9
Effect of Hydrotreatment on Polar and Heteroatom Content of VGO

Properties	Test Method	VGO	HT VGO
Distillation	ASTM D 2887		
0.5% (IBP), °C		343	121
50%, °C		499	393
99.5% (FBP), °C		641	546
Hydrocarbon types	HPLC		
Saturates, wt%		13.3	28.9
Aromatics, wt%		68.1	63.4
Polars, wt%		18.6	7.7
Total S, wt%	CHNS Analyzer	2.67	<0.1
Total N, wt%	CHNS Analyzer	0.22	0.10
Asphaltenes (C_7 insolubles)		0	0

Source: From Ali, M.A., *Petroleum Science and Technology*, 19, 1063, 2001. Reproduced by permission of Taylor & Francis Group, LLC, http://www.taylorandfrancis.com.

The typical 100SN solvent refined base stock was reported to have a VI of 101, a TAN of 0.01 mg KOH/g, and a flash point of 212°C. Another 100SN solvent refined base stock, having a higher VI of 107, was reported to have a higher TAN of 0.02 mg KOH/g and a lower flash point of 193°C. Higher viscosity 150SN base stocks, having a higher VI of 111, were also reported to

TABLE 8.10
Effect of Processing Severity on VI and TAN of 100SN and 150SN Base Oils

Properties	100SN (Typical)	100SN	150SN
Density at 20°C, g/cc	0.865	0.8595	0.8664
Kin. Viscosity, cSt			
At 40°C	20	21.82	28.73
At 100°C	4	4.23	5.19
VI	101	107	111
TAN, mg KOH/g	0.01	0.02	0.02
Flash point, °C	212	193	210
Carbon residue, wt%	0.01	0.01	0.02

Source: From Men, G. et al., *Petroleum Science and Technology*, 19, 1251, 2001. Reproduced by permission of Taylor & Francis Group, LLC, http://www. taylorandfrancis.com.

TABLE 8.11
Effect of Hydroisomerization on Air Release Time
of Different SWI 6 Base Stocks

Hydroisomerization	SWI 6 # 1	SWI 6 # 2
Kin. Viscosity, cSt		
At 40°C	29.38	31.35
At 100°C	5.77	5.96
VI	143	138
Saturates, wt%	>99.5	>99.5
Aromatics, wt%	<0.5	<0.5
Sulfur, ppm	<1	17
Nitrogen, ppm	<1	2
Surface tension, mN/m	32.7	32.8
ASTM D 892 Foam Test		
Seq. I at 24°C		
Foaming tendency, mL	145	180
Foam stability, mL	0	0
Seq. II at 93.5°C		
Foaming tendency, mL	10	25
Foam stability, mL	0	0
ASTM D 3427 50°C		
Air release time, min	1.1	1.2

Source: From Pillon, L.Z. and Asselin, A.E., *Antifoaming Agents for Lubricating Oils*, US Patent 5,766,513, 1998.

have a higher TAN of 0.02 mg KOH/g indicating an increase in the presence of acidic molecules. With an increase in viscosity of base stocks from 100SN to 150SN, their flash point increased to 210°C and some increase in CCR from 0.01 to 0.02 wt% was reported. The usual components of TAN are organic soaps, soaps of heavy metals, organic nitrates, other compounds used as additives, and oxidation by-products. The effect of hydroisomerization on the foaming characteristics and the air release time of different SWI 6 base stocks is shown in Table 8.11.

SWI 6 base stock # 1 was found to contain less than 0.5 wt% of aromatics and practically no presence of S and N compounds. When compared to similar viscosity 150N mineral base stocks, it was found to have a lower ST of 32.7 mN/m, a lower Seq. I foaming tendency of 145 mL, measured at 24°C, and a shorter air release time of only 1.1 min, measured at 50°C. Another SWI 6 base stock # 2 also containing less than 0.5 wt% of aromatics was found to contain a trace of 17 ppm of S and 2 ppm of N compounds. With a small increase in viscosity at 40°C, a small increase in ST, an increase in Seq. I and Seq. II foaming tendency, and a small increase in air release time are observed. The use of

TABLE 8.12
Trace Polar Content of Hydrofined SWI 6 Base Stocks

Hydroisomerization Temperature	HF SWI 6, Low	HF SWI 6, High
Daylight stability, days	9	3
DMSO/CHCl₃ Extract		
FTIR	1–2 Ring aromatics	1–2 Ring aromatics
	Ketones, acids, esters	Ketones, acids, esters

Source: From Pillon, L.Z., *Petroleum Science and Technology*, 19, 1263, 2001. Reproduced by permission of Taylor & Francis Group, LLC, http://www.taylorandfrancis.com.

hydroprocessing technology, such as slack wax hydroisomerization, was found to improve the air separation properties of base stocks despite the fact that, after hydrofining, the presence of polar contaminants affecting their daylight stability was found. The FTIR analysis of DMSO/CHCl₃ extracts from hydrofined SWI 6 base stocks, hydroisomerized at different temperatures, is shown in Table 8.12.

Hydrofined slack wax isomerate base stock, hydroisomerized under low temperature conditions, was found to have a daylight stability of 9 days. The FTIR analysis of the DMSO/CHCl₃ extract indicated the presence of 1–2 ring aromatic compounds, ketones, acids, and esters. Another hydrofined slack wax isomerate base stock, hydroisomerized under high temperature conditions, was found to have a daylight stability of only 3 days indicating a higher content of polar contaminants. The FTIR analysis of the DMSO/CHCl₃ extract also indicated the presence of 1–2 ring aromatic compounds, ketones, acids, and esters. The effect of oxidation on the ST of different hydrocarbons is shown in Table 8.13.

The oxidation process of aliphatic hydrocarbons, such as heptane, increases the ST from 17.19 to 23.48 mN/m for ketones and further to 25.64 mN/m for

TABLE 8.13
Effect of Oxidation on ST of Aliphatic and Aromatic Hydrocarbons

ST, mN/m	Formula	25°C	50°C	75°C	100°C
Heptane	C_7H_{16}	19.66	17.19	14.73	—
2-Heptanone	$C_7H_{14}O$	26.12	23.48	—	—
Heptanoic acid	$C_7H_{14}O_2$	27.76	25.64	—	—
Ethylbenzene	C_8H_{10}	28.75	26.01	23.28	20.54
Acetophenone	C_8H_8O	39.04	36.15	33.27	—
Methyl benzoate	$C_8H_8O_2$	37.17	34.25	31.32	—
Benzyl benzoate	$C_{14}H_{12}O_2$	42.82	40.06	37.31	34.55

Source: From *CRC Handbook of Chemistry and Physics*, 86th ed., Lide, D.R., Editor-in-Chief, CRC Press, Boca Raton, 2005. Reproduced by permission of Routledge/Taylor & Francis Group, LLC.

TABLE 8.14
Oxidation By-Products of Mineral Oil

Mineral Oil/Fe Surface	Volatility, %	High MW Products, %	Unreacted, %
10 min at 225°C	<2	14	58
20 min at 225°C	13	27	24
30 min at 225°C	25	31	16

Source: Reprinted with permission from Naidu, S.K., Klaus, E.E., and Duda, J.L., *Industrial and Engineering Chemistry Product Research and Development*, 23, 613, 1984. Copyright American Chemical Society.

acids, measured at 50°C. The oxidation process of 1-ring aromatic hydrocarbons, such as ethyl benzene, increases the ST from 26.01 to 36.15 mN/m for ketones and to 34.25 mN/m for esters, measured at 50°C. With an increase in the aromatic ring number, their ST further increases and the ST of 2-ring aromatic ester, benzyl benzoate, was found to further increase to 40.06 mN/m at 50°C. Since the polar contaminants found in SWI 6 were oxidized 1–2 ring aromatics, their ST is higher than the ST of base stocks and no adsorption at the oil/air interface will take place. The oxidation by-products of mineral oil formed in the presence of low carbon steel are shown in Table 8.14.

With an increase in the oxidation time at 225°C from 10 to 30 min, an increase in volatility and high molecular weight oxidation by-products was observed. The primary low molecular weight oxidation by-products were reported to contain ketones, aldehydes, alcohols, and acids. These products can polymerize to form viscous high molecular by-products in the form of sludge and varnish (Naidu et al., 1984). The formation of various oxidation by-products was reported to be affected by the presence of the type of metal surface, ranging from iron to aluminum and other metals (Naidu et al., 1984). Before the oxidation, the UV absorbance in the range of 180–340 nm with the max absorbance at 230 nm was reported. After oxidation, the UV absorbance in the range of 180–320 nm with the max absorbance at 230 and 285 nm was reported (Naidu et al., 1984). The literature reported that in lubricating oils, under low temperature oxidation conditions, peroxides, alcohols, aldehydes, ketones, and water are formed. Under high temperature oxidation conditions, acids are formed (Bardasz and Lamb, 2003). Oxidation by-products were reported to be the primary cause of metal corrosion, viscosity increase, sludge, and varnish formation in lubricant products.

8.3 EFFECT OF ANTIFOAMING AGENTS

The petroleum oils foam and the use of antifoaming agents is required. The literature reported that silicone antifoaming agents degrade the air separation properties of lubricating oils (Pirro and Wessol, 2001). To prevent air entrainment, some lubricating oils require the use of nonsilicone type antifoaming agents. The literature reported that polymethacrylate is more effective as an antifoaming agent than other polymers such as PVC, polypropylene glycol, or polyesters (Si et al., 1990). Other nonsilicone materials, such as alkoxylated aliphatic acids, polyethers such as polyethylene glycols, polyvinylethers, and polyalkoxyamines, have been reported to prevent foaming in lubricating oils (Crawford et al., 1997). Many lubricating oils, such as turbine oils, require resistance to foaming and good air separation properties. To prevent foaming, antifoaming agents need to be used at their effective treat rate to reduce the foaming tendency below 50 mL and meet the air release time specifications. The polyacrylate polymers, such as emulsions in diluent solvents, are used as antifoaming agents in applications where good air separation properties are required. The early literature reported on difficulties to study the oil/air interface because the use of ST measurements is not sensitive to the presence of trace content of surface active molecules, including highly surface active silicone antifoaming agents. The theoretical calculations of the effect of surface active molecules on the ST of mineral oil are shown in Table 8.15.

The use of highly surface active polydimethylsiloxane, even when used at a high treat rate of 10 ppm, depressed the ST of mineral oil by only 0.72 mN/m. Span 20 contains fatty acids and is a mixture of about 50% of lauric acid (C12:0), myristic acid (C14:0), palmitic acid (C16:0), and linolenic acid (C18:3) (Aldrich, 2007). The addition of Span 20, even at such a high treat rate as 400 ppm, depressed the ST of mineral oil by only 0.65 mN/m (Ross, 1987). The early literature reported that the surface active molecules decrease the ST of petroleum oils, which can be measured, when used at very high concentrations (Bondi, 1951). The effect of surface active molecules, at high concentrations, and temperature on ST of mineral lubricating oil is shown in Table 8.16.

At a low temperature of 27°C, the addition of 1 wt% of lauryl sulfonic acid decreased the ST of mineral lubricating oil from about 31–32 mN/m to

TABLE 8.15
Effect of Surface Active Molecules on ST of Mineral Oil

Surface Active Molecules	Concentration in Mineral Oil, ppm	ST Depression (Calculated), mN/m
Polydimethylsiloxane	0.5	0.09
(PDMS 1000 cSt)	3	0.20
	10	0.72
Span 20	55	0.09
(Fatty acid mixture)	150	0.18
	400	0.65

Source: From Ross, S., *Interfacial Phenomena in Apolar Media*, Eicke, H.F. and Parfitt, G.D., Eds., Marcel Dekker, Inc., New York, 1987. Reproduced by permission of Routledge/Taylor & Francis Group, LLC.

TABLE 8.16
Effect of Surface Active Molecules and Temperature on ST of Lubricating Oil

Surface Active Molecules in Mineral Lubricating Oil	Liquid ST, mN/m	Content, wt%	Solution ST, mN/m
Lubricating oil at 25°C	31–32	0	—
Lubricating oil at 100°C	25–26	0	—
Lauryl sulfonic acid at 27°C	33.0	1.0	27.8
Petroleum resin at 100°C	27.0	—	26.3
Sodium stearate at 112°C	25.2	1.0	20.8

Source: From Bondi, A., *Physical Chemistry of Lubricating Oils*, Reinhold Publishing Corporation, New York, 1951.

about 28–29 mN/m. At a high temperature of 100°C, the literature reported that petroleum resins did not decrease the ST of mineral lubricating oil. However, the addition of 1 wt% of sodium stearate at 112°C was found to decrease the ST of mineral lubricating oil from about 25–26 mN/m to 21–22 mN/m. Silicone antifoaming agents are effective in preventing foaming of mineral base stocks at low treat rates of only 1–3 ppm; however, even at these low treat rates, a severe degradation in their air separation properties is observed. The typical polyacrylate antifoaming agent, a 40 wt% emulsion in kerosene PA (40 wt%), is used in some industrial oils. During the actual application, the diluent oil is present and it is not clear if the diluent solvent is actually entering the oil/air interface and affecting the surface activity of the antifoaming agent. More experimental data are needed to understand the surface activity of polymer dispersions. The effect of different chemistry antifoaming agents on the foam inhibition and the air release time of 150N mineral base stock is shown in Table 8.17.

TABLE 8.17
Effect of Different Antifoaming Agents on Air Entrainment of 150N Mineral Oil

Antifoaming Agents	Treat Rate, ppm (Active)	Seq. I at 24°C, Foaming, mL	Seq. II at 93.5°C, Foaming, mL	Air Release Time, min at 50°C
HF 150N oil	0	340/0	25/0	1.6
Si 12500	3 (3)	10/0	35/0	6.3
FSi 300	3 (3)	25/0	15/0	5.5
PA (40 wt%)	100 (40)	5/0	20/0	2.2

Source: From Pillon, L.Z. and Asselin, A.E., *Antifoaming Agents for Lubricating Oils*, US Patent 5,766,513, 1998.

The use of 3 ppm of nondiluted Si 12500 was found effective in preventing foaming of 150N mineral base stock but a significant increase in air release time from 1.6 to 6.3 min is observed. The use of 3 ppm of nondiluted FSi 300, while less effective in preventing foaming, was found to increase the air release time to 5.5 min at 50°C. The use of 100 ppm of PA (40 wt%), 40 ppm of active ingredient, was required to prevent foaming of 150N mineral base stock; however, no significant increase in the air release time is observed. While the max foaming tendency of 150N mineral base stock is observed at 24°C, the air release time is measured at an increased temperature of 50°C. With an increase in the temperature from 24°C to 93.5°C, some decrease in ST of silicones is observed. The ST of low viscosity Si 12500 was found to decrease by 3.8 mN/m. The ST of FSi 300 was found to decrease by 5.3 mN/m. The effect of different viscosity silicones on the air release time of 150N mineral base stock is shown in Table 8.18.

With an increase in the treat rate of low viscosity Si 350, from 1 to 3 ppm, the Seq. I foaming tendency of 150N mineral base stock decreased from 280 to 5–35 mL, and the air release time was found to increase from 2.1 to 5.2–6.3 min, measured at 50°C. With an increase in the treat rate of Si 12500, from 1 to 3 ppm, the Seq. I foaming tendency of 150N mineral base stock decreased from 280 to 0–45 mL, and the air release time was found to increase from 2.1 to 4.2–5.5 min, measured at 50°C. With an increase in the treat rate of high viscosity Si 60000, from 1 to 3 ppm, the Seq. I foaming tendency of 150N mineral base stock decreased from 280 to 5–65 mL, and the air release time was found to increase from 2.1 to 3.2–5.5 min, measured at 50°C. Silicones are the most effective antifoaming agents; however, their use needs to be strictly controlled to minimize an increase in air entrainment. The effect of different viscosity silicones on the air release time of 150N mineral base stock is shown in Figure 8.1.

Different viscosity silicones were found to increase the air release time of base stocks; however, the use of high viscosity Si 60000 was found to have less effect in degrading the air separation properties of 150N mineral base stock. Fluorosilicones, having a higher ST, were found less effective than silicones in

---

TABLE 8.18
Effect of Different Viscosity Silicones on Air Release Time of 150N Mineral Oil

150N Oil at 50°C, Silicones ppm	Air Release Time, min, Si 350	Air Release Time, min, Si 12500	Air Release Time, min, Si 60000
0	2.1	2.1	2.1
1	5.2	4.2	3.2
2	6.2	5.0	4.2
3	6.3	5.5	5.5

Source: From Pillon, L.Z. and Asselin, A.E., *Antifoaming Agents for Lubricating Oils*, US Patent 5,766,513, 1998.

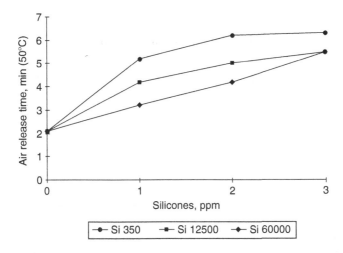

FIGURE 8.1 Effect of different viscosity silicones on the air release time of 150N mineral base stock.

preventing foaming of base stocks and higher treat rates were required. The effect of FSi 300 on the air release time of different 150N mineral base stocks, measured at 50°C, is shown in Table 8.19.

While fluorosilicones were found less effective in preventing foaming and higher treat rates are required, their use was found to have a significantly less effect on the air separation properties of base stocks. At low treat rates of only 1–3 ppm, low viscosity Si 350 was found to increase the air release time of 150N base stock from 2.1 to 5.2–6.3 min. A higher treat rate of FSi 300, up to 10 ppm, was required to decrease the Seq. I foaming tendency of 150N base stock from 370 to 40 mL. Despite a high treat rate of FSi 300, the air release time of 150N base stock was found to increase from 1.7 to only 5.1 min. The use of less surface

TABLE 8.19
Effect of FSi 300 on the Air Release Time of Different 150N Mineral Base Stocks

Mineral Base Stocks FSi 300, ppm	HF 150N (Phenol), Air Release Time, min	HF 150N (NMP), Air Release Time, min
0	1.6	2.1
3	3.7	4.3
5	4.8	4.8
10	5.1	5.1

Source: From Pillon, L.Z. and Asselin, A.E., *Antifoaming Agents for Lubricating Oils*, US Patent 5,766,513, 1998.

TABLE 8.20
Effect of Si 12500 and FSi 300 on Foaming and Air Release
Time of SWI 6 Oil

SWI 6 Base Stock Antifoaming Agents	Seq. I at 24°C, Foaming, mL	Seq. II at 93.5°C, Foaming, mL	Air Release Time, min at 50°C
Neat SWI 6	175/0	25/0	1.2
Si 12500 (3 ppm)	0/0	0/0	2.8
FSi 300 (3 ppm)	35/0	0/0	1.6

Source: From Pillon, L.Z., Asselin, A.E., and Vernon, P.D.F., *Method for Reducing Foaming of Lubricating Oils*, US Patent 6,090,758, 2000.

active fluorosilicones requires higher treat rates, but was less effective in degrading the air separation properties of base stocks. The foaming tendency, foam stability, and air release time of mineral base stocks increase with an increase in their viscosity. While low viscosity FSi 300 was found to be more effective in preventing foaming tendency of higher viscosity 600N mineral base stocks due to their higher ST, their air release time was found to drastically increase from 8.9 to 23.2 min. An increase in surface activity of FSi 300 in higher viscosity mineral base stocks leads to an increase in air entrainment. The effect of Si 12500 and FSi 300 on the foaming tendency and the air release time of SWI 6 base stock is shown in Table 8.20.

At low treat rates of 3 ppm, Si 12500, while effective in preventing foaming, was found to increase the air release time of SWI 6 base stock from 1.2 to 2.8 min, measured at 50°C. At the same treat rate of 3 ppm, FSi 300, while less effective in preventing foaming, was found to increase the air release time of SWI 6 base stock from 1.2 to only 1.6 min. Fluorosilicone overtreatment, such as the use of 20 ppm of FSi 300, was found to increase the air release time of SWI 6 base stock to only 1.9 min. While fluorosilicones were found less effective in preventing foaming and higher treat rates were required, their use was found to have a significantly less effect on the air separation properties of lower viscosity 150N mineral and SWI 6 base stocks. The effect of the same viscosity Si 12500 on air release time of different base stocks is shown in Figure 8.2.

The same viscosity silicones, such as Si 12500, were found effective in preventing the foaming tendency of 150N mineral and SWI 6 base stocks at low treat rates of 1–3 ppm. The addition of 1–3 ppm of Si 12500 was found to increase the air release time of 150N NMP extracted base stock from 2.1 to 4.2–5.5 min. The addition of 1–3 ppm of Si 12500 was found to increase the air release time of SWI 6 base stock from 1.1 to only 1.7–2.8 min. The effect of PA (40 wt%) dispersion on foaming tendency and air release time in different base stocks is shown in Table 8.21.

At the high treat rate of 100 ppm of PA (40 wt%), 40 ppm of an active ingredient, the Seq. I foaming tendency of HF 150N phenol extracted base stock was found to decrease from 345 to 5 mL, and a small increase in the air release

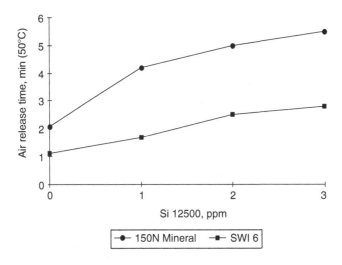

FIGURE 8.2 Effect of the same viscosity Si 12500 on air release time of different base stocks.

time from 1.6 to 2.2 min was observed. At a very high treat rate of 200 ppm of PA (40 wt%), 80 ppm of an active ingredient, the Seq. I foaming tendency of HF 150N NMP extracted base stock was found to be totally eliminated and some increase in air release time from 2.1 to 3.2 min was observed. The use of only 20 ppm of PA (40 wt%), 8 ppm of active ingredient, was found sufficient to prevent Seq. I foaming tendency of 600N NMP extracted base stock indicating an increase in surface activity due to an increase in the ST of higher viscosity mineral base stock. However, despite a significant decrease in the effective treat rate, a significant increase in air release time in higher viscosity mineral base stocks from

TABLE 8.21
Effect of PA (40%) on Foaming and Air Release Time of Different Base Stocks

Antifoaming Agents	PA (40 wt%), ppm (Active)	Seq. I at 24°C, Foaming, mL	Seq. II at 93.5°C, Foaming, mL	Air Release Time, min at 50°C
150N mineral oil	0	345/0	30/0	1.6
	100 (40)	5/0	30/0	2.2
150N mineral oil	0	280/0	20/0	2.1
	200 (80)	0/0	20/0	3.2
SWI 6 base stock	0	170/0	15/0	1
	150 (60)	135/0	15/0	1

Source: From Pillon, L.Z. and Asselin, A.E., *Antifoaming Agents for Lubricating Oils*, US Patent 5,766,513, 1998.

TABLE 8.22
Effect of Temperature on Spreading of Antifoaming Agents in 150N Mineral Oil

150N Mineral Base Stock	Antifoaming Agents, ST, mN/m at 24°C	Antifoaming Agents, S, mN/m at 24°C	Antifoaming Agents, S, mN/m at 93.5°C
Si 12500	24.0	8.2	5.5
FSi 300	26.2	1.7	3.9
PA (40 wt%)	27.2	5	−1.1

Source: From Pillon, L.Z. and Asselin, A.E., *Antifoaming Agents for Lubricating Oils*, US Patent 5,766,513, 1998.

9 to 16 min is observed. With an increase in the ST and an increase in adsorption of PA (40%) at the oil/air interface, an increase in air entrainment is observed. The use of PA (40 wt%) in SWI 6 base stock confirmed the need for higher treat rates. The effect of temperature on the efficiency of different antifoaming agents to spread (*S*) on the oil surface of 150N mineral base stock, having an ST of 33.7 mN/m, is shown in Table 8.22.

With an increase in the temperature from 24°C to 93.5°C, the efficiency of Si 12500 to spread on the oil surface of 150N mineral base stock was found to decrease from 8.2 to 5.5 mN/m. However, with an increase in temperature, some increase in the efficiency of FSi 300 to spread on the oil surface of 150N base stock from 1.7 to 3.9 mN/m is observed. The polyacrylate antifoaming agent, while effective in preventing foaming when used at high treat rates, was found not effective in spreading on the oil surface at 93.5°C. The air release time measured at 50°C indicated a significant increase when Si 12500 was used, less degradation when lower surface activity FSi 300 was used, and practically no increase when PA (40%) was used. The effect of temperature on the efficiency of different antifoaming agents to spread (*S*) on the oil surface of SWI 6 base stock, having an ST of 32.8 mN/m, is shown in Table 8.23.

TABLE 8.23
Effect of Temperature on Spreading of Antifoaming Agents in SWI 6 Base Stock

SWI 6 Base Stock	Antifoaming Agents, ST, mN/m at 24°C	Antifoaming Agents, S, mN/m at 24°C	Antifoaming Agents, S, mN/m at 93.5°C
Si 12500	24.0	7.0	1.9
FSi 300	26.2	−0.2	1.8
PA (40 wt%)	27.2	−5.3	−2.0

Source: From Pillon, L.Z., Asselin, A.E., and Vernon, P.D.F., *Method for Reducing Foaming of Lubricating Oils*, US Patent 6,090,758, 2000.

TABLE 8.24

Air Release Time and Storage Stability of Antifoaming Agents in SWI 6 Oil

SWI 6 Antifoaming Agents	Treat Rate, ppm (Active)	Air Release Time, min at 50°C	Storage Stability, Days at RT
Si 12500	3 (3)	2.8	>35
PA (40 wt%)	200 (80)	1.1	1–2

Source: From Pillon, L.Z., Asselin, A.E., and Vernon, P.D.F., *Method for Reducing Foaming of Lubricating Oils*, US Patent 6,090,758, 2000.

With an increase in the temperature from 24°C to 93.5°C, the efficiency of Si 12500 to spread on the oil surface of SWI 6 was found to drastically decrease from 7 to 1.9 mN/m. However, with an increase in temperature, some increase in the efficiency of FSi 300 to spread on the oil surface of SWI 6 base stock from −0.2 to 1.8 mN/m is observed. The polyacrylate antifoaming agent, while effective in preventing foaming when used at very high treat rates, was found not effective in spreading on the oil surface at 24°C and 93.5°C. The air release time measured at 50°C indicated an increase when Si 12500 was used and less degradation when lower surface activity FSi 300 was used and no increase when PA (40%) was used. The effect of Si 12500 and PA (40%) on air release time and the storage stability of SWI 6 base stock is shown in Table 8.24.

The use of 3 ppm of Si 12500 is effective in preventing the foaming of SWI 6 base stock; however, a significant increase in air release time from 1.1 to 2.8 min at 50°C is observed. SWI 6 base stock, containing 3 ppm of Si 12500, was found to have an excellent storage stability and no increase in the foaming tendency after 35 days was observed. SWI 6 base stock requires a high treat rate of 200 ppm of PA (40%), 80 ppm of active ingredient, to prevent foaming. Despite a high treat rate of polyacrylate antifoaming agent, no significant increase in the air release time is observed. After only 1–2 days of storage at room temperature, SWI 6 base stock, containing 200 ppm PA (40%), was found to foam. A high treat rate of poly-acrylate does not cause air entrainment but a poor solvency of SWI 6 base stocks can lead to poor storage stability. There are no additives to prevent air entrainment. Antifoaming agents and their treat rates need to be carefully selected to assure good air separation properties and a good storage stability of petroleum products.

8.4 USE OF ANTIFOAMING AGENT BLENDS

Many lubricating oils are complex formulations which require a resistance to foaming and good air separation properties. The literature reported that there are two modes of action by which antifoaming agents prevent foam formation. After entering the oil/air interface, an antifoaming agent will either spread over the surface or form mixed layers with stabilizing surface active molecules. Both

TABLE 8.25

Use of FSi 300/PA (40 wt%) Blend in HF 150N Phenol Extracted Base Stock

Antifoaming Agents	Treat Rate, ppm	Seq. I at 24°C, Foaming, mL	Seq. II at 93.5°C, Foaming, mL	Air Release Time, min at 50°C
Neat 150N oil	0	345/0	30/0	1.6
Si 12500	2	0/0	10/0	5.0
FSi 300	3	340/0	25/0	3.7
PA (40%)	100	5/0	30/0	2.2
Si 12500/FSi 300	2/3	10/0	5/0	5.3
FSi 300/PA (40%)	3/100	0/0	15/0	2.6

Source: From Pillon, L.Z. and Asselin, A.E., *Antifoaming Agents for Lubricating Oils*, US Patent 5,766,513, 1998.

approaches, entering and spreading on the oil surface, will enable the antifoaming agent to alter some of the stabilizing factors like film elasticity, surface viscosity, gas diffusion, or double layer repulsion which will lead to foam collapse (Rome, 2001). To prevent foaming, an effective antifoaming agent needs to be efficient in entering the oil/air interface and spreading on the oil surface. To prevent air entrainment, an effective antifoaming agent needs to be efficient in entering the oil/air interface without spreading on the oil surface. In some lubricant applications, there is a need for antifoaming agents which will be effective in preventing foaming at low and high temperatures without an increase in air entrainment. The literature reported that the polymethacrylate polymer is more effective in preventing foaming at lower temperatures, below 93.5°C (Si et al., 1990). The effect of Si 1200/FSi 300 and FSi 300/PA (40%) blends on foaming tendency and the air release time of 150N mineral base stock is shown in Table 8.25.

At the treat rate of 2 ppm, Si 12500 was found effective in preventing the foaming tendency of 150N base stock; however, an increase in the air release time from 1.6 to 5.0 min at 50°C was observed. At the treat rate of 3 ppm, FSi 300 was found not effective in preventing the foaming tendency of 150N phenol extracted base stock; however, an increase in the air release time from 1.6 to 3.7 min at 50°C was observed. While FSi 300 is not effective at low temperature, at 24°C, in preventing foaming due to its higher ST, it becomes more surface active at the higher temperature of 50°C and an increase in air entrainment is observed. At the treat rate of 100 ppm, PA (40%) was found effective in preventing Seq. I foaming of 150N phenol extracted base stock and the air release time was found to increase to only 2.2 min. Blending of silicone and fluorosilicone was found not effective in reducing the air entrainment of 150N mineral base stock. The use of an FSi 300/PA (40 wt%) blend was found effective in preventing the foaming tendency of 150N phenol extracted base stock but no improvement in the air release time was observed. The effect of Si 12500/PA (40%) and FSi 300/PA (40%) blends on foaming tendency and the air release time of another HF 150N mineral base stock is shown in Table 8.26.

TABLE 8.26
Use of Si 12500/PA (40%) and FSi 300/PA (40%) Blends in 150N Mineral Oil

Antifoaming Agents	Treat Rate, ppm	Seq. I at 24°C, Foaming, mL	Seq. II at 93.5°C, Foaming, mL	Air Release Time, min at 50°C
Neat 150N oil	0	280/0	20/0	2.1
Si 12500	2	0/0	10/0	5.0
FSi 300	3	300/0	0/0	4.3
PA (40%)	200	0/0	20/0	3.1
Si 12500/PA (40%)	2/200	0/0	10/0	4.4
FSi 300/PA (40%)	3/200	0/0	0/0	3.1

Source: From Pillon, L.Z. and Asselin, A.E., *Antifoaming Agents for Lubricating Oils*, US Patent 5,766,513, 1998.

At a low treat rate of 2 ppm, Si 12500 was found effective in preventing the Seq. I foaming tendency of 150N mineral base stock but the air release time was found to increase from 2.1 to 5 min at 50°C. At a treat rate of 3 ppm, FSi 300 was found not effective in preventing the Seq. I foaming tendency of 150N mineral base stock but the air release time was found to increase from 2.1 to 4.3 min at 50°C. The results confirmed that FSi 300 is more surface active at higher temperatures. At a high treat rate of 200 ppm, PA (40%) was found effective in preventing the Seq. I foaming tendency of 150N mineral base stock and the air release time was found to increase to only 3.1 min at 50°C. The use of Si 12500/PA (40%) and FSi 300/PA (40%) blends was found effective in preventing the foaming tendency of 150N mineral base stock but no improvement in the air release time was observed. The effect of Si 12500/PA (40%) and FSi 300/PA (40%) blends on foaming tendency and the air release time of SWI 6 base stock is shown in Table 8.27.

At the treat rate of 3 ppm, Si 12500 was found effective in preventing foaming of SWI 6 base stock but the air release time was found to increase from 1 to 2.8 min at 50°C. At the treat rate of 3 ppm, FSi 300 was found only effective in decreasing the Seq. I foaming tendency of SWI 6 base stock from 170 to 35 mL; however, no significant increase in air release time was observed. At the treat rate of 150 ppm of PA (40%), only some decrease in the Seq. I foaming tendency of SWI 6 base stock from 170 to 135 mL was observed with no increase in the air release time. At the higher treat rate of 200 ppm of PA (40%), poor storage stability is observed. The use of an Si 12500/PA (40 wt%) blend, containing 3 ppm of silicone and 150 ppm of PA (40%), while effective in preventing foaming, was found to increase the air release time of SWI 6 base stock to 2.5 min at 50°C. The use of an FSi 300/PA (40%) blend, containing 3 ppm of FSi 300 and 150 ppm of PA (40%), was found effective in preventing foaming without an increase in the air release time. The patent literature reported on the effect of some additives on the air release time of lubricating fluids, containing a blend of silicone and

TABLE 8.27
Use of Si 12500/PA (40%) and FSi 300/PA (40%) Blends in SWI 6
Base Stock

Antifoaming Agents	Treat Rate, ppm	Seq. I at 24°C, Foaming, mL	Seq. II at 93.5°C, Foaming, mL	Air Release Time, min at 50°C
Neat SWI 6 oil	0	170/0	25/0	1.0
Si 12500	3	0/0	0/0	2.8
FSi 300	3	35/0	0/0	1.1
PA (40%)	150	135/0	15/0	1.0
Si 12500/PA (40%)	3/150	20/0	5/0	2.5
FSi 300/PA (40%)	3/150	0/0	0/0	1.0

Source: From Pillon, L.Z. and Asselin, A.E., *Antifoaming Agents for Lubricating Oils*, US Patent 5,766,513, 1998.

TABLE 8.28
Use of Silicone/Fluorosilicone Blend and the Air Release Time
of Lubricating Fluids

Composition, wt%	Lubricating Fluid # 1	Lubricating Fluid # 2
Base stock	PAO blend	PAO blend
Succinimide dispersant	2.4	2.4
Borated epoxide	0.2	0.2
S-containing antioxidant	0.5	0.5
Di(para-nonyl phenyl) amine	0.65	0.65
Dialkyl hydrogen phosphite	0.2	0.2
Alkyl naphthalene	0.2	0.2
Diluent oil	0.34	0.34
Sulfolane seal swell	0.6	0.6
Alkylthiodimercaptothiadiazole	0.5	—
Dimercaptothiadiazole	—	0.5
Silicone/fluorosilicone blend	40 ppm	40 ppm
Red dye	250 ppm	250 ppm
Kin. viscosity at 100°C	4.63	4.71
ASTM D 3427, 50°C	—	—
Air release time, min	3.5	4.9

Source: From Ward, W.C. Jr., Tipton, C.D., and Murray, K.A., *Lubrication Fluids for Reduced Air Entrainment and Improved Gear protection*, CA Patent 2,184,969, 1996.

fluorosilicone antifoaming agents (Ward et al., 1996). The use of a silicone/fluorosilicone blend and the air release time of lubricating fluids are shown in Table 8.28.

Pure synthetic PAO base stocks have resistance to foaming and do not require the use of antifoaming agents. Many additives used in lubricating oils are surface active and designed to protect the metal surface and prevent sediment formation. Despite the fact that there are hundreds of new additives being developed and designed to adsorb at the specific interface, they are usually surface active toward any surface and competing to adsorb at the interface. Since there are no antifoaming agents effective in preventing foaming at the low and high temperatures without an increase in air entrainment, the use of their blends might be of benefit. The use of the most surface active silicones might be of benefit at low temperatures due to their lower ST and the presence of fluorosilicones might be of benefit at high temperatures due to their higher ST. The presence of fluorosilicones, having a significantly lower tendency to spread on the oil surface, will lead to less air entrainment and a shorter air release time of lubricating oils. Foam inhibition at low and high temperatures, without air entrainment, is a complex issue and all additives and their treat rates need to be carefully selected. Many different additives used in some lubricant products might be effective in improving a specific performance area while degrading other properties.

REFERENCES

Aldrich Handbook of Fine Chemicals, Advancing Science, Sigma-Aldrich, 2007–2008.
Ali, M.A., *Petroleum Science and Technology*, 19(9&10), 1063, 2001.
Bardasz, E.A. and Lamb, G.D., Additives for crankcase lubricant applications, in *Lubricant Additives Chemistry and Applications*, Rudnick, L.R., Ed., Marcel Dekker, Inc., New York, 2003.
Basu, B. et al., *ASLE Transactions*, 28(3), 313, 1985.
Bondi, A., *Physical Chemistry of Lubricating Oils*, Reinhold Publishing Corporation, New York, 1951.
Crawford, J., Psaila, A., and Orszulik, S.T., Miscellaneous additives and vegetable oils, in *Chemistry and Technology of Lubricants*, 2nd ed., Mortier, R.M. and Orszulik, S.T., Eds., Blackie Academic & Professional, London, 1997.
CRC Handbook of Chemistry and Physics, 86th ed., Lide, D.R., Editor-in-Chief, CRC Press, Boca Raton, 2005.
Gary, J.H. and Handwerk, G.E., *Petroleum Refining: Technology and Economics*, 4th ed., Marcel Dekker, Inc., New York, 2001.
Ivanov, A.V., Lazarev, N.P., and Yaushev, R.G., *Chemistry and Technology of Fuels and Oils*, 36(5), 352, 2000.
Klingsberg, E., *Heterocyclic Compounds—Pyridine and Derivatives*, Interscience Publishers Inc., New York, 1969.
Men, G. et al., *Petroleum Science and Technology*, 19(9&10), 1251, 2001.
Naidu, S.K., Klaus, E.E., and Duda, J.L., *Industrial and Engineering Chemistry Product Research and Development*, 23, 613, 1984.
Pillon, L.Z., *Petroleum Science and Technology*, 19(9&10), 1263, 2001.

Pillon, L.Z. and Asselin, A.E., *Antifoaming Agents for Lubricating Oils*, US Patent 5,766,513, 1998.

Pillon, L.Z., Reid, L.E., and Asselin, A.E., *Lubricating Oil for Inhibiting Rust Formation*, US Patent 5,397,487, 1995.

Pillon, L.Z., Asselin, A.E., and Vernon, P.D.F., *Method for Reducing Foaming of Lubricating Oils*, US Patent 6,090,758, 2000.

Pirro, D.M. and Wessol, A.A., *Lubrication Fundamentals*, 2nd ed., Marcel Dekker, Inc., New York, 2001.

Prince, R.J., Base oils from petroleum, in *Chemistry and Technology of Lubricants*, 2nd ed., Mortier, R.M. and Orszulik, S.T., Eds., Blackie Academic & Professional, London, 1997.

Rome, F.C., *Hydrocarbon Asia*, 42, May/June 2001.

Ross, S., Foaminess and capilarity in apolar solutions, in *Interfacial Phenomena in Apolar Media*, Eicke, H.F. and Parfitt, G.D., Eds., Marcel Dekker, Inc., New York, 1987.

Sequeira, A., *Pre-Prints Division of Petroleum Chemistry*, ACS, 37(4), 1286, 1992.

Sequeira, A. Jr., *Lubricant Base Oil and Wax Processing*, Marcel Dekker, Inc., New York, 1994.

Si, M.H. et al., *Chieh Mien Ko Hsueh Hui Chih*, 13(2), 23, 1990.

Speight, J.G., *The Chemistry and Technology of Petroleum*, 3rd ed., Marcel Dekker, Inc., New York, 1999.

Ward, W.C. Jr., Tipton, C.D., and Murray, K.A., *Lubrication Fluids for Reduced Air Entrainment and Improved Gear Protection*, CA Patent 2,184,969, 1996.

Zaki, N.N., Poindexter, M.K., and Kilpatrick, P.K., *Energy and Fuels*, 16, 711, 2002.

9 Water Contamination

9.1 SPREADING COEFFICIENT OF CLAY TREATED JET FUEL ON WATER

Crude oils are often contaminated with water during their recovery. The literature reported on an asphaltenic crude oil, having an API gravity of 41 and a kinematic viscosity of 223 cSt at 60°F, containing 7.18 wt% of asphaltenes, 3.1 wt% of wax, and 0.5 wt% of water (Al-Sabagh et al., 2002). The formation of water-in-oil emulsions, their stabilization by solids, rheology, and conductivity was reviewed (Becher, 1987). Different petroleum products are also contaminated with water during their processing, storage, or their use. The early literature reported that fuels can become contaminated with entrained aqueous droplets by the condensation of atmospheric moisture in fuel storage tanks (Hughes, 1969). The literature also reported that, without any agitation, the suspended droplets of water in oil can coalesce. According to Stoke's equation, for the chosen oil, the density and the viscosity are constant, so the rate of coalescence will depend on the size of the water droplet. For the same viscosity oils, a decrease in oil/water interfacial tension (IFT) will decrease the water droplet size and their rate of coalescence (Hughes, 1969).

The ASTM D 971 test method for IFT of oil against water by the ring method covers the measurement of the IFT between mineral oil and water at room temperature (RT). The literature reported on the use of a processor tensiometer K12 to study the effect of asphaltenes, resins, and paraffin waxes on crude oil emulsions (Zaki et al., 2000). The standard processor tensiometer K12 allows a constant temperature to be maintained in the sample vessel from $-10°C$ to $+100°C$. It can also repeat the measurements until the equilibrium at the oil/water interface is established. The effect of molecular weight and chemical structure on the surface tension (ST) and the liquid/water tension (IFT) of some liquid molecules, measured at 20°C, is shown in Table 9.1.

The saturated hydrocarbons, such as n-octane, dodecane, and hexadecane, have a liquid/vapor tension of 21.7–27.5 mN/m and a high liquid/water tension of 51.7–53.8 mN/m. The liquid butanol and octanol have a liquid/vapor tension of 24.6–27.5 mN/m and a significantly lower liquid/water tension of only 1.6–8.5 mN/m. The liquid diethyl ether was reported to have a liquid/vapor tension of 17 mN/m and a liquid/water tension of 10.7 mN/m. An aromatic amine,

TABLE 9.1

Surface Tension and Liquid/Water Tension of Some Organic Liquids

Liquid Compounds	Liquid/Vapor Tension, ST, mN/m at 20°C	Liquid/Water Tension, IFT, mN/m at 20°C
Mercury	484	426
Water	72.75	—
Octane	21.69	51.68
Dodecane	25.44	52.90
Hexadecane	27.46	53.77
Benzene	28.88	35.0
Tetrachloromethane	26.77	45.0
Octanol	27.53	8.5
Butanol	24.6	1.6
Aminobenzene	42.9	5.9
Diethylether	17.0	10.7
Ethyl ethanoate	23.9	3

Source: From Jaycock, M.J. and Parfitt, G.D., *Chemistry of Interfaces*, Ellis Horwood Ltd., Chichester, 1981. Copyright John Wiley & Sons Limited. Reproduced with permission.

aminobenzene, was reported to have a significantly higher liquid/vapor tension of 42.9 mN/m and a drastically lower liquid/water tension of only 5.9 mN/m.

The early literature reported on the effect of chemical structure on the efficiency of different molecules to spread (S) on the water surface. Saturated hydrocarbons were reported to have a low or negative spreading coefficient while aromatic benzene and its derivatives were reported to have a positive spreading coefficient. Aliphatic acids, amines, and alcohols were reported to have a high spreading coefficient. Purified mineral oil, known as Nujol, was reported to have a negative spreading coefficient (S) of −11.2 mN/m (Bondi, 1951). The effect of chemical structure on the spreading coefficient of different molecules on the water surface is shown in Table 9.2.

The water contamination of jet fuels was reported to be a frequent problem and jet fuels were required to have good water separation properties. Attapulgite clay has been known for years to be effective in improving the thermal–oxidation stability and the water separation properties of jet fuels. The literature reported that clay treating of jet fuel is used to remove surface active molecules which adversely affect their water separator index specifications (Gary and Handwerk, 2001). Attapulgite clay is activated by heat and the degree of activity is determined by the amount of water of hydration retained after the thermal treatment (Sequeira, 1994).

The two main components of Attapulgite clay are silica oxide and aluminum oxide and the salt drier is used before the clay treatment of the Merox treated–caustic washed jet fuels to lower their water content. The basic nitrogen (BN) type

TABLE 9.2

Spreading Coefficient of Different Molecules on the Water Surface

Chemical Structure	Spreading Coefficient (S), mN/m
Hexane	3.2
Octane	0.6
n-Tetradecane	−7.6
Benzene	9.4
Toluene	8.0
m-Xylene	6.0
Carbon tetrachloride	1.8
Ethylene bromide	3.6
Ethyl iodide	4.6
Diamyl amine	35
Octyl alcohol	37
Propyl alcohol	48.6
Oleic acid	25
Petroleum mineral oil	−11.2

Source: From Bondi, A., *Physical Chemistry of Lubricating Oils*, Reinhold Publishing Corporation, New York, 1951.

molecules can be titrated with perchloric acid in a solution of glacial acetic acid. According to the ASTM D 2896 test method, basic constituents of petroleum products include not only lower molecular weight BN compounds but also other organic and inorganic bases, amino compounds, salts of weak acids (soaps), basic salts of polyacidic bases, and salts of heavy metals. Before the clay treatment, the Merox treated–caustic washed jet fuel, containing 31 ppm of water, was found to have poor thermal–oxidation stability. The stability of jet fuels is tested following the ASTM D 3241 procedure for thermal–oxidation stability of aviation turbine fuels (JFTOT), usually at 275°C. The effect of fresh Attapulgite clay treatment on composition, oil/water IFT, and spreading coefficient (S) of jet fuel on the water surface is shown in Table 9.3.

After the fresh Attapulgite clay treatment, under the column adsorption conditions and using 1 L of jet fuel per 15 g of clay at RT, no change in S content of the jet fuel was observed. However, the N content decreased from 15 to 1 ppm, and the BN content decreased from 12 to 3 ppm, and jet fuel passed the JFTOT test. The IFT of clay treated jet fuel significantly increased from 32.9 to 42.6 mN/m measured at 25°C, indicating a decrease in the surface active content capable of adsorbing at the jet fuel/water interface. The spreading coefficient (S) of the clay treated jet fuel decreased from about 10 mN/m to 0, indicating a decrease in the content of molecules having a high tendency to spread on the water surface.

TABLE 9.3
Effect of Fresh Clay on Spreading Coefficient of Jet Fuel on Water Surface

Fresh Attapulgite Clay Column (1 L jet fuel/15 g clay)	Jet Fuel before Clay Treatment	Jet Fuel after Clay Treatment
Density at 15°C, g/cc	0.8276	—
Sulfur, wt%	0.23	0.23
Nitrogen, ppm	15	1
Cu, ppb	<10	—
Basic nitrogen (BN), ppm	12	3
Water, ppm	31	—
ASTM D 3241 (JFTOT)	Fail	Pass
IFT, mN/m at 25°C	32.9	42.6
Spreading (S), mN/m	10	0

Source: From Pillon, L.Z., *Petroleum Science and Technology*, 19, 1109, 2001. Reproduced by permission of Taylor & Francis Group, LLC, http://www.taylorandfrancis.com.

The Attapulgite clay has been known for years to be effective in adsorbing polar inorganic and organic contaminants. While the metal contamination of jet fuel can occur, the water contamination was reported to be a more frequent problem. The untreated "dry" jet fuel, containing 31 ppm of water, was contaminated with additional water to increase its water level to 94 ppm. The "wet" jet fuel, containing 94 ppm of water, was passed through column adsorption containing 15 g of fresh Attapulgite clay. The Attapulgite clay was reported to contain 1 wt% of water. The quality of naturally occurring clays, such as fuller's earth, is often difficult to define with any degree of precision from batch to batch. The

TABLE 9.4
Effect of Water Contamination on the BN Content of the Effluent

Effluent Volume, L	Dry Jet Fuel Sulfur, wt%	Dry Jet Fuel BN, ppm	Wet Jet Fuel BN, ppm	Wet Jet Fuel Water, ppm
1	0.23	3	2	35
2	—	9	14	58
3	0.23	10	12	64
4	—	11	13	87
5	0.23	12	12	89

Source: From Pillon, L.Z., *Petroleum Science and Technology*, 19, 1109, 2001. Reproduced by permission of Taylor & Francis Group, LLC, http://www.taylorandfrancis.com.

effect of water content of untreated jet fuel on the Attapulgite clay capacity to retain BN molecules is shown in Table 9.4.

The dry jet fuel, containing 31 ppm of water, and the wet jet fuel, containing 94 ppm of water, were passed through the adsorption column containing 15 g of fresh Attapulgite clay. With an increase in the volume of effluent dry jet fuel from 1 to 5 L, no change in S content was observed. However, the polar BN content of the effluent was found to gradually increase from 3 to 12 ppm, thus reaching the level present in untreated jet fuel. With an increase in the volume of wet jet fuel from 1 to 5 L, a gradual increase in the BN content from 2 to 12 ppm was also observed. With an increase in the volume of wet jet fuel, the water content also gradually increased from 35 to 89 ppm, indicating a decrease in clay capacity to retain anymore water.

After passing 5 L of the wet jet fuel, the water content of effluent jet fuel increased to the level indicating that clay was practically saturated and lost its capacity to retain BN molecules. At this point, the wet spent Attapulgite clay was analyzed. The wet spent clay was carefully mixed to assure its uniformity and washed with n-heptane to remove any retained hydrocarbon molecules. The GC/MS analysis of the wet spent Attapulgite clay was carried out in a He purge gas in three steps. In the first step, the clay sample was heated from 50°C to 200°C and analyzed for any desorption of organic molecules. In the second step, the clay sample was heated from 50°C to 400°C and, in the third step, from 50°C to 600°C. The GC/MS analysis of fresh and wet spent Attapulgite clays is shown in Table 9.5.

In the temperature range of 50°C–600°C, the GC/MS analysis of fresh clay shows only desorption of water. At high temperatures above 200°C, desorption of water indicated that the water molecules were held inside the crystalline structure of the fresh Attapulgite clay. In the temperature range of 50°C–200°C, GC–MS analysis of the wet spent Attapulgite clay indicated only desorption of water and C_9–C_{15} jet fuel hydrocarbons. At a higher temperature of 200°C–400°C, the

TABLE 9.5
GC/MS Analysis of Fresh and Wet Spent Attapulgite Clay

GC/MS Analysis, Attapulgite Clay	Temperature 50°C–200°C	Temperature 200°C–400°C	Temperature 400°C–600°C
Fresh clay	Water	Water	Water
Wet spent clay	Water	Water	Water
	C_9–C_{15} Hydrocarbons	C_{11}–C_{15} Hydrocarbons	C_{11}–C_{13} Hydrocarbons
		Quinolines	Pyridines
			Quinolines
			Other N compounds

Source: From Pillon, L.Z., *Petroleum Science and Technology*, 19, 1109, 2001. Reproduced by permission of Taylor & Francis Group, LLC, http://www.taylorandfrancis.com.

analysis indicated desorption of water, C_{11}–C_{15} hydrocarbons, and different chemistry quinolines. At the highest temperature of 400°C–600°C, the analysis indicated desorption of water, C_{11}–C_{13} hydrocarbons, different chemistry quinolines, only one peak indicating the presence of pyridines, and some other unidentified N molecules.

Different molecules, found in jet fuel, adsorb on Attapulgite clay at low temperatures and desorb at high temperatures. The use of the GC–MS technique can only detect the presence of molecules having a high volatility. The analysis of the organic content of wet spent clay indicated the presence of some sulfur so the absence of sulfur compounds during GC/MS analysis would indicate their low volatility. Some commercial Merox treated–caustic washed jet fuels, containing increased water content, were also found to have a slight yellow color and form haze. The effect of clay treatment on IFT, spreading coefficient (S) of jet fuel on the water surface, and the haze formation is shown in Table 9.6.

Untreated jet fuel, containing 10 ppm of BN, no Cu contamination and 75 ppm of water, was found to have a slight yellow color, form haze at 4°C and have poor thermal–oxidation stability. Haze number, based on the colonial pipeline haze test, puts a value on the "haziness" of the sample caused by water suspension. A sample, in round clear bottle, is placed in water bath at 4°C and allowed to reach equilibrium with the bath temperature. The haze number is reported as the nearest comparison between the sample and the chart. After the

TABLE 9.6
Effect of Clay Treatment on Spreading Coefficient of Jet Fuel on Water and Haze

Merox Treated–Caustic Washed	Before Clay Treatment	After Clay Treatment	After Fresh Clay Treatment
Density at 15°C, g/cc	0.8195	0.8195	—
Saturates, wt%	77.1	77.2	—
Aromatics, wt%	22.9	22.8	—
Sulfur, wt%	0.17	0.16	—
Nitrogen, ppm	11	10	—
BN, ppm	10	9	0
Cu, ppb	<10	<10	—
Water, ppm	75	74	41
IFT, mN/m at 25°C	36.8	39.4	46.2
Spreading (S), mN/m	6	3	−3.4
Color	Slight yellow	Slight yellow	Clear
Haze at 4°C	Yes	Yes	No

Source: From Pillon, L.Z., *Petroleum Science and Technology*, 19, 1109, 2001. Reproduced by permission of Taylor & Francis Group, LLC, http://www.taylorandfrancis.com.

commercial clay treatment, the jet fuel continued to have a high water content, slight yellow color, failed JFTOT, and formed haze. While the IFT of clay treated jet fuel increased from 36.8 to 39.4 mN/m, no improvement in performance was observed indicating a need for clay replacement. After the fresh clay treatment, the BN content was totally eliminated and a clear color was observed. At 4°C, no haze was formed and the IFT significantly increased to 46.2 mN/m.

Many different molecules can be surface active and adsorb at the oil/water interface; however, they must also have the emulsifying properties to form haze and degrade the water separation properties of petroleum oils. The use of additives having surfactant properties, or their formation during processing of crude oils, can lead to an increase in the spreading coefficient of jet fuel on the water surface. These surfactant molecules might be present in many products and not affect their performance if no water present. While the waste clay disposal is expensive and an environmental problem, the use of inefficient clay treatment will lead to haze and poor water separation properties when jet fuel is contaminated with water.

9.2 OIL/WATER INTERFACIAL TENSION OF WHITE OILS

In many other applications, petroleum products might become contaminated with water and are required to have good water separation properties. Mineral lube oil distillates contain paraffins, naphthenes, aromatics, and heteroatom containing molecules. Olefins are normally not present but they can be present as a result of cracking. The literature reported that the vacuum distillates and deasphalted oils contain many undesirable molecules which cause darkening, oxidation, and sludging of lubricant products. The vacuum distillate from Dagang oil was reported to have a negative VI of -5 and a high total acid number (TAN) of 1.7 mg KOH/g (Wang et al., 2001). The TAN value of a petroleum product is the weight of potassium hydroxide (KOH) required to neutralize 1 g of oil. The usual components of TAN are not only naphthenic acids but also organic soaps, soaps of heavy metals, organic nitrates, other compounds used as additives, and oxidation by-products. The literature reported on Bohai crude oil, from China, having a high TAN of 3.34 and causing severe corrosion (Qi et al., 2004). The literature reported that the TAN of lube oil distillates can increase with an increase in the deep exploitation of oil fields and in viscosity of lube oil distillate (Wang et al., 2001). The properties and TAN of Bohai crude oil are shown in Table 9.7.

The acids present in Bohai crude oil distillates, boiling in the range of 200°C–370°C and 370°C–550°C, were isolated by aqueous alcoholic sodium hydroxide and converted into methyl esters with BF_3 as catalyst. The analysis indicated the presence of 1–3 ring naphthenic acids and carboxylic acids in lower boiling diesel distillate while higher boiling lubricant distillate was found to contain 1–6 ring naphthenic acids and carboxylic acids (Qi et al., 2004). The effect of boiling point on the acid content of diesel and lube oil distillates from Bohai crude oil is shown in Table 9.8.

The presence of acidic components in petroleum feedstocks is of concern because it leads to corrosion and strong acids can react with alumina leading to

TABLE 9.7

Properties and TAN of Bohai Crude Oil

Bohai Crude Oil	Properties
Density at 20°C, g/cm^3	0.95
Viscosity at 50°C, mm^2/s	478.3
Solidifying point, °C	−6
Sulfur, wt%	0.36
Nitrogen, wt%	0.26
TAN, mg KOH/g	3.34

Source: From Qi et al., *Petroleum Science and Technology*, 22, 463, 2004. Reproduced by permission of Taylor & Francis Group, LLC, http://www.taylorandfrancis.com.

its hydrolysis. Natural clays contain alumina and hydroprocessing catalysts often use alumina support. Clays are natural compounds of silica and alumina, also containing oxides of sodium, potassium, magnesium, calcium, and other alkaline earth metals. The literature described the Attapulgite clay as magnesium–aluminum silicate while porocel is described as hydrated aluminum oxide, known as bauxite (Sequeira, 1994).

Attapulgite clay is effective in decolorizing and neutralizing any petroleum oil. It is less effective in removing odorous compounds and metals and does not adsorb aromatics (Sequeira, 1994). Porocel clay, composed primarily of hydrated aluminum oxide (bauxite) and activated by heat, was reported effective in decolorization, reducing organic acidity, effective deodorizer, and also adsorbing aromatic type molecules and polar compounds containing sulfur and nitrogen

TABLE 9.8

Acid Content of Diesel and Lube Oil Distillates from Bohai Crude Oil

Petroleum Fractions Boiling Range, °C	Diesel Distillate 200–370	Lube Oil Distillate 370–550
TAN, mg KOH/g	1.1	2.7
Acids, wt%	1.47	3.53
Naphthenic acids	1–3 ring	1–6 ring
Carboxylic acids	$C_{16}H_{27}O_2SN$	$C_{34}H_{51}O_2SN$
Molecular weight	253	489

Source: From Qi et al., *Petroleum Science and Technology*, 22, 463, 2004. Reproduced by permission of Taylor & Francis Group, LLC, http://www.taylorandfrancis.com.

TABLE 9.9

Main Components of Attapulgite and Porocel Clays

Composition	Attapulgite Clay	Porocel Clay
Main content	SiO_2	Al_2O_3
Minor content	Al_2O_3	SiO_2
	Fe_2O_3	Titania
	CaO	Kaolinite
	MgO	Hematite
	K_2O	

Source: From Sequeira, A. Jr., *Lubricant Base Oil and Wax Processing*, Marcel Dekker, Inc., New York, 1994; USGS, *Clays Statistics and Information*, 2007.

(Sequeira, 1994). Bauxite is a naturally occurring and the bulk of world bauxite production is used as a feed for the manufacture of alumina via a wet chemical caustic leach method (USGS, 2007). The composition of different clays is shown in Table 9.9.

The early chemical refining process was based on sulfuric acid and clay treatment. The action of sulfuric acid on lube oil distillates was reported to be complex and varied depending on the acid concentration, reaction temperature, and the residence time. Acid–alkali refining, called wet refining, is a chemical process in which lube oils are contacted with sulfuric acid followed by neutralization with aqueous or alcoholic alkali (Sequeira, 1994). Paraffins and naphthenes are relatively nonreactive but aromatics and olefins are readily attacked by sulfuric acid. The S compounds are mainly thiophenes and converted to thiophene-sulfonic acid and removed as acid sludge. Nitrogen compounds consist of amines, amides, and minor amounts of aminoacids which are converted to salts and removed with acid sludge. The oxygen compounds are primary acids with minor amounts of aldehydes and alcohols which are oxidized to acids and removed during neutralization (Sequeira, 1994). The early literature reported on the different effects of alkali and sulfuric acid–clay treatment on IFT of lube oil distillates (Bondi, 1951).The effect of alkali and acid–clay treatments on IFT of lube oil distillate is shown in Table 9.10.

After the NaOH treatment, in the presence of distilled water, the IFT decreased from 22.9 to 19.9 mN/m. In the presence of aqueous KCl, no significant decrease in the IFT is observed (Bondi, 1951). Caustic washing is mostly used when refining gasoline and diesel fuel. Caustic washing can remove most of the naphthenic acids of lube oil distillates but it may also cause stable emulsions which are difficult to break (Wang et al., 2001). After the acid–clay treatments in the presence of distilled water, the IFT significantly increased from 22.9 to 30.7–34.7 mN/m. In the presence of aqueous KCl, a significant increase in IFT from 23.9 to 31.6–39.4 mN/m is also observed. Resins and naphthenic acids are

TABLE 9.10
Effect of Alkali and Acid–Clay Treatments on IFT of Lube Oil Distillate

Lube Oil Distillate Treatments	Distilled Water IFT, mN/m	Aqueous KCl IFT, mN/m
None	22.9	23.9
NaOH treated	19.9	23.2
3% H_2SO_4/3% clay	30.7	31.6
5% H_2SO_4/15% clay	34.7	39.4

Source: From Bondi, A., *Physical Chemistry of Lubricating Oils*, Reinhold Publishing Corporation, New York, 1951.

present in crude oils. The effect of resins and naphthenic acid on IFT of white oil is shown in Table 9.11.

The addition of resins and naphthenic acid leads to a significant decrease in the IFT of white oil indicating that they adsorb at the oil/water interface. Water and brine are present in crude oils. The use of an aqueous KCl solution leads to a further decrease in the IFT indicating that additional chemical reactions are taking place leading to an increase in adsorption of surface active molecules at the oil/aqueous KCl interface. The sodium and potassium salts of fatty acids are soluble in water. The ones with long alkyl chains are also soluble in oil which gives them their cleaning power as soaps. Commercial soaps are mixtures of sodium salts of long chain fatty acids having from C_{12} to C_{18} carbon atoms.

The demulsibility properties of lube oil base stocks are usually tested following the ASTM D 1401 method for water separability of petroleum oils and synthetic fluids. The ASTM D 1401 procedure requires the mixing of 40 mL of oil with 40 mL of distilled water. Both liquids are stirred for 5 min at 54°C or 82°C, depending on their viscosity. The time required for the separation of the

TABLE 9.11
Effect of Resins and Naphthenic Acid on IFT of White Oil

White Oil/Water Interface at 20°C	Oil/Water IFT, mN/m	Oil/Aqueous KCl IFT, mN/m
Neat white oil	52.0	52.2
White oil (1.4% resins)	19.8	7.8
White oil (1.4% naphthenic acid)	24.5	11.8
White oil (0.5% resin/1% naphthenic acid)	22.9	9.9

Source: From Bondi, A., *Physical Chemistry of Lubricating Oils*, Reinhold Publishing Corporation, New York, 1951.

oil/water/emulsion and the volumes of remaining oil/water/emulsion (O/W/E) are recorded. The surface activity at the oil/water interface and the demulsibility properties of hydrorefined white oil and other highly saturated base stocks, having a similar viscosity, are shown in Table 9.12.

Despite similar saturate contents, above 99 wt% and no presence of sulfur and nitrogen content, the IFT values of the saturated base stocks were different. The use of hydrorefining produces white oils, having a high saturate content, above 99 wt%, and practically no sulfur and nitrogen. In the case of the white oil, a significant decrease in the IFT with an increase in the equilibrium time was observed. After 1 min on standing, 17 mL of emulsion was present. Hydrorefined white oil was found to require 3 min to separate the emulsion. The use of slack wax hydroisomerization technology results in lube oil base stocks, having a high saturate content, above 99.5 wt%, and practically no sulfur and nitrogen. After 1 min on standing, only 9 mL of emulsion was present. SWI 6 base stock was found to require only 2 min to separate the emulsion. The synthetic PAO 6 base stock, having 100% saturate content and no S and N, was found to require only 1 min to separate the emulsion and to have a negative S of -1.5 mN/m thus confirming the effect of different chemistry hydrocarbons on their surface activity at the oil/water interface.

TABLE 9.12
Surface Activity and Demulsibility Properties of Hydrorefined White Oil

Base Stock Properties	Hydrorefined White Oil	Hydroisomerized SWI 6	Synthetic PAO 6
Kinematic Viscosity			
At 40°C, cSt	32.71	29.38	30.43
At 100°C, cSt	5.55	5.77	5.77
VI	106	143	134
Aromatics, wt%	0.9	<0.5	0
Sulfur, wt%	<1	<1	0
Nitrogen, ppm	<1	<1	0
Equilibrium Time at 24°C			
IFT, mN/m (30 min)	39.0	45.6	41.9
Spreading (*S*), mN/m	1.3	−9.1	−1.5
ASTM D 1401 (54°C)			
O/W/E, mL	24/39/17	32/39/9	40/40/0
Time, min	1	1	1
O/W/E, mL	40/40/0	40/40/0	—
Time, min	3	2	—

Source: From Pillon, L.Z., Reid, L.E., and Asselin, A.E., *Lubricating Oil for Inhibiting Rust Formation*, US Patent 5,397,487, 1995; Pillon, L.Z. and Asselin, A.E., *Method for Improving the Demulsibility of Base Oils*, US Patent 5,282,960, 1994.

9.3 SURFACE ACTIVITY OF DEMULSIFIERS

Lube oil base stocks, used in lubricating oils, are required to have good demulsibility properties by which is meant the ability to separate the water. Lubricating oils, such as turbine oils and engine oils, are contaminated with water during their use. The dispersion of seawater in marine lubricating systems and the condensation of water in the crankcases of internal combustion engines operating intermittently at temperatures below normal running can lead to water contamination (Hughes, 1969). In engine oils, the water is formed as a by-product of the combustion and the oil must resist the emulsion formation (Prince, 1997). Many lubricating oils require good water separation properties and need to use lube oil base stocks having good demulsibility properties. After solvent extraction and dewaxing, finishing of mineral base stocks includes adsorbent clay treatment which removes some undesirable molecules, such as residual solvents, or catalytic hydrofinishing which can saturate some aromatics.

Some refiners use clay treatment to finish lube oil base stocks. Porocel clay has been known for years to be a good refining agent for some lubricating oils, such as turbine and transformer oils (Sequeira, 1994). Clay refining can be used to remove acids from lube oil base stocks if the TAN is below 0.1 mg KOH/g. At a higher TAN, a high volume of clay is required and the waste clay disposal problem makes the use of clay not practical (Wang et al., 2001). Spent clay disposal problems and other operating restrictions have caused the clay treating processes to be replaced by hydrofinishing. The majority of refineries use solvent refining and hydrofining, as the last step, to improve the color, stability and remove impurities introduced during the various processing steps. According to the literature, the most hydrofining operations are operated at a severity set by the color improvement needed (Sequeira, 1994). The effect of hydrogen finishing on the properties of mineral base stocks is shown in Table 9.13.

TABLE 9.13
Effect of Hydrogen Finishing on Base Stock Properties and Stability

Hydrogen Finishing, Effect on Properties	Hydrogen Finishing, Effect on Stability
Decreases specific gravity	Increases pour point
Increases wax content	Improves color
May decrease aromatics	Improves color stability
Decreases sulfur content	Improves thermal stability
Decreases nitrogen content	Decreases carbon content
Decreases resin content	Improves demulsibility

Source: From Sequeira, A. Jr., *Lubricant Base Oil and Wax Processing*, Marcel Dekker, Inc., New York, 1994. Reproduced by permission of Routledge/Taylor & Francis Group, LLC.

The early literature reported that hydrofining was effective in improving VI, color, neutralization number, and the demulsibility properties of mineral base stocks (Folkins et al., 1960). The early literature also reported that hydrofining was more effective than clay treatment in improving the demulsibility properties and the oxidation stability of base stocks (Golbshtein et al., 1966). According to more recent literature, an increase in the hydrogen pressure or temperature will usually improve neutralization, desulfurization, denitrification, color, and stability. An increase in the temperature above a certain maximum might degrade the color, oxidation stability, and other properties (Sequeira, 1994).

Naphthenic base stocks, having a kinematic viscosity of only 8–9 cSt at 40°C and containing aromatic, sulfur, and nitrogen components, separated the oil/water/emulsions in less than 5 min at 54°C indicating good demulsibility properties. However, with an increase in the viscosity of hydrofined paraffinic base stocks, an increase in the emulsion separation time with a significant variation in the water separation properties is observed. The IFT of mineral base stocks is usually measured after 30 min of the equilibrium time. The effect of viscosity on surface activity at the oil/water interface and the demulsibility properties of hydrofined phenol extracted base stocks is shown in Table 9.14.

After a typical equilibrium time of 30 min at 24°C, hydrofined 150N phenol extracted base stock was found to have an IFT of 43.1 mN/m and a negative

TABLE 9.14
Effect of Viscosity on Demulsibility Properties of HF Phenol Extracted Base Stocks

Mineral Base Stocks (Extraction Solvent)	HF 150N (Phenol)	HF 600N (Phenol)	HF 1400N (Phenol)
Kinematic Viscosity			
At 40°C, cSt	29.48	105.91	301.71
At 100°C, cSt	5.03	11.32	22.01
VI	94	92	89
Aromatics, wt%	16.1	19.5	28.2
Sulfur, wt%	0.09	0.12	0.19
Nitrogen, ppm	8	30	141
BN, ppm	4	16	51
Equilibrium Time at 24°C			
IFT, mN/m (30 min)	43.1	42.8	40.8
Spreading coefficient (S), mN/m	−4	−4.7	−3
ASTM D 1401 (54°C)			
O/W/E, mL	40/40/0	41/28/12	44/36/2
Time, min	2	1	45
		(needed 30 min)	

Source: From Pillon, L.Z. and Asselin, A.E., *Method for Improving the Demulsibility of Base Oils*, US Patent 5,282,960, 1994.

spreading coefficient (S) of −4 mN/m. To meet the quality requirements of lubricating oils, the emulsion needs to be separated in less than 30 min. HF 150N phenol extracted base stock was found to require only 2 min to separate the emulsion indicating excellent water separation properties. After a typical equilibrium time of 30 min at 24°C, hydrofined 600N phenol extracted base stock was found to have an IFT of 42.8 mN/m and a negative spreading coefficient (S) of −4.7 mN/m. After 1 min standing time, higher viscosity HF 600N phenol extracted base stock was found to contain 41 ml of oil, 28 mL of water, and 12 mL of emulsion and needed 30 min to separate the emulsion indicating poor water separation properties. After a typical equilibrium time of 30 min at 24°C, hydrofined 1400N phenol extracted base stock was found to have a lower IFT of 40.8 mN/m and a negative spreading coefficient (S) of −3 mN/m. High viscosity HF 1400N phenol extracted base stock was found to require 45 min to separate the emulsion and not meeting the quality requirements.

The ASTM D 1401 method for water separability of petroleum oils and synthetic fluids requires the mixing of oil with distilled water. In some cases, deionized water is used. The level of alkali metals, such as Na, Ca, and Mg, was found to increase with an increase in the pH of the deionized water. The deionized water, having a pH of 6, was found to contain only 1 ppm of Na while the deionized water, having a pH of about 8, was found to contain 11 ppm of Ca, 4 ppm of Na, and 4 ppm of Mg. When using the deionized water, having the pH of 6, the hydrofined 600N phenol extracted base stock required 45 min to decrease the volume of emulsion from 12 to 3 mL. When using the deionized water, having a higher pH of 8, the same hydrofined 600N phenol extracted base stock was found to require only 20 min to decrease the volume of emulsion to 3 mL. The IFT and demulsibility properties of hydrofined mineral base stocks were found to vary from batch to batch. The use of N-methylpyrrolidone (NMP) as an extraction solvent leads to an increase in N and BN content of hydrofined base stocks. The surface activity at the oil/water interface and demulsibility properties of different batches of HF NMP extracted base stocks are shown in Table 9.15.

After a typical equilibrium time of 30 min at 24°C, hydrofined 150N NMP extracted base stock # 1 was found to have an IFT of 33.4 mN/m and a positive spreading coefficient (S) of 5.6 mN/m. HF 150N NMP extracted base stock, batch # 1, was found to have very poor demulsibility properties. After 1 min on standing, 77 mL of stable emulsion was present. After an additional 60 min on standing, 14 mL of stable emulsion was still present. Another batch of HF 150N NMP extracted base stock, batch # 2, was found to have good demulsibility properties. After 10 min on standing, no emulsion was present. After a typical equilibrium time of 30 min at 24°C, hydrofined 600N NMP extracted base stock was found to have an IFT of 39.4 mN/m and a negative spreading coefficient of −1.4 mN/m. After only 5 min on standing, the emulsion was separated indicating excellent demulsibility properties. The early literature reported that the storage of mineral base stocks caused degradation in their demulsibility properties and that hydrofining followed by the clay treatment was effective in improving their demulsibility properties (Watanabe et al., 1976).

TABLE 9.15
Demulsibility Properties of Different HF NMP Extracted Base Stocks

Mineral Base Stocks (Extraction Solvent)	HF 150N # 1 (NMP)	HF 150N # 2 (NMP)	HF 600N (NMP)
Kinematic Viscosity			
At 40°C, cSt	29.71	29.6	111.4
At 100°C, cSt	5.07	4.99	11.6
VI	96	90	89
Aromatics, wt%	14.0	20.1	19.6
Sulfur, wt%	0.06	0.12	0.11
Nitrogen, ppm	36	66	100
BN, ppm	33	53	88
Equilibrium Time at 24°C			
IFT, mN/m (30 min)	33.4	—	39.4
Spreading (*S*), mN/m	5.6	—	−1.4
ASTM D 1401 (54°C)			
O/W/E, mL	3/0/77	40/40/0	39/40/1
Time, min	1	10	5
O/W/E, mL	41/25/14	—	—
Time, min	60	—	—

Source: From Pillon, L.Z. and Asselin, A.E., *Method for Improving the Demulsibility of Base Oils*, US Patent 5,282,960, 1994; Pillon, L.Z. and Asselin, A.E., *Antifoaming Agents for Lubricating Oils*, US Patent 5,766,513, 1998.

Due to a variation in demulsibility properties of HF mineral base stocks, surfactants, known as demulsifiers, are used in lubricating oils to assure good water separation properties. The ethoxylated alcohols are one of the most widely used classes of nonionic surfactants and their adsorption at the oil/water interface was reported to decrease with an increase in the degree of ethoxylation and to increase with an increase in hydrophobe size (Rulison and Falberg, 1997). The effect of hydrocarbon solvents on the surfactant aggregation number is shown in Table 9.16.

With an increase in molecular weight of *n*-paraffinic hydrocarbons from heptane to decane at the same concentration, the aggregation number of surfactant molecule increases. The use of cyclohexane, as a solvent, decreases the surfactant aggregation number indicating its higher solvency. It is known that naphthenes have increased solvency over paraffinic solvents. The same chemistry surfactant, used at the same concentration, will form different aggregates in petroleum products having different hydrocarbon content which might lead to a variation in their performance. The effect of temperature on the surfactant aggregation number in hydrocarbon solvent is shown in Table 9.17.

With an increase in temperature from 10°C to 45°C for the same concentration of the same chemistry surfactant, its aggregation number decreases. At 45°C, the

TABLE 9.16
Effect of Hydrocarbon Solvents on Surfactant Aggregation Number

Hydrocarbon Solvents at 20°C	Hydrocarbon Formula	$C_{12}(EO)_4$, wt%	$C_{12}(EO)_4$ Aggregation Number
Heptane	C_7H_{16}	10	5
		15	6
		20	8
Decane	$C_{10}H_{22}$	3	1–2
		7	9
		11	9
		15	10
Cyclohexane	C_6H_{12}	2	1
		4	2
		6	2
		10	3
		15	4

Source: From Friberg, S.E., *Interfacial Phenomena in Apolar Media*, Eicke, H.F. and Parfitt, G.D., Eds., Marcel Dekker, Inc., New York, 1987. Reproduced by permission of Routledge/Taylor & Francis Group, LLC.

TABLE 9.17
Effect of Temperature on Surfactant Aggregation Number in Hydrocarbon Solvent

Hydrocarbon Solvent	Temperature,°C	$C_{12}(EO)_6$, wt%	$C_{12}(EO)_6$, Aggregation Number
Decane	10	3	22
$C_{10}H_{22}$		5	25
		7	28
		11	30
	20	6	23
		8	23
		10	23
		12	28
	45	4	10
		6	10
		8	10
		10	10

Source: From Friberg, S.E., *Interfacial Phenomena in Apolar Media*, Eicke, H.F. and Parfitt, G.D., Eds., Marcel Dekker, Inc., New York, 1987. Reproduced by permission of Routledge/Taylor & Francis Group, LLC.

aggregation number was reported as constant even when the surfactant concentration was increased. According to the literature, demulsifiers usually contain alkylene oxide block-copolymers, hydroxyalkylated alkylphenol formaldehyde resins, and polyesters (Klimova et al., 1999). Typical demulsifiers used to improve demulsibility properties of mineral lube oil base stocks are ethoxylated and stearified phenolic resins and glycerol derivative of phenolic resins. In a chemical demulsification process, the water phase needs to be separated from the oil phase through the oil/water interface destabilization, the droplet flocculation, and coalescence (Rome, 2001). According to the industry standards, the treat rate of 0.005 wt% should be sufficient for an effective demulsifier used in solvent refined mineral base stocks.

For a demulsifier, the first challenge in the chemical demulsification is to reach the oil/water interface. The second challenge is the displacement of the surface active molecules adsorbed at the oil/water interface by a more surface active demulsifier (Rome, 2001). Some additives, such as demulsifiers, can have detrimental side effects if the dosage is excessive. An overtreatment needs to be avoided to prevent an increase in the emulsion stability. Molecules used as demulsifiers need to concentrate at the oil/water interface and form weak intermolecular interactions. In some cases, by strongly adsorbing at the oil/water interface, demulsifiers can build up the layer around droplets of water and delay their flocculation (Rome, 2001). The surface activity of a typical demulsifier at the oil/water interface of different viscosity hydrofined phenol extracted base stocks is shown in Table 9.18.

HF 150N phenol extracted base stock, having an ST of 33.7 mN/m and a negative S of -4 mN/m, was found to have excellent demulsibility properties and required only 2 min to separate the emulsion. At the treat rate of only 0.005 wt%, the addition of demulsifier decreased the IFT of base stock indicating its adsorption at the oil/water interface. With a decrease in the IFT, the spreading (S) of

TABLE 9.18
Surface Activity of Demulsifier in HF Phenol Extracted Base Stocks

HF Base Stocks (Phenol Extracted)	Demulsifier, wt%	Oil/Water IFT, mN/m	Spreading (S), mN/m	Separation Time, min
Neat 150N	0	43.1	-4.0	2
Demulsifier	0.005	Decrease	Increase	Increase
Neat 600N	0	42.8	-4.7	45
Demulsifier	0.005	Decrease	Increase	Decrease
Demulsifier	0.015	Decrease	Increase	Decrease
Neat 1400N	0	40.8	-3.0	40
Demulsifier	0.005	Decrease	Increase	Decrease
Demulsifier	0.015	Decrease	Increase	Increase

Source: From Pillon, L.Z., *Petroleum Science and Technology*, 21, 1469, 2003.

base stock on the water surface increased and the emulsion separation time increased. Higher viscosity HF 600N phenol extracted base stock, having an ST of 34.7 mN/m and a negative spreading (S) of −4.7 mN/m, was found to have poor demulsibility properties since it required 45 min to separate the emulsion. At the treat rate of only 0.005 wt%, the addition of demulsifier decreased the IFT of base stock indicating its adsorption at the oil/water interface. With a decrease in the IFT, the S increased and the emulsion separation time decreased. High viscosity HF 1400N phenol extracted base stock, having an ST of 35 mN/m and a negative S of −3 mN/m, was found to require 40 min to separate the emulsion. At the treat rate of only 0.005 wt%, the addition of demulsifier decreased the IFT of base stock indicating its adsorption at the oil/water interface. With a decrease in the IFT, the S increased and the emulsion separation time decreased. With an increase in the treat rate of demulsifier from 0.005 to 0.015 wt%, the IFT further decreased and the spreading of the base stock on the water surface increased leading to an increase in the emulsion stability.

The same chemistry demulsifier was found to act as an emulsifier in lower viscosity 150N base stock but it was found to have excellent properties as a demulsifier in higher viscosity 600N base stock. Many industrial oils use lower and higher viscosity base stocks to meet the viscosity requirements and need to be carefully tested to select the most effective treat rate and to avoid overtreatment. The use of phenol extraction might lead to an increase in acidic content caused by the presence of residual phenol solvent and affect the trace contaminants found in some batches of the same base stocks. The surface activity of the same chemistry demulsifier at the oil/water interface of different viscosity hydrofined NMP extracted base stocks is shown in Table 9.19.

Hydrofined 150N NMP extracted base stock, batch # 1, was found to have very poor demulsibility properties and required over 60 min to separate the emulsion. At the treat rate of 0.005 wt%, the addition of demulsifier decreased the IFT indicating its adsorption at the oil/water interface; however, no significant

TABLE 9.19
Surface Activity of Demulsifier in HF NMP Extracted Base Stocks

HF Base Stocks (NMP Extracted)	Demulsifier, wt%	Oil/Water IFT, mN/m	Spreading (S), mN/m	Separation Time, min
Neat 150N	0	33.4	5.6	>60
Demulsifier	0.005	Decrease	Increase	No change
Demulsifier	0.015	Decrease	Increase	No change
Neat 600N	0	39.4	−1.4	5
Demulsifier	0.005	Decrease	Increase	No change
Demulsifier	0.015	Decrease	Increase	No change

Source: From Pillon, L.Z., *Petroleum Science and Technology*, 21, 1469, 2003.

decrease in the emulsion stability was observed. With an increase in the treat rate of demulsifier, the IFT further decreased; however, no significant decrease in the emulsion stability was observed. Higher viscosity hydrofined 600N NMP extracted base stock was found to have excellent demulsibility properties and required only 5 min to separate the emulsion. At the treat rate of 0.005 wt%, the addition of demulsifier decreased the IFT indicating its adsorption at the oil/water interface; however, no change in the emulsion stability was observed. With an increase in the treat rate of demulsifier from 0.005 to 0.015 wt%, the IFT further decreased; however, no change in the emulsion stability was observed. Depending on the extraction solvent, different hydrofined mineral base stocks might have a different surface active content at the oil/water interface. Hydrofined NMP extracted base stocks contain a higher content of nitrogen, including BN type compounds, when compared to phenol extracted base stocks. The surface activity of the same chemistry demulsifier at the oil/water interface of highly saturated base stocks, having excellent demulsibility properties, is shown in Table 9.20.

Highly saturated base stock, such as SWI 6, has very good water separation properties and requires 2 min to separate the emulsion. At a treat rate of 0.005 wt%, the addition of demulsifier was found to have no significant effect on the stability of the oil/water interface. The synthetic PAO 6 base stock has excellent water separation properties and requires only 1 min to separate the emulsion. At the same treat rate of only 0.005 wt%, the addition of the same demulsifier was found to increase the time required to separate the emulsion from 1 to 5 min. The use of demulsifiers, while usually effective in destabilizing the oil/water interface, in some cases can lead to an increase in the stability of emulsions. The adsorption of the demulsifier at the oil/water interface decreases the IFT which also leads to an increase in the spreading coefficient of petroleum products on the water surface which, in some cases, leads to an increase in the stability of the oil/water interface.

An effective demulsifier needs to have a low ST and a low IFT against water to have a high spreading coefficient on the water surface. After adsorbing at the oil/water interface, the efficiency of the same demulsifier in different base stocks

TABLE 9.20
Surface Activity of Demulsifier in Highly Saturated Base Stocks

Highly Saturated Base Stocks	Demulsifier, wt%	Oil/Water/Emulsion Volume, mL	Separation Time, min
Neat SWI 6	0	40/40/0	2
Demulsifier	0.005	40/40/0	2
Syntehtic PAO 6	0	40/40/0	1
Demulsifier	0.005	40/40/0	5

Source: From Pillon, L.Z., *Petroleum Science and Technology*, 21, 1469, 2003.

in destabilizing the oil/water interface was found to vary indicating the variation in interactions with other trace contaminants. The literature reported on synthesis of novel demulsifiers prepared by reaction of oleic acid with maleic anhydride to form oleic acid–maleic anhydride adduct which reacted with different compounds, such as triethanolamine, triglycerol, cetylamine, triethanolamine with lauryl alcohol, triethanolamine with glycol, polyethylene glycol, and triethanolamine with polyethylene glycol to produce different chemistry demulsifiers (Al-Sabagh et al., 2002). The early literature reported on the high efficiency of oleic acid to spread on the water surface and its effect on petroleum white oil (Bondi, 1951). The surface activity of oleic acid at the oil/water interface of white oil is shown in Table 9.21.

An effective demulsifier needs to have a low ST and a low IFT against water to have a very high spreading coefficient on the water surface. A fatty acid, oleic acid, is a surface active molecule which has a high spreading coefficient of 25 mN/m on the water surface. High purity saturated petroleum oil, such as white oil, contains practically no aromatics and heteroatom molecules and has a high negative spreading coefficient of −11.7 mNm on the water surface. The addition of 0.9 wt% of oleic acid changes the properties of white oil at the oil/water interface. The spreading coefficient of white oil was reported to increase from a negative of −11.7 to 7.4 mN/m (Bondi, 1951). An increase in spreading of petroleum oils on the water surface might lead to an increase in emulsion stability.

Silicone antifoaming agents, such as polydimethylsiloxanes, are not effective as demulsifiers despite a low ST, due to a high siloxane/water IFT and high solubility in petroleum oils. Modified silicones, such as polyethers and silicone glycol, have a low ST and a lower IFT. The silicone copolymers, used as demulsifiers, have an "ABA" or a "rake" structure. Silicone polyethers can be used efficiently on their own or, in some applications, they are combined with other organic demulsifiers to improve their performance (Rome, 2001). Silicone polyethers, used as demulsifiers, are considered a more environmentally acceptable alternative to ethoxylated phenolic/formaldehyde resins or oxyalkylated phenols. The use of silicone polyethers was reported to be effective in producing

---

TABLE 9.21
Effect of Oleic Acid on Spreading of White Oil on Water Surface

Oil/Water Interface at 20°C	Chemistry	Spreading (S), mN/m
Oleic acid ($C_{18}H_{34}O_2$)/water	Cis-9-octadecenoic acid	25.0
Neat white oil/water	Saturated petroleum oil	−11.7
White oil (0.9% oleic acid)/water	Petroleum oil/surfactant	7.4

Source: From Bondi, A., *Physical Chemistry of Lubricating Oils*, Reinhold Publishing Corporation, New York, 1951.

a clean crude oil/water interface, to be cost efficient, and to reduce sensitivity to temperature reduction (Rome, 2001). The literature reported that the demulsification of water-in-bitumen emulsion was studied using 52 nonionic demulsifiers for which their relative solubility number (RSN) and molecular weights were determined. The results indicated that there was no correlation between RSN and their demulsification properties (Wu et al., 2003). The performance of demulsifiers can vary in different petroleum products depending on their solubility and interactions with other surface active contaminants also capable of adsorbing at the oil/water interface. An increase in the surface active contaminants of petroleum products will lead to competition at the oil/water interface which makes some additives ineffective.

9.4   AQUEOUS FOAMS

Petroleum products are frequently contaminated with water and the early literature reported that the contamination with water, gasoline, or gasket-sealing compounds may affect the foaming characteristics of oil (Tourret and White, 1952). Nonaqueous and aqueous foams are formed during processing of crude oils and the literature reported on the profiles of foams formed from model aqueous and nonaqueous foaming solutions using the optical transmission foam meter (Callaghan et al., 1986). Foams and emulsions have much in common, both being examples of two immiscible phases. During the demulsification process, a liquid phase needs to be separated from another liquid phase. In a defoaming process, a gas phase needs to be separated from a liquid phase (Rome, 2001). Pure saturate hydrocarbons have low ST but high IFT and negative spreading coefficient on the water surface. They are incompatible when together which allows separation of water from hydrocarbon mixtures. However, the properties of aromatic hydrocarbons, such as benzene, are different. Benzene has a higher ST but a significantly lower IFT and a positive spreading coefficient on the water surface which will increase the water solubility. Hydrofined lube oil base stocks have only traces of water contamination, usually below 50 ppm, and silicone antifoaming agents are highly effective in spreading on the oil surface and preventing foaming. The spreading coefficient (S) of silicone antifoaming agent, Si 12500, on the oil surface of hydrofined base stocks is shown in Table 9.22.

Despite a relatively low ST of petroleum oils, silicones have a high efficiency to spread due to a low oil/silicone IFT. In hydrofined 150N mineral base stock, not contaminated with water, Si 12500 has a high spreading coefficient of 8.2 mN/m measured at 24°C. Silicone antifoaming agents are also effective in hydroprocessed base stocks despite some decrease in their ST and some increase in their IFT. In hydroprocessed SWI 6 base stocks, not contaminated with water, Si 12500 has a still high spreading coefficient of 6.6 mN/m measured at 24°C. Silicone antifoaming agents, such as polydimethylsiloxanes, prevent foaming by spreading on the oil surface. Silicone antifoaming agents are also effective in preventing foaming of aqueous solutions due to their high spreading coefficient on the water surface. The literature reported on the effect of polymer chemistry on

TABLE 9.22
Spreading Coefficient of Si 12500 on Oil Surface of Hydrofined Base Stocks

Base Stock/Si 12500 Interface at 24°C	Lube Oil ST, mN/m	Si 12500 ST, mN/m	Oil/Si 12500 IFT, mN/m	Si 12500 (S), mN/m
150N mineral oil	33.7	24.0	1.5	8.2
SWI 6 base stock	32.8	24.0	2.2	6.6

Source: From Pillon, L.Z. and Asselin, A.E., *Antifoaming Agents for Lubricating Oils*, US Patent 5,766,513, 1998.

their spreading coefficient on the water surface, including octamethyltrisiloxane (Owen, 1980). The spreading coefficient (*S*) of different low molecular weight polymers on the water surface is shown in Table 9.23.

Low molecular weight octamethyltrisiloxane polymer has a low ST of 17.0 mN/m and a high water/polymer IFT of 42.5 mN/m. Despite a high IFT against water, octamethyltrisiloxane has a high spreading coefficient of 12.5 mN/m on the water surface due to the high ST of water. Another low molecular weight polymer, *n*-perfluorononane, has a lower ST of 14.3 mN/m but a low spreading coefficient of only 1.3 mN/m due to a higher IFT against water of 56.4 mN/m. However, the spreading coefficient of octamethyltrisiloxane and *n*-perfluorononane can further decrease or increase if there are other surface active molecules present capable of changing the ST or IFT of the aqueous solution. The early literature reported that even a high content of surface active molecules, including silicone antifoaming agents, has a very small effect on the ST of liquids which is difficult to measure. The effect of surface active molecules, at varying treat rates, on the ST of mineral oil is shown in Table 9.24.

While there is some difference in the temperature of ST measurements, except for one surface active molecule, lauryl sulfonic acid, no significant decrease in ST of mineral oil was reported. Even when used at high treat rate of 9 wt%,

TABLE 9.23
Spreading Coefficient of Different Polymers on Water Surface

Water/Polymer Interface at 20°C	Polymers bp, °C	Polymer ST, mN/m	Water/Polymer IFT, mN/m	Polymer (S), mN/m
Octamethyltrisiloxane	153	17.0	42.5	12.5
n-Perfluorononane	125	14.3	56.4	1.3

Source: Reprinted from Owen, M.J., *Industrial and Engineering Chemistry Product Research and Development*, 19, 97, 1980. Copyright American Chemical Society. With permission.

TABLE 9.24
Effect of Surface Active Molecules on ST of Mineral Oil

Mineral Oil/Air Interface Surface Active Molecules	Temperature, °C	Content, wt%	Liquid ST, mN/m	Solution ST, mN/m
Neat mineral oil	20	0	31.0	31.0
Mineral oil (stearic acid)	27	0.7	33.5	32.5
Mineral oil (lauryl sulfonic acid)	27	1.0	33.5	25.9
Mineral oil (lauric acid)	27	3.0	33.5	31.5
Mineral oil (Triethanolamine oleate)	27	9.0	33.5	31.5

Source: From Bondi, A., *Physical Chemistry of Lubricating Oils*, Reinhold Publishing Corporation, New York, 1951.

the triethanolamine oleate was found to have no significant effect on the ST of mineral oil. Depending on the chemistry and the content of surface active molecules, the water solubility in petroleum oils might increase leading to haze and increased emulsion stability. The early literature reported that IFT is very sensitive to even traces of hydrophilic impurities and will greatly vary for the same technical materials (Bondi, 1951). The literature also reported that immiscible liquids, such as benzene and water, might become miscible in the presence of a cosolvent, such as ethanol (Ross, 1987). The effect of surface active alcohols on the spreading coefficient of petroleum white oil on the water surface is shown in Table 9.25.

Purified mineral oil, containing saturate, aromatic, and heteroatom containing molecules, was reported to have a negative spreading coefficient of -11.2 mN/m on the water surface. At the treat rate of 0.9 wt%, oleyl alcohol increased the spreading coefficient of white oil from negative -11.7 to 14.4 mN/m. At a lower treat rate of 0.7 wt%, glycerol monooleate drastically increased the spreading

TABLE 9.25
Effect of Surface Active Alcohols on Spreading of White Oil on Water Surface

Petroleum Oil/Water Interface at 20°C	Chemistry	(S), mN/m
Neat mineral oil	Hydrocarbon mixture	-11.2
Neat white oil	Saturated hydrocarbons	-11.7
White oil (0.9% oleyl alcohol)	*Cis*-9-octadecen-1-ol	14.4
White oil (0.7% glycerol monooleate)	$C_3H_8O_3$-oleate	37.8

Source: From Bondi, A., *Physical Chemistry of Lubricating Oils*, Reinhold Publishing Corporation, New York, 1951.

coefficient of white oil from negative -11.7 to 37.8 mN/m. Surface active molecules, such as oleyl alcohol and glycerol monooleate, drastically increase the spreading coefficient of white oil on the water surface by adsorbing at the oil/water interface and decreasing the IFT. Calcium ions contained in "hard" water will precipitate soap from solution and build deposits. The literature reported that in the presence of calcium ions or other heavy metals, the spreading of the fatty alcohols will not take place and a rigid film at the oil/water interface will form (Bondi, 1951).

The literature also reported that the equilibrium at the interface might take a fraction of a second or even a few days because the adsorption process involves the diffusion of the solute molecule from the bulk phase to the surface first (Ross, 1987). The standard processor tensiometer K12 can repeat the measurements until the equilibrium at the oil/water interface is established. Some base stocks require significantly more time to reach the equilibrium IFT indicating the presence of surface active contaminants adsorbing at the oil/water interface. Hydroprocessed SWI 6 base stock, containing practically 100% saturate content, quickly reaches the equilibrium IFT in less than 5 min while many other mineral base stocks require 1–2 h. The effect of equilibrium time and different chemistry rust and corrosion inhibitors on IFT of SWI 6 base stock is shown in Table 9.26.

In crude oils and other petroleum products, containing water, there is usually a separate oil/water and oil/air interface where silicone antifoaming agents can adsorb. Water and brine are often present in crude oils. Rust and corrosion inhibitors are surfactant type molecules which are used during crude oil recovery and transportation to protect the metal surface. The IFT between oil and water provides an indication of compounds in the oil that have an affinity for water

TABLE 9.26
Effect of Equilibrium Time and Rust and Corrosion Inhibitors on IFT of SWI 6

SWI 6/Water Interface Inhibitors (0.15 wt%)	After 5 min IFT, mN/m	After 30 min IFT, mN/m	After 60 min IFT, mN/m	After 60 min (S), mN/m
Neat SWI 6 base stock	54.7	55.4	54.8	-14.8
Alkyl succinic acid	8.3	8.3	8.1	31.9
Dodecyl succinic acid	7.8	7.1	7.4	32.6
Esterified alkyl succinic acid	7.5	7.2	7.1	32.9
Sodium sulfonate	5.9	6.1	6.1	33.9
Calcium sulfonate	3.8	4.9	4.7	35.3
Polyamine	4.4	3.6	3.7	36.3
Succinic anhydride amine	2.6	2.5	2.2	37.8

Source: From Pillon, L.Z., Asselin, A.E., and MacAlpine, G.A., *Lubricating Oil Having an Average Ring Number of Less Than 1.5 Per Mole Containing Succinic Anhydride Amine Rust Inhibitor*, US Patent 5,225,094, 1993.

(Speight, 1999). Highly surface active rust and corrosion inhibitors, dissolved in SWI 6 base stock, quickly reach the oil/water interface and a drastic decrease in the IFT is observed. The spreading coefficient of SWI 6 base stock also drastically increases from a negative -14.8 to 31.9–37.8 mN/m, indicating a significant change in base stock properties. Many other surfactant molecules are used during crude oil recovery and processing and adsorb at the oil/water interface.

A decrease in IFT and an increased water content of petroleum products might increase their solvency properties and affect the solubility properties of surface active additives, such as demulsifiers and antifoaming agents, and make them ineffective. Increased solubility of surface active additives decreases their surface activity. Refining of crude oils, under varying conditions, leads to a significant variation in composition of water and surface active content which can decrease the surface activity of antifoaming agents. In some petroleum applications, to prevent foaming of aqueous foams, emulsions of silicone containing silica are used. These emulsions, containing silica, cannot be used in lubricating oils. Water contamination and foaming of some lubricant products, during their use, can be a serious problem. The oxidation by-products of petroleum are known to be surface active and adsorb at the oil/water interface (Speight, 1999). In some cases, an increase in the treat rate of antifoaming agent can be effective but it might also lead to a severe air entrainment.

REFERENCES

Al-Sabagh, A.M., Badawi, A.M., and Noor-El-Den, M.R., *Petroleum Science and Technology*, 20(9&10), 887, 2002.

Becher, P., Water-in-oil emulsions, in *Interfacial Phenomena in Apolar Media*, Eicke, H.F. and Parfitt, G.D., Eds., Marcel Dekker, Inc., New York, 1987.

Bondi, A., *Physical Chemistry of Lubricating Oils*, Reinhold Publishing Corporation, New York, 1951.

Callaghan, I.C., Lawrence, F.T., and Melton, P.M., *Colloid and Polymer Science*, 264(5), 423, 1986.

Folkins, H.O., Miller, E.L., and O'Malley, C.T., *Hydrocatalytic Refining of Lubricating Oils*, US Patent 2,921,024, 1960.

Friberg, S.E., Stabilization of inverse micelles by non-ionic surfactants, in *Interfacial Phenomena in Apolar Media*, Eicke, H.F. and Parfitt, G.D., Eds., Marcel Dekker, Inc., New York, 1987.

Gary, J.H. and Handwerk, G.E., *Petroleum Refining Technology and Economics*, 4th ed., Marcel Dekker, Inc., New York, 2001.

Golbshtein, D.L. et al., *Khim. Tekhnol. Topl. Masel*, 11(7), 18, 1966.

Hughes, R.I., *Corrosion Science*, 9, 535, 1969.

Jaycock, M.J. and Parfitt, G.D., *Chemistry of Interfaces*, Ellis Horwood Ltd., Chichester, 1981.

Klimova, L.Z. et al., *Petroleum Chemistry*, 39, 201, 1999.

Owen, M.J., *Industrial and Engineering Chemistry Product Research and Development*, 19, 97, 1980.

Pillon, L.Z., *Petroleum Science and Technology*, 19(9&10), 1109, 2001.

Pillon, L.Z., *Petroleum Science and Technology*, 21(9&10), 1469, 2003.

Pillon, L.Z. and Asselin, A.E., *Method for Improving the Demulsibility of Base Oils*, US Patent 5,282,960, 1994.

Pillon, L.Z. and Asselin, A.E., *Antifoaming Agents for Lubricating Oils*, US Patent 5,766,513, 1998.

Pillon, L.Z., Asselin, A.E., and MacAlpine, G.A., *Lubricating Oil Having an Average Ring Number of Less Than 1.5 Per Mole Containing Succinic Anhydride Amine Rust Inhibitor*, US Patent 5,225,094, 1993.

Pillon, L.Z., Reid, L.E., and Asselin, A.E., *Lubricating Oil for Inhibiting Rust Formation*, US Patent 5,397,487, 1995.

Prince, R.J., Base oils from petroleum, in *Chemistry and Technology of Lubricants*, 2nd ed., Mortier, R.M. and Orszulik, S.T., Eds., Blackie Academic & Professional, London, 1997.

Qi, B. et al., *Petroleum Science and Technology*, 22(3&4), 463, 2004.

Rome, F.C., *Hydrocarbon Asia*, 42, May/June, 2001.

Ross, S., Foaminess and capillarity in apolar solutions, in *Interfacial Phenomena in Apolar Media*, Eicke, H.F. and Parfitt, G.D., Eds., Marcel Dekker Inc., New York, 1987.

Rulison, C. and Falberg, D., *Book of Abstracts, 214th ACS National Meeting*, Las Vegas, NV, 1997.

Sequeira, A. Jr., *Lubricant Base Oil and Wax Processing*, Marcel Dekker, Inc., New York, 1994.

Speight, J.G., *The Chemistry and Technology of Petroleum*, 3rd ed., Marcel Dekker, Inc., New York, 1999.

Tourret, R. and White, N., *Aircraft Engineering*, 24(279), 122, 1952.

USGS, *Clays Statistics and Information*, http://minerals.er.usgs.gov (accessed February 2007).

Wang, Y. et al., *Petroleum Science and Technology*, 19(7&8), 923, 2001.

Watanabe, J., Fukushima, T., and Nose, Y., *Proceedings of the JSLE-ASLE International Lubrication Conference*, 641, 1976.

Wu, J. et al., *Energy and Fuels*, 17, 1554, 2003.

Zaki, N., Schorling, P.C., and Rahimian, I., *Petroleum Science and Technology*, 18(7&8), 945, 2000.

10 Rust and Corrosion

10.1 EFFECT OF DISTILLED WATER

The literature reported that the combined action of water and oxygen causes rust while the presence of acids in water causes corrosion. The early literature reported that many fractions of lube oil base stocks can be strongly adsorbed on cast iron, and that this is connected with the presence of molecules having long normal hydrocarbon chains (Groszek, 1962). The more recent literature reported that a severe hydrotreatment and a further reduction in aromatic and sulfur contents of fuels might lead to a decrease in the fuel inherent lubricity causing an increase in the wear of engine components (Hughes et al., 2002). The literature reported on the use of different testing procedures to study the rust and corrosion formation on the metal surface. A comprehensive review of analytical techniques used to analyze solids and their surfaces, including x-ray photoemission spectroscopy (XPS), was published (Yacobi et al., 1994). Solid surfaces are irradiated with x-rays and the emitted energy of photoelectrons is analyzed. The difference between the x-ray energy and the energy of photoelectrons gives the binding energy of the core level electrons. The energy of photoelectrons depends on the excited atoms and therefore gives elemental information. According to the literature, the XPS technique can be used to measure the surface composition, elemental distributions, and interfacial chemical reactions (Nelson, 1994). The early literature reported on the visual standard corrosion ratings which varied from "0," indicating no rusting, to "8" indicating that the entire metal surface is covered with rust (Hughes, 1969). The more recent literature reported on the similar rusting severity based on the surface coverage (Pillon, 2003a). The rating system for visual evaluation of severity of rusting based on rusting severity or metal surface coverage is shown in Table 10.1.

The early literature also reported that the mineral lube oil base stocks, without water, are not corrosive towards the metal surface (Hughes, 1969). The more recent literature reported on the effect of distilled water on the rust preventive properties of different uninhibited lube oil base stocks. Without the presence of water, mineral base stocks, white oil, SWI 6, and synthetic PAO 6 base stocks did not cause rusting, indicating the absence of any contaminants which could lead to rust formation on the metal surface (Pillon, 2003a). The early

TABLE 10.1
Visual Rating System for Evaluation of Rusting Severity on the Metal Surface

Visual Rating	Rusting Severity	Metal Surface Coverage
0	No rusting	None
1	Trace rusting	1–2 Rust spots
2	Slight rusting	3–5 Rust spots
3	Light rusting	>6 Rust spots
4	Moderate rusting	1%–2% Surface covered
5	Fairly severe rusting	2%–3% Surface covered
6	Severe rusting	3%–5% Surface covered
7	Very severe rusting	5%–75% Surface covered
8	Entire surface rust covered	100% Surface covered

Source: From Hughes, R.I., *Corrosion Science*, 9, 535, 1969; Pillon, L.Z., *Petroleum Science and Technology*, 21, 1453, 2003a.

literature reported on the effect of temperature on the severity of rusting of the metal surface immersed in different base stocks, containing drops of water. The rust preventive properties of medicinal white oil and mineral base stocks were studied at the temperatures of 25°C and 40°C. At 25°C, the corrosion of the metal surface occurred only in the presence of the white oil. At 40°C, the extent of corrosion increased and the rusting of the metal surface also occurred in the presence of mineral base stocks (Hughes, 1969). The effect of temperature on rusting severity on the metal surface immersed in different base stocks, containing drops of water, is shown in Table 10.2.

At 25°C, no rusting in mineral base stocks was reported. The advancing contact angle measurements indicated that water drops did not appear to wet the

TABLE 10.2
Effect of Temperature on Rusting Severity in Base Stocks Containing Drops of Water

Rusting Conditions	White Oil	Mineral Oil # 1	Mineral Oil # 2
25°C/1 h	Severe rusting	No rusting	No rusting
25°C/18 h	Severe rusting	No rusting	No rusting
40°C/1 h	Severe rusting	Slight rusting	Slight rusting
40°C/18 h	Severe rusting	Slight rusting	Slight rusting

Source: Reprinted from Hughes, R.I., *Corrosion Science*, 9, 535, 1969. With permission from Elsevier.

metal surface and no rusting was observed. However, an increase in temperature from 25°C to 40°C was reported to cause rust formation on the metal surface immersed in different mineral base stocks (Hughes, 1969). The presence of aromatic, S and N molecules in mineral base stocks was found to have no significant effect in improving their rust preventive properties and protecting the metal surface. According to the early literature, for the same viscosity base stocks, a decrease in the oil/water IFT will decrease the water droplet size which will decrease the rate of water approaching the metal surface and thus delay the wetting process (Hughes, 1969). The effect of distilled water on rusting severity on the metal surface immersed in different viscosity mineral base stocks is shown in Table 10.3.

The uninhibited hydrofined mineral base stocks, not containing any rust inhibitors, were stirred at 60°C with a cylindrical steel test rod completely immersed therein. After 24 h, the test rod was observed for signs of rusting. While some base stocks might contain trace water content, up to 50 ppm, no water was added. After 24 h of testing at 60°C, without the addition of water, no sign of rusting was observed. The rust test was carried out for an additional 12 days and no rusting was observed. After addition of only 1 mL of distilled water, the rust formation was observed. After the first 10 min of the rust test, the metal pin immersed in 150N phenol extracted base stock showed a slight rust formation while no rust formation was observed in high viscosity 1400N phenol and 600N

TABLE 10.3
Effect of Distilled Water on Rusting Severity in Mineral Base Stocks

Mineral Base Stocks, Extraction Solvent	150N Phenol	1400N Phenol	600N NMP
Kin. Viscosity			
At 40°C, cSt	29.48	301.71	111.53
At 100°C, cSt	5.03	22.01	11.55
VI	94	89	89
Aromatics, wt%	17.2	31.7	19.6
Sulfur, wt%	0.09	0.19	0.12
Total N (TN), ppm	8	141	100
Basic N (BN), ppm	4	51	88
IFT, mN/m (30 min)	43.1	40.8	39.4
Rusting Severity			
No water/60°C/24 h	No rusting	No rusting	No rusting
No water/60°C/12 days	No rusting	No rusting	No rusting
1 mL Water/60°C/10 min	Slight rusting	No rusting	No rusting
30 mL Water/25°C/25 min	Slight rusting	Slight rusting	Slight rusting

Source: From Pillon, L.Z., *Petroleum Science and Technology*, 21, 1453, 2003a.

NMP extracted base stocks. The rust preventive properties of uninhibited phenol extracted base stocks were found to improve with an increase in their viscosity. With an increase in viscosity from 150N to 1400N, their IFT was found to decrease from 43.1 to 40.8 mN/m. After an increase in distilled water content from 1 to 30 mL, and even a decrease in temperature from 60°C to 25°C, all different mineral base stocks showed the same rusting severity.

The rust preventive properties of uninhibited mineral base stocks were also found to vary depending on the extraction solvent. The uninhibited 600N NMP extracted base stock was found to have improved rust preventive properties and also a lower IFT of 39.4 mN/m. The NMP extraction process was developed as a replacement for the phenol extraction process mostly due to the safety, health, and environmental problems associated with the use of phenol. The NMP extracted base stocks contain a higher N and basic N (BN) contents. The nitrogen compounds found in petroleum are classified as basic or nonbasic, depending on whether they can be titrated with perchloric acid in a solution of glacial acetic acid. Basic constituents of petroleum include not only BN compounds but also other compounds, such as salts of weak acids (soaps), basic salts of polyacidic bases, and salts of heavy metals. The effect of extraction solvent on BN/TN ratio of different hydrofined mineral base stocks is shown in Figure 10.1.

The BN/TN ratio of phenol extracted base stocks was found to be in the range of 0.4–0.5. The BN/TN ratio of NMP extracted base stocks was found to be significantly higher in the range of 0.9. The literature reported that the lubricating oil cuts extracted with NMP solvent were found to have a lower IFT than the

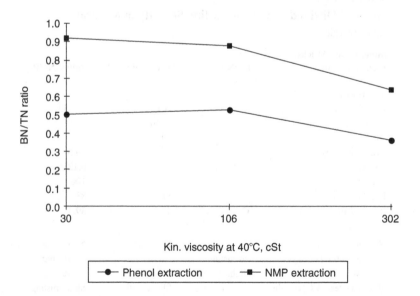

FIGURE 10.1 The effect of extraction solvent on BN/TN ratio of different hydrofined mineral base stocks.

lubricating oil cuts extracted with phenol or furfural (Vyazovkin et al., 1990). The actual BN content of mineral base stocks can be extracted with diluted mineral acids and analyzed. FTIR analysis of BN molecules separated from HF NMP extracted base stocks indicated the presence of amines, pyridines, and acids. The BN type molecules, separated from the NMP extracted base stocks, were added to phenol extracted base stocks and their effect on IFT was tested. After the addition of trace BN molecules, 30–80 ppm, the IFT of different viscosity phenol extracted base stocks drastically decreased by 10–15 mN/m. Despite the same viscosity and practically the same composition, the IFT of highly saturated base stocks was found to vary indicating the presence of some other surface active contaminants capable of adsorbing at the oil/water interface and decreasing the IFT of base stocks. The effect of distilled water on rusting severity on the metal surface immersed in different hydroprocessed base stocks, having a similar viscosity at 100°C, is shown in Table 10.4.

The use of hydrorefining and hydroisomerization produces base stocks which contain practically no S, N, and BN contents. The uninhibited base stocks, not containing any rust inhibitors, were stirred at 60°C with a cylindrical steel test rod completely immersed therein. After 24 h of testing at 60°C, without the presence of water, no sign of rusting was observed. The rust test was carried out for an additional 12 days and no rusting was observed. After addition of 1 mL of distilled water, the rust formation was observed. After the first 10 min of the

TABLE 10.4
Effect of Distilled Water on Rusting Severity in Hydroprocessed Base Stocks

Hydroprocessed Base Stocks	Hydrorefined White Oil	Hydroisomerized SWI 6	Synthetic PAO 6
Kin. Viscosity			
At 40°C, cSt	32.71	29.38	30.43
At 100°C, cSt	5.55	5.77	5.77
VI	106	143	134
Aromatics, wt%	0.9	<0.5	0
Sulfur, ppm	<1	<1	0
Total N (TN), ppm	<1	<1	0
Basic N (BN), ppm	0	0	0
IFT, mN/m (30 min)	39.1	45.6	41.8
Rusting Severity			
No water/60°C/24 h	No rusting	No rusting	No rusting
No water/60°C/12 days	No rusting	No rusting	No rusting
1 mL Water/60°C/10 min	Light rusting	Light rusting	Light rusting
30 mL Water/25°C/25 min	Light rusting	Light rusting	Light rusting

Source: From Pillon, L.Z., *Petroleum Science and Technology*, 21, 1453, 2003a.

TABLE 10.5

Effect of Distilled Water on Rusting Severity in Different PAO Base Stocks

Synthetic Base Stocks	PAO 4	PAO 6	PAO 8
Kin. Viscosity			
At 40°C, cSt	16.68	30.89	46.3
At 100°C, cSt	3.84	5.98	7.74
VI	124	134	136
Aromatics, wt%	0	0	0
Sulfur, ppm	0	0	0
Total N (TN), ppm	0	0	0
Basic N (BN), ppm	0	0	0
IFT, mN/m (30 min)	46.3	43.7	35.1
Rusting Severity			
1 mL Water/60°C/30 min	Severe	Moderate	Light

Source: From Pillon, L.Z., *Petroleum Science and Technology*, 21, 1453, 2003a.

rust test, the metal pin immersed in white oil and SWI 6 showed increased rust formation rated "light rusting" and less rust formation rated "slight rusting" in PAO 6 base stock. After an increase in distilled water content from 1 to 30 mL and a decrease in temperature from 60°C to 25°C, all different saturated base stocks showed the same light rusting severity. The synthetic PAO base stocks, having 100% saturate content, are commercially available in different viscosity grades (Rudnick and Shubkin, 1999). The effect of distilled water on rusting severity on the metal surface immersed in different viscosity PAO base stocks is shown in Table 10.5.

With an increase in viscosity of PAO base stocks, a decrease in the IFT and the rusting severity was observed. To understand the significance of the IFT in the rust protective mechanism, it is necessary to determine how readily an entrained droplet of water can make contact with the metal surface in the presence of the base stock. To delay the wetting of the metal surface, the droplets of water need to be small and a requirement for small size droplets is that the tension at the oil/water interface is low. Without any presence of water, different 600N uninhibited mineral and PAO 6 base stocks were exposed to the iron surface, for 30 min at 60°C. After being exposed, their IFT was measured. In the presence of the iron surface, no change in the IFT of 600N phenol extracted base stocks and synthetic PAO 6 base stock was observed. However, in the presence of the iron surface, an increase in the IFT of 600N NMP extracted base stock was observed indicating a desorption of surface active molecules from the oil/water interface. Some mineral base stocks contain surface active molecules capable of adsorbing at the oil/water and oil/metal interfaces. While there is no test method to evaluate the rust preventive properties of uninhibited base stocks, a variation in the distilled water content, the temperature, and the testing time was found to affect the severity of rusting.

10.2 RUST INHIBITION

According to the early literature, rust inhibitors are polar oil-soluble compounds capable of adsorbing on the metal surface and forming a protective layer. The polar groups need to attach themselves to the metal surface in an oriented fashion called chemisorption and their hydrocarbon chains need to extend into the oil phase. These hydrocarbon chains were reported to form a hydrophobic film on the metal surface which is easily wetted by oil but not water (Von Fuchs, 1956). The early literature reported on alkylsuccinic acid, known as ASA, forming "film" on the metal surface immersed in hydrocarbon solutions, containing water, and preventing rusting (Hughes, 1969).

$HOOC–CH_2–CH_2–COOH$
Butanedioic acid (succinic acid)

Many different alkyl derivatives of succinic acid, such as dodecenyl succinic acid (DSA), are used as rust inhibitors.

$HOOC–CH_2–CHR–COOH$
Alkylsuccinic acid (ASA)

The patent literature reported on the partially esterified alkylsuccinic acid (EASA) which is a mixture of about 75 wt% of unreacted tetrapropenylsuccinic acid and about 25 wt% of a partially esterified tetrapropenylsuccinic acid as a rust inhibitor (Pillon and Asselin, 1993).

$C_{12}H_{23}–CH(COOH)–CH_2–COOH$
Tetrapropenylsuccinic acid (~75 wt%)
$C_{12}H_{23}–CH(COOH)–CH_2–COO(CH_2)_3OH$
Partially esterified tetrapropenylsuccinic acid (~25 wt%)

The rust preventive properties of lubricating oils are usually tested following the ASTM D 665 method for rust preventing characteristics of inhibited mineral oils in the presence of water. A mixture of 300 mL of oil is stirred with 30 mL of distilled water (method A) or 30 mL of synthetic seawater (method B) at 60°C with a cylindrical steel test rod completely immersed therein. After 24 h, the test rod is observed for signs of rusting. Usually, the evaluation pass/fail is used and the oil passes the test only if no rusting is observed, not even one rust spot. The effect of the testing severity and the presence of rust inhibitor on the rust preventive properties of base stocks are shown in Table 10.6.

The uninhibited mineral and hydroprocessed base stocks are not capable of preventing rust formation, even in the presence of distilled water, and very severe

TABLE 10.6
Effect of Severity of ASTM D 665 Rust Test on Rust Inhibitor Performance

ASTM D 665 Rust Test	Distilled Water, A	Synthetic Water, B
Uninhibited white oil	Fail	Fail
Inhibited white oil (0.015% dodecenyl succinic acid)	Pass	Fail
Uninhibited 150N mineral base stock	Fail	Fail
Inhibited 150N oil (0.04% succinic acid derivative)	Pass	Pass

Source: From ASTM D 665-06 Standard Test Method for Rust Preventing Characteristics of Inhibited Mineral Oil in the Presence of Water, *Annual Book of ASTM Standards*, 2006; Pillon, L.Z., *Petroleum Science and Technology*, 21, 1461, 2003b.

rusting covering over 75% of the metal surface is observed. The XPS analysis of iron surface immersed in uninhibited mineral base stock indicated the presence of N and O but no presence of S despite the fact that the base stocks contained an S content over 0.1 wt% and only a trace of N content of less than 50 ppm. The addition of 2,4,6-trimethyl pyridine, known as collidine, was found to further increase the N content on the iron surface. Mineral base stocks contain sulfur molecules which are mostly dibenzothiophenes and clearly they are not surface active. According to the ASTM D 665 test method, the inhibited white oil, containing 0.015% of DSA, can prevent rust formation in the presence of distilled water but not in the presence of synthetic water. An increased treat rate of rust inhibitor, 0.04 wt% of succinic acid derivative in mineral base stock, is required to pass the less severe A test and the more severe B test. The composition of synthetic seawater, used in ASTM D 665 B rust test, is shown in Table 10.7.

The use of different inorganic salts at low concentrations, below 5 wt%, has a small effect on the ST of water. At 20°C, an increase from 72.75 to 73–74 mN/m is observed. The effect of inorganic salts, at low concentrations, on the ST of water is shown in Table 10.8.

While some lubricant applications allow the use of a less severe ASTM D 665 A procedure, the majority of lubricating oils, such as turbine oils, require the use of the more severe ASTM D 665 B rust test. The presence of salts in water can lead to chemical interactions with rust inhibitors and other components of lubricating oils. The effect of surface active molecules and aqueous inorganic salt solutions on IFT of different petroleum oils is shown in Table 10.9.

The addition of naphthenic acid decreased the IFT of white oil from 52.0 to 24.5 mN/m, which further decreased to a low of 11.8 mN/m in the presence of aqueous KCl solution. The addition of sodium alkyl sulfate decreased the IFT of mineral oil, from 53.0 to only 51.5 mN/m, which further decreased to only 50.4 mN/m in the presence of aqueous 0.2% NaCl solution (Bondi, 1951).

TABLE 10.7
Composition of Synthetic Seawater Used in ASTM D 665 B Rust Test

Salts	Content, g/L
NaCl	24.54
$MgCl_2 \cdot 6H_2O$	11.10
Na_2SO_4	4.09
$CaCl_2$	1.16
KCl	0.69
$NaHCO_3$	0.20
KBr	0.10
H_3BO_3	0.03
$SrCl_2 \cdot 6H_2O$	0.04
NaF	0.003

Source: Reproduced with permission from ASTM D 665-06 Standard Test Method for Rust Preventing Characteristics of Inhibited Mineral Oil in the Presence of Water, *Annual Book of ASTM Standards*, 2006. Copyright ASTM International.

The XPS analysis was used to study the adsorption of the ASA derivatives on the iron surface in the presence of distilled water and synthetic seawater. The acidic group of the ASA derivatives was found to bond directly with the iron surface and, under more severe conditions of the ASTM D 665 B test, less coverage of the

TABLE 10.8
Effect of Inorganic Salts at Low Concentrations on ST of Water

Inorganic Salts	Content in Water, wt%	Temp., °C	Solution ST, mN/m
Water	0	20	72.75
NaCl	2.84	20	73.75
$MgCl_2$	0.94	20	73.07
Na_2SO_4	2.76	20	73.25
KCl	0.74	20	72.99
Na_2CO_3	2.58	20	73.45
NaBr	4.89	20	73.45

Source: From *CRC Handbook of Chemistry and Physics*, 86th ed., Lide, D.R., Editor-in-Chief, CRC Press, Boca Raton, 2005. Reproduced by permission of Routledge/Taylor & Francis Group, LLC.

TABLE 10.9
Effect of Surface Active Molecules and Aqueous Salts on IFT of Petroleum Oils

Oil/Water Interface at 20°C	Distilled Water IFT, mN/m	Aqueous KCl IFT, mN/m	Aqueous NaCl IFT, mN/m
Neat white oil	52.0	52.2	—
White oil (1.4% naphthenic acid)	24.5	11.8	—
Neat mineral oil	53.0	—	—
Mineral oil (sodium alkyl sulfate)	51.5	—	50.4

Source: From Bondi, A., *Physical Chemistry of Lubricating Oils*, Reinhold Publishing Corporation, New York, 1951.

iron surface by the rust inhibitor was observed. Water is present during the ASTM D 665 rust test and the amount of the rust inhibitor adsorbed at the oil/water interface will decrease the amount available for adsorption on the metal surface. The effect of rust inhibitor, EASA, on the IFT and the rust preventive properties of hydrorefined white oil are shown in Table 10.10.

With an increase in the treat rate of the rust inhibitor EASA from 0 to 0.1 wt %, the IFT of white oil gradually decreased from 45.1 to 9.3 mN/m, indicating its adsorption at the oil/water interface. The severity of rusting can be

TABLE 10.10
Effect of EASA on IFT and Rust Preventive Properties of White Oil

White Oil EASA, wt%	Oil/Distilled Water IFT, mN/m	ASTM D 665 B Rusting	ASTM D 665 B Rating
0	45.1	Severe	Fail
0.03	19.0	Severe	Fail
0.04	16.0	Severe	Fail
0.05	13.5	Severe	Fail
0.07	11.9	Moderate	Fail
0.08	11.4	Moderate	Fail
0.09	10.9	Light	Fail
0.10	9.3	None	Pass

Source: From Pillon, L.Z. and Asselin, A.E., *Lubricating Oil Having Improved Rust Inhibition and Demulsibility*, US Patent 5,227,082, 1993.

TABLE 10.11
Effect of 2,4,6-Trimethylpyridine on Performance of EASA in White Oil

ASTM D 665 B Test, Pyridines in White Oil	Pyridines, ppm	EASA, 0.03 wt%	EASA, 0.04 wt%	EASA, 0.05 wt%	EASA, 0.1 wt%
None	0	Fail	Fail	Fail	Pass
2,4,6-Trimethylpyridine	5	Fail	Pass		
2,6-Di-*tert*-butylpyridine	7	Fail	Fail	Fail	

Source: From Pillon, L.Z. and Asselin, A.E., *Lubricating Oil Having Improved Rust Inhibition and Demulsibility*, US Patent 5,227,082, 1993.

described as light, moderate, or severe. Light rusting indicates no more than six rust spots. Moderate rusting indicates more than six rust spots but confined to less than 5% of the surface of the test rod. Severe rusting indicates more than 5% of the surface is covered with rust. In order to pass the test, not even one rust spot can be present on the surface of the test rod. With a decrease in the IFT, a visual evaluation of the iron pin showed a gradual decrease in the rusting severity, from severe to moderate and light. Inhibited white oil, containing 0.1 wt% of EASA and having a significantly lower IFT of only 9.3 mN/m, was found effective in preventing rust formation and passed the severe ASTM D 665 B rust test. The patent literature reported on the effect of a trace amount of 2,4,6-trimethylpyridine, collidine, on increasing the effectiveness of the rust inhibitor EASA in preventing rust formation in different base stocks, including white oil (Pillon and Asselin, 1993). The effect of 2,4,6-trimethylpyridine on the performance of the rust inhibitor EASA in white oil is shown in Table 10.11.

The hydrorefined white oil required 0.1 wt% of EASA to prevent the rust formation and pass the more severe ASTM D 665 B rust test. In the presence of a trace amount of 2,4,6-trimethylpyridine, only 5 ppm, the effectiveness of the rust inhibitor significantly improved and only 0.04 wt% was needed to prevent the rust formation. The min required treat rate of the rust inhibitor decreased by 60% from 0.1 to 0.04 wt%. Without the presence of the rust inhibitor, the addition of collidine at the treat rate of 50 ppm and an even higher treat rate of 100 ppm did not improve the rust preventive properties of white oil. Another commercial alkylated pyridine, 2,6-di-tert-butylpyridine, was tested but it was found not effective in improving the performance of the rust inhibitor EASA. A different chemistry or stereochemistry of 2,6-di-*tert*-butylpyridine prevented it from interacting with the rust inhibitor or the metal surface. The patent literature reported that the presence of 2,4,6-trimethylpyridine, without the presence of rust inhibitor EASA, had no effect on the IFT and the rust preventive properties of hydrorefined white oil (Pillon and Asselin, 1993). The IFT and spreading coefficient (*S*) of different hydrorefined white oils, containing 2,4,6-trimethylpyridine, are shown in Table 10.12.

While some batches of hydrorefined oils, having the same composition, were found to have a different IFT indicating the presence of some surface active

TABLE 10.12

IFT of Different Hydrorefined White Oils Containing
2,4,6-Trimethylpyridine

Oil/Distilled Water Interface at 24°C	2,4,6-Trimethylpyridine, ppm	IFT, mN/m	(S), mN/m
Hydrorefined white oil # 1	0	45.1	−4.8
	5	45.1	−4.8
	11	45.1	−4.8
Hydrorefined white oil # 2	0	39.0	1.3
	20	39.0	1.3

Source: From Pillon, L.Z. and Asselin, A.E., *Lubricating Oil Having Improved Rust Inhibition and Demulsibility*, US Patent 5,227,082, 1993.

contaminants, the presence of trace content of 2,4,6-trimethylpyridine had no effect on their IFT indicating no adsorption at the oil/water interface. The addition of 20 ppm of 2,4,6-trimethylpyridine was also found to have no effect on IFT of synthetic PAO 6 base stock. At a higher treat rate of 50 ppm, 2,4,6-trimethyl-pyridine had no significant effect on the IFT of SWI 6 base stocks thus confirming no surface activity towards the oil/water interface of base stocks. While the addition of collidine did not affect the IFT of hydrofined phenol extracted base stocks, it was found to increase the IFT of hydrofined NMP extracted base stocks. The addition of 20 ppm of 2,4,6-trimethylpyridine was found to increase the IFT of hydrofined 600N NMP extracted base stock from about 33 to 35 mN/m, measured at 24°C. Depending on the extraction solvent, the trace polar contaminants of hydrofined mineral base stocks can vary. The hydrofinished mineral base stocks may contain residual phenol solvent or higher N and BN contents. The effect of 2,4,6-trimethylpyridine on the IFT and the performance of the rust inhibitor EASA in 150N phenol extracted base stock is shown in Table 10.13.

At the treat rate of 0.03 wt% of EASA, the IFT of 150N phenol extracted base stock was found to decrease from 43.1 to 16.6 mN/m, indicating its adsorption at the oil/water interface and the rust inhibitor was ineffective in preventing the rust formation. At a higher treat rate of 0.04 wt% of EASA, the IFT of 150N phenol extracted base stock was found to further decrease from 16.6 to 15.8 mN/m, indicating its increased adsorption at the oil/water interface and the rust inhibitor was effective in preventing the rust formation. After the addition of only 22 ppm of collidine, at the treat rate of only 0.03 wt% of EASA, the IFT of 150N phenol extracted base stock actually increased from 16.6 to 20.7 mN/m, and the rust inhibitor was found effective in preventing the rust formation. The effect of 2,4, 6-trimethylpyridine on the IFT and the performance of the rust inhibitor EASA in 600N phenol extracted base stock is shown in Table 10.14.

At the treat rate of 0.04 wt% of EASA, the IFT of 600N phenol extracted base stock was found to decrease from 42.8 to 14.5 mN/m, indicating its adsorption at

TABLE 10.13
Effect of 2,4,6-Trimethylpyridine on IFT and Performance
of EASA in 150N Mineral Oil

150N (Phenol) EASA, wt%	2,4,6-Trimethylpyridine, ppm	Oil/Distilled Water IFT, mN/m	ASTM D 665 B Rating
0	0	43.1	Fail
0.03	0	16.6	Fail
0.04	0	15.8	Pass
0.03	22	20.7	Pass

Source: From Pillon, L.Z. and Asselin, A.E., *Lubricating Oil Having Improved Rust Inhibition and Demulsibility*, US Patent 5,227,082, 1993.

the oil/water interface and the rust formation was observed. At an increased treat rate of 0.05 wt%, the IFT further decreased from 14.5 to 13.4 mN/m, and no rust formation was observed. In the presence of only 5 ppm of collidine, the IFT of 600N phenol extracted base stock, containing only 0.04 wt% of EASA, increased from 14.5 to 20.9 mN/m, and no rust formation was observed. The performance of rust inhibitor, such as EASA, was found to be affected by the composition of base stocks, mainly by its N content, and other surface active contaminants. In the presence of collidine, the effectiveness of the rust inhibitor EASA in preventing rust formation in white oil and phenol extracted base stocks was found to increase. An increase in the IFT would indicate an interaction between collidine and the rust inhibitor leading to a decrease in its adsorption at the oil/water interface and thus an increase in its concentration at the oil/metal interface. The effect of 2,4,6-trimethylpyridine on the performance of the rust inhibitor EASA in synthetic PAO 6 base stock is shown in Table 10.15.

TABLE 10.14
Effect of 2,4,6-Trimethylpyridine on IFT and Performance
of EASA in 600N Mineral Oil

600N (Phenol) EASA, wt%	2,4,6-Trimethylpyridine, ppm	Oil/Distilled Water IFT, mN/m	ASTM D 665 B Rating
0	0	42.8	Fail
0.04	0	14.5	Fail
0.05	0	13.4	Pass
0.04	5	20.9	Pass

Source: From Pillon, L.Z. and Asselin, A.E., *Lubricating Oil Having Improved Rust Inhibition and Demulsibility*, US Patent 5,227,082, 1993.

TABLE 10.15

Effect of 2,4,6-Trimethylpyridine on Performance of EASA in Synthetic PAO 6 Oil

Synthetic PAO 6 EASA, wt%	2,4,6-Trimethylpyridine, ppm	ASTM D 665 B Rating
0	0	Fail
0.05	0	Fail
0.10	0	Fail
0.15	0	Fail
0.05	10	Pass
0.05	30	Fail
0.05	50	Fail

Source: From Pillon, L.Z. and Asselin, A.E., *Lubricating Oil Having Improved Rust Inhibition and Demulsibility*, US Patent 5,227,082, 1993.

At the treat rate of 0.05 wt%, the partially EASA was found not effective in PAO 6 base stock and a very severe rusting of the metal pin was observed. With an increase in the treat rate from 0.05 to as high as 0.15 wt%, no improvement in the rust preventive properties was observed. After the addition of only 10 ppm of collidine, 0.05 wt% of the rust inhibitor EASA was found effective in preventing the rust formation and no rusting of the metal pin was observed. However, an increase in the collidine content to 30–50 ppm was found not effective in improving the performance of the rust inhibitor EASA and a severe rusting was observed. The effect of the extraction solvent and the viscosity on the performance of the rust inhibitor EASA in different mineral base stocks is shown in Table 10.16.

TABLE 10.16

Effect of Extraction Solvent and Viscosity on EASA in Mineral Base Stocks

ASTM D 665 B Test	EASA, 0.03 wt%	EASA, 0.04 wt%	EASA, 0.05 wt%	EASA, 0.15 wt%
150N (Phenol)	Fail	Pass	—	—
150N (NMP) # 1	Fail	Pass	—	—
150N (NMP) # 2	Pass	—	—	—
600N (Phenol)	Fail	Fail	Pass	—
600N (NMP)	Pass	—	—	—
1400N (Phenol)	Fail	Fail	Fail	Fail

Source: From Pillon, L.Z. and Asselin, A.E., *Lubricating Oil Having Improved Rust Inhibition and Demulsibility*, US Patent 5,227,082, 1993.

150N phenol extracted base stocks were found to require a min 0.04 wt% of the rust inhibitor EASA to prevent rust formation and pass the more severe ASTM D 665 B rust test. In many cases, 150N NMP extracted base stocks, containing higher N and BN contents, were found to require lower treat rates of 0.03 wt% of the same rust inhibitor to prevent rusting. Higher viscosity 600N phenol extracted base stock was found to require a min 0.05 wt% of the rust inhibitor EASA to prevent rust formation and pass the more severe ASTM D 665 B rust test. The same viscosity 600N NMP extracted base stocks, containing higher N and BN contents, were found to require lower treat rates of 0.03 wt% of the same rust inhibitor to prevent rusting. The performance of rust inhibitor EASA was found to decrease with an increase in viscosity of mineral base stocks and, even at such a high treat rate of 0.15 wt%, was found not effective in high viscosity 1400N mineral base stocks. NMP extracted base stocks contain higher N and BN contents which consist of amines, pyridines, and acids. Amines and pyridines are soluble in hydrocarbons and water and they can form salts in the presence of acids. The solubility of amine and pyridine salts in hydrocarbons will vary depending on their chemistry and the solvency of base stocks.

10.3   CORROSION INHIBITION

In petroleum refining, the equipment rust and corrosion is a serious problem and a costly economic loss. Rusting and corrosion weaken the metal surface and lead to enhanced wear. The literature reported on some crude oils causing corrosion in the crude unit heater, the crude heater transfer line, and some parts of vacuum columns. The crude oils from California, Romania, Russia, Texas, and the Arabian Peninsula, having a total acid number (TAN) above 0.5 mg KOH/g, were reported to contain naphthenic acids which were found corrosive at temperatures above 220°C. Their low molecular weight decomposition products, such as propionic, butyric, and formic acids, were found in petroleum fractions. Additionally, the hydrolysis and decomposition of magnesium chloride to produce hydrochloric acid was reported to start at around 120°C (Oyekunle and Dosunmu, 2003). Most Nigerian crude oils were reported to be relatively light and have a low sulfur content and, without a residue upgrading process, the total yield of distillates is above 70%. High yields of light products, such as gasoline, kerosene, and gas oils, were obtained. Nigerian crude oils were also reported to have a high naphthenic acid content. Their TAN was reported to vary from 0.14 mg KOH/g for Brass blend to 0.34 mg KOH/g for Escravos crude oil which was reported to cause severe corrosion (Oyekunle and Dosunmu, 2003). The effect of crude oils on TAN of kerosene and gas oil fractions is shown in Table 10.17.

The use of light Nigerian crude oil increases the TAN from 0.012 mg KOH/g of kerosene to 0.04 mg KOH/g of gas oil. The use of heavy Nigerian crude oil further increases the TAN from 0.021 mg KOH/g of kerosene to 0.052 mg KOH/g of gas oil. The use of foreign crude oil also increases the TAN from 0.010 mg KOH/g of kerosene to 0.045 mg KOH/g of gas oil. The TAN of petroleum fractions increases with the use of heavy crude oils and an increase in

TABLE 10.17
Effect of Heavy Crude Oil on TAN of Kerosene and Gas Oil Fractions

Nigeria Crude Oils	Light Crude Oil	Heavy Crude Oil	Foreign Crude Oil
Kerosene			
Density, g/cc	0.804	0.794	0.804
TAN, mg KOH/g	0.012	0.021	0.010
Gas Oil			
Density, g/cc	0.846	0.836	0.853
TAN, mg KOH/g	0.040	0.052	0.045

Source: From Oyekunle, L.O. and Dosunmu, A.J., *Petroleum Science and Technology*, 21, 1499, 2003. Reproduced by permission of Taylor & Francis Group, LLC, http://www.taylorandfrancis.com.

the density of petroleum fractions. An increase in the temperature was reported to further increase the corrosive properties of kerosene and gas oil on iron and galvanized steel surfaces. The kerosene fraction, having lower density and a lower TAN, was found to be more corrosive. The corrosion growth rate in kerosene and gas oil fractions, produced from Nigerian crude oils, is shown in Table 10.18.

The literature reported on the rate of corrosion of iron and galvanized steel immersed in kerosene and gas oil at the temperature varying from 200°C to 260°C. With an increase in temperature, the rate of corrosion drastically increased from 1.80 to 4.70 for kerosene and 1.50 to 4.29 for gas oil, indicating that

TABLE 10.18
Corrosion on Iron and Galvanized Steel in Kerosene and Gas Oil Fractions

Corrosion Growth Rate Weight Loss %	Nigeria Light Crude Oil	Nigeria Heavy Crude Oil	Foreign Crude Oil
Kerosene			
Iron	1.80	2.30	4.40
Galvanized steel	4.20	2.90	4.70
Gas Oil			
Iron	1.50	2.70	3.20
Galvanized steel	4.20	3.20	2.60

Source: From Oyekunle, L.O. and Dosunmu, A.J., *Petroleum Science and Technology*, 21, 1499, 2003. Reproduced by permission of Taylor & Francis Group, LLC, http://www.taylorandfrancis.com.

TABLE 10.19
Effect of Basic Type Corrosion Inhibitors on Corrosion of Carbon Steel

Tetraethylene Glycol with 10% Water	Basic Type Inhibitor	Liquid Phase Corrosion Rate, mm/year	Vapor Phase Corrosion Rate, mm/year
Carbon steel at 120°C	None	0.21	0.73
Carbon steel at 120°C	MEA/200	0.001	0.438
Carbon steel at 120°C	HBSI/200	0.0252	0.4013
Carbon steel at 120°C	TBSI/200	0.0420	0.4526

Source: From Khidr, T.T. and Mohamed, M.S., *Petroleum Science and Technology*, 19, 547, 2001. Reproduced by permission of Taylor & Francis Group, LLC, http://www.taylorandfrancis.com.

kerosene was more corrosive (Oyekunle and Dosunmu, 2003). The usual components of TAN are acids but also organic soaps, soaps of heavy metals, organic nitrates, other compounds used as additives, and oxidation by-products. The literature also reported that the presence of organic acids within the aqueous tetraethylene glycol solvent causes corrosion of the steel material and the basic type amine compounds are added to neutralize acids which produce amine salts and amides. The effect of corrosion inhibitors, such as monoethanolamine (MEA/200) and synthesized corrosion inhibitors, such as hexadecene bis-succinimides (HBSI) and tetradecene bis-succinimides (TBSI), was reported (Khidr and Mohamed, 2001). The effect of basic type corrosion inhibitors on liquid and vapor corrosion growth rate on carbon steel at 120°C is shown in Table 10.19.

The use of a basic amine, such as MEA, was reported effective in decreasing the corrosion in the liquid phase but not in the vapor phase. The use of the synthesized basic type derivatives, such as HBSI and TBSI, was reported less effective as corrosion inhibitors but was reported effective in decreasing the pour points (Khidr and Mohamed, 2001). The literature reported on an extensive coupon corrosion testing program to study the effect of various heat stable salts on methyldiethanolamine (MDEA) corrosivity to various steels at different temperatures (Rooney et al., 1997). Each application, where petroleum products are used, will be affected by the variation in chemistry of acids and the composition of water which might include many other acidic type contaminants. Different electrochemical processes were established to study corrosion of carbon steel similar to corrosion damage in a catalytic oil refinery plant. The corrosion process was studied in a medium similar to sour refinery waters and containing $(NH_4)_2S$, NaCN, and having a pH of 8.8 (Sosa et al., 2002). The effect of organic and inorganic liquids, including acids, on the ST of water is shown in Table 10.20.

TABLE 10.20

Effect of Different Organic and Inorganic Compounds on ST of Water

Compounds	Content, wt%	Temp., °C	Solution ST, mN/m
Water	0	25	72.0
Acetic acid	10.01	30	54.6
n-Butyric acid	8.60	25	33.0
Aminobutyric acid	9.34	25	71.67
Propionic acid	9.80	25	44.0
o-Aminobenzoic acid	12.35	25	71.96
Diethanolamine	10.0	25	66.8
HCl	12.81	20	71.85
H_2SO_4	12.18	25	72.8
HNO_3	8.64	20	71.7
NaOH	5.66	18	75.9
NaOH	16.66	18	83.1
KOH	10.08	18	76.6
NH_4OH	9.51	18	67.9

Source: From *CRC Handbook of Chemistry and Physics*, 86th ed., Lide, D.R., Editor-in-Chief, CRC
Press, Boca Raton, 2005. Reproduced by permission of Routledge/Taylor & Francis Group, LLC.

The presence of about 10 wt% of low molecular weight organic acids, such as acetic, butyric, and propionic, decreases the ST of water. The presence of organic acids, containing an amino group, does not decrease the ST of water. Acidic molecules and amines adsorb at the oil/water interface and decrease the IFT of petroleum oils. While the presence of organic acids might increase the spreading coefficient of petroleum oil on water, the presence of amine derivatives which do not decrease the ST of water but decrease the IFT of petroleum oil against water might lead to water emulsification and decrease the wetting of the metal surface. An increase in temperature leads to rust and corrosion in burners, heat exchangers, storage tanks, and other equipment. At high temperatures, the presence of mercaptans from crude oils, and alkali chlorides, after desalting, will lead to corrosive acids, such as H_2S and HCl, and cause corrosion. Inorganic acids do not decrease the ST of water and the effect of hydrochloric acid will be different from organic acids. The early literature reported on the effect of different aqueous phases and their effect on the rust and corrosion formation on the metal surface immersed in different base stocks (Hughes, 1969). The effect of water pH on rusting and corrosion severity on the metal surface immersed in different base stocks is shown in Table 10.21.

In white oil, under the specific test conditions of 32 h at 22°C with agitation, the same severity rusting and corrosion was reported. The presence of distilled water, having a neutral pH of 6–7, aqueous H_2SO_4, having a strong acidic pH of 3, and aqueous NaOH, having a strong basic pH of 11, caused the same rusting

TABLE 10.21
Effect of Water pH on Rusting and Corrosion Severity in Different Base Stocks

22°C/32 h/Agitation Rusting Rating	White Oil Severity (0–8)	Mineral Oil # 1 Severity (0–8)	Mineral Oil # 2 Severity (0–8)
Distilled water (pH 6–7)	4–5	2–3	2–3
Aqueous H₂SO₄ (pH 3)	4–5	4	4
Aqueous NaOH (pH 11)	4–5	0	0

Source: From Hughes, R.I., *Corrosion Science*, 9, 535, 1969. Reprinted with permission from Elsevier.

and corrosion rated 4–5. The visual rating 4–5 indicates moderate to fairly severe rusting which means that 1%–3% of the metal surface is covered by rust and corrosion. In mineral base stocks, under the same test conditions and in the presence of distilled water, the severity of rusting decreased to 2–3. In the presence of aqueous sulfuric acid, no significant change in corrosion severity was observed but in the presence of highly basic aqueous NaOH, no rust or corrosion was observed.

The effect of water pH on the emulsion stability in mineral base stocks was studied. Using aqueous HCl, the pH of deionized water was adjusted from 6 to a strongly acidic pH of 2–4, and both hydrofined phenol and NMP extracted base stocks showed an increase in emulsion stability. Using aqueous NaOH, the pH of deionized water was adjusted to a strongly basic pH of 11–12 and both hydrofined phenol and NMP extracted base stocks showed an increase in emulsion stability. In petroleum products, an increased water-in-oil emulsion stability would indicate the formation of smaller water droplets and delay in the wetting of the metal surface. In mineral base stocks containing basic and acidic trace polar contaminants, water emulsification can delay the rusting and corrosion of the metal surface. It takes time to form rust or corrosion on the metal surface and different base stocks require different equilibrium times to achieve the constant IFT. The effect of equilibrium time and aqueous H₂SO₄ on the IFT and the corrosion severity on the metal surface immersed in different base stocks is shown in Table 10.22.

White oil, containing practically 100% saturate content, was reported to have a high IFT of 46 mN/m which decreased to only 45 mN/m after a long equilibrium time of 2 h. Mineral oils, containing saturate, aromatic, and heteroatom molecules, were reported to have a lower IFT of 35 mN/m which gradually decreased to 25–26 mN/m after the equilibrium time of 2 h indicating that it takes time for surface active molecules to adsorb. The presence of H₂SO₄ acid in water was reported to produce etched and pitted scars on the metal surface and the most extensive corrosion was observed in white oil, having the highest oil-aqueous H₂SO₄ IFT (Hughes, 1969). Corrosion is a more severe form of rusting caused by the presence of acid compounds in water. Similarly to rust formation in

TABLE 10.22

Effect of Time and Aqueous H_2SO_4 on IFT and Corrosion Severity in Base Stocks

Oil/Aqueous H_2SO_4 Interface at 25°C	White Oil IFT, mN/m	Mineral Oil # 1 IFT, mN/m	Mineral Oil # 2 IFT, mN/m
Initial	46	35	35
After 1 h equilibrium	45	28	26
After 2 h equilibrium	45	26	25
Metal/Aqueous H_2SO_4 Interface at 25°C/18 h	Severe corrosion	Corrosion	Corrosion

Source: From Hughes, R.I., *Corrosion Science*, 9, 535, 1969. Reprinted with permission from Elsevier.

the presence of distilled water, a lower IFT leads to a smaller size dispersed aqueous H_2SO_4 water droplets and their rate of approaching the metal surface is lower which delays the wetting and corrosion formation.

10.4 SURFACE ACTIVITY OF RUST AND CORROSION INHIBITORS

Rust inhibitors typically contain polar functional groups attached to the hydrocarbon. According to the early literature, to prevent rust formation, surface active rust inhibitor molecules need to adsorb and form a protective layer on the iron surface. The more strongly these polar compounds are attached to the metal surface, the better is the resulting rust protection (Von Fuchs, 1956). The adsorption mechanism can involve a physical or chemical interaction between the rust inhibitors and the metal surface but the rust inhibitor must be soluble in oil and in the monomer in order for interaction to take place. In many instances, water can become mixed with the lubricating oil and rusting of metal parts can occur. The use of effective rust inhibitors is required. The EASA was found to be an effective rust inhibitor in many mineral base stocks but not in highly saturated base stocks, such as SWI 6 (Pillon and Asselin, 1993). The literature reported on the use of alkylated succinic acids, their partial esters, and half amides, as effective rust inhibitors in lubricating oils (Mang, 2001). The literature also reported on the variation in the performance of rust inhibitors in different solvent refined mineral and hydroprocessed base stocks (Pillon, 2003b). The early literature reported on ASA in preventing rusting but not effective in preventing corrosion. Several other monocarboxylic acids, having chain length in a range of C_6–C_{18}, were studied but were found not as effective as ASA in preventing rust and corrosion (Hughes, 1969). The effect of acid chemistry on the rust and corrosion preventive properties of white oil is shown in Table 10.23.

Tested at 22°C and in the presence of distilled water, only ASA, having two acidic groups, and stearic acid, having a long hydrophobic chain, were effective in

TABLE 10.23
Effect of ASA and Other Acids on Rust and Corrosion Formation in White Oil

Dynamic Corrosion Test at 22°C/ White Oil (0.1 wt% Additive)	Acid Formula	Distilled Water Rating (0–8)	Aq H_2SO_4 Rating (0–8)
Alkyl succinic acid (ASA)	$HOOCCH_2CHRCOOH$	0	3
Octadecanoic acid (stearic acid)	$CH_3(CH_2)_{16}COOH$	0	6–7
Dodecanoic acid (lauric acid)	$CH_3(CH_2)_{10}COOH$	1	5
Octanoic acid (caproic acid)	$CH_3(CH_2)_6COOH$	3	6

Source: From Hughes, R.I., Corrosion Science, 9, 535, 1969. Reprinted with permission from Elsevier.

preventing rust formation in white oil. Other monocarboxylic acids, having a shorter hydrophobic chain, were not effective. At 22°C and in the presence of aqueous H_2SO_4, none of the acids, including ASA, were effective in preventing corrosion in white oil. In the presence of ASA, the severity of corrosion was rated "3" while more severe, rated "5–7" in the presence of other acids. The patent literature reported on the effect of acidic rust inhibitors, such as partially esterified alkylsuccinic acid (EASA), and basic corrosion inhibitors, such as polyamine (PA) and succinic anhydride amine (SAA), on the IFT of base stocks (Pillon et al., 1993). The performance of rust and corrosion inhibitors in different viscosity mineral base stocks is shown in Table 10.24.

TABLE 10.24
Performance of EASA and SAA in Different Viscosity Mineral Base Stocks

Mineral Base Stocks ASTM D 665 B Test	Inhibitor, 0	Inhibitor, 0.04 wt%	Inhibitor, 0.05 wt%	Inhibitor, 0.1 wt%	Inhibitor, 0.15 wt%
Neat 150N (phenol)	Fail				
150N/EASA	Fail	Pass			
150N/SAA	Fail	Fail	Fail	Pass	
Neat 600N (phenol)	Fail				
600N/EASA	Fail	Fail	Pass		
600N/SAA	Fail	Fail	Fail	Pass	
Neat 1400N (phenol)	Fail				
1400N/EASA	Fail	Fail	Fail	Fail	Fail
1400N/SAA	Fail	Fail	Fail	Fail	Pass

Source: From Pillon, L.Z., Reid, L.E., and Asselin, A.E., Lubricating Oil for Inhibiting Rust Formation, US Patent 5,397,487, 1995.

In comparison to the partially esterified alkylsuccinic derivative, the use of SAA in 150N and 600N mineral base stocks requires significantly higher treat rates. 150N phenol extracted base stock was found to require 0.1 wt% of the SAA to prevent the rust formation. 600N phenol extracted base stock was found to require 0.1 wt% of the SAA to prevent the rust formation. High viscosity 1400N phenol extracted base stock, containing up to 0.15 wt% of the partially EASA, did not prevent rust formation. Only the use of a highly basic corrosion inhibitor, such as SAA, at a high treat rate of 0.15 wt% was effective in preventing rust formation in high viscosity 1400N mineral base stock. Commercial rust inhibitors were developed for conventionally processed mineral base stocks which contain significant amounts of aromatic and polar compounds and have good solvency properties. The rust inhibitor EASA was found not effective in SWI 6 base stock and synthetic PAO 6, containing 100% saturate content, in preventing rust formation. The performance of rust inhibitor EASA and corrosion inhibitor SAA in different highly saturated base stocks is shown in Table 10.25.

The use of highly polar rust inhibitors in highly saturated base stocks, having poor solvency properties, will lead to additive precipitation or micelle formation. Hydrorefined white oil has only decreased solvency properties, due to a decrease in aromatic content but contains a high content of naphthenic molecules which are known to have good solvency properties. Highly saturated base stocks, such as SWI 6, contain over 99 wt% of the saturate content which is mostly iso-paraffinic indicating a decrease in solvency of SWI 6 when compared to white oil. In SWI 6, EASA was found ineffective while highly basic corrosion inhibitor SAA was found effective in preventing rust formation. Synthetic PAO 6 contains 100% of saturate content which has a different chemistry iso-paraffinic content and EASA

TABLE 10.25

Performance of EASA and SAA in Highly Saturated Base Stocks

Saturated Base Stocks, ASTM D 665 B Test	Inhibitor, 0	Inhibitor, 0.05 wt%	Inhibitor, 0.06 wt%	Inhibitor, 0.1 wt%	Inhibitor, 0.15 wt%
Neat white oil	Fail				
White oil/EASA	Fail	Fail	Fail	Pass	
White oil/SAA	Fail	Fail	Fail	Fail	Pass
Neat SWI 6	Fail				
SWI 6/EASA	Fail	Fail	Fail	Fail	Fail
SWI 6/SAA	Fail	Fail	Pass		
Neat PAO 6	Fail				
PAO 6/EASA	Fail	Fail	Fail	Fail	Fail
PAO 6/SAA	Fail	Fail	Pass		

Source: From Pillon, L.Z., Asselin, A.E., and MacAlpine, G.A., *Lubricating Oil Having an Average Ring Number of Less Than 1.5 Per Mole Containing Succinic Anhydride Amine Rust Inhibitor*, US Patent 5,225,094, 1993.

TABLE 10.26
Effect of Rust and Corrosion Inhibitors on IFT and Rust Formation in SWI 6 Oil

Rust Inhibitors in SWI 6	Treat Rate, wt%	IFT, mN/m	ASTM D 665 B Rating
Neat SWI 6 base stock	0	55.4	Fail
Alkyl succinic acid (ASA)	0.05	—	Fail
	0.10	—	Fail
	0.15	8.3	Fail
Dodecyl succinic acid (DSA)	0.05	—	Fail
	0.15	7.1	Fail
	0.25	—	Fail
Esterified alkyl succinic acid (EASA)	0.05	—	Fail
	0.10	—	Fail
	0.15	7.2	Fail
Sodium sulfonate	0.15	6.1	Fail
	0.30	—	Fail
	0.50	—	Fail
	0.70	—	Fail
Calcium sulfonate	0.15	4.9	Fail
	0.30	—	Fail
	0.50	—	Fail
	0.70	—	Fail
Polyamine (PA)	0.05	—	Fail
	0.10	—	Fail
	0.15	3.6	Fail
Succinic anhydride amine (SAA)	0.06	—	Pass
	0.10	—	Pass
	0.15	2.5	Pass

Source: From Pillon, L.Z., Asselin, A.E., and MacAlpine, G.A., *Lubricating Oil Having an Average Ring Number of Less Than 1.5 Per Mole Containing Succinic Anhydride Amine Rust Inhibitor*, US Patent 5,225,094, 1993.

was found ineffective while highly basic corrosion inhibitor SAA was found effective in preventing rust formation. The effect of different chemistry rust and corrosion inhibitors on IFT and rust preventive properties of SWI 6 base stock is shown in Table 10.26.

Highly polar additives, such as rust inhibitors, are known to have limited solubility in nonpolar highly saturated base stocks. At a high treat rate of 0.15 wt%, a visible precipitation of ASA from SWI 6 base stock was observed. The poor performance of ASA and DSA in SWI 6 base stock can be related to their low solubility in highly saturated base stocks. The use of rust inhibitor EASA, in place of ASA, should increase its solubility in highly saturated base stocks but poor performance was observed. The use of sodium and calcium sulfonates was also found ineffective in preventing rust formation in SWI 6 base stock. The use of a basic type inhibitor, such as PA, was also ineffective and only the use of a highly basic corrosion inhibitor, such as SAA, was found effective in preventing rust formation in SWI 6 base stock. At the same treat rate of 0.15 wt%, highly basic SAA was also found most effective in decreasing the IFT of SWI 6 base stock indicating a different mechanism of rust prevention. While typical acidic rust inhibitors are expected to adsorb on the metal surface, basic type corrosion inhibitors drastically decrease the IFT and delay the wetting of the metal surface.

10.5 EFFECT OF RUST AND CORROSION INHIBITORS ON EMULSION STABILITY

In industrial equipment, rusting is the most common type of corrosion and additives, such as rust and corrosion inhibitors, are used to protect the metal surface. The literature reported that the concentration of rust inhibitors needs to be carefully selected to give adequate protection under a variety of mild-to-severe rust regimes but not too high (Mang, 2001). The main concern with the use of rust and corrosion inhibitors is the fact that many lubricating oils, such as turbine oils, are required to possess the ability to protect the metal surface against rust formation and have good water separation properties. The adsorption of rust and corrosion inhibitors at the oil/water interface can lead to stable emulsions. The rust preventive properties of lubricating oils are usually tested following the more severe ASTM D 665 B rust test. The ASTM D 1401 test method for water separability of petroleum oils and synthetic fluids is usually used and low viscosity oils are tested at 54°C. The effect of the rust inhibitor EASA on IFT, the emulsion stability, and the rust preventive properties of white oil is shown in Table 10.27.

Without the rust inhibitor, testing of water separation properties of uninhibited white oil at 54°C indicated the presence of 17 mL of emulsion after standing

TABLE 10.27

Effect of EASA on IFT, Demulsibility, and Rust Preventive Properties of White Oil

White Oil EASA, wt%	IFT at 24°C, mN/m	Oil/Water/Emulsion, mL (after 1 min)	Rust Test, after 24 h
0	45.1	24/39/17	Fail
0.05	13.5	15/23/42	Fail
0.10	9.3	3/7/70	Pass

Source: From Pillon, L.Z. and Asselin, A.E., *Lubricating Oil Having Improved Rust Inhibition and Demulsibility*, US Patent 5,227,082, 1993.

TABLE 10.28
Effect of Rust and Corrosion Inhibitors on Demulsibility Properties
of Base Stocks

Base Stocks Inhibitor, 0.05%	EASA/Emulsion Separation Time, min	PA/Emulsion Separation Time, min	SAA/Emulsion Separation Time, min
Neat 150N mineral	10	10	10
150N (inhibitor)	No change	Increase	Drastic increase
Neat SWI 6	4	4	4
SWI 6 (inhibitor)	Decrease	No change	Increase

Source: From Pillon, L.Z., *Petroleum Science and Technology*, 21, 1461, 2003b.

for only 1 min. After the addition of 0.05 wt% of the rust inhibitor EASA, the IFT of white oil was found to decrease from 45.1 to 13.5 mN/m, indicating its adsorption at the oil/water interface. While inhibited white oil was found to have poor rust preventive properties, the volume of emulsion, tested after standing for 1 min, increased from 17 to 42 mL. To prevent rust formation, white oil requires 0.1 wt% of the partially EASA to pass the ASTM D 665 B rust test. With an increase in the treat rate of the rust inhibitor from 0.05 to 0.1 wt%, the IFT of white oil was found to further decrease to 9.3 mN/m indicating a further increase in adsorption at the oil/water interface and the volume of the emulsion, tested after standing for 1 min, was also found to further increase to 70 mL. Lubricating oils, containing rust inhibitors and other additives, are required to separate the emulsion in less than 30 min. The effect of a typical acidic rust inhibitor, such as EASA, and basic type corrosion inhibitors, such as PA and SAA, on demulsibility properties of base stocks is shown in Table 10.28.

All different chemistry rust inhibitors and corrosion inhibitors were found effective in preventing rust formation in 150N mineral base stock but their effect on the demulsibility properties was found to vary. The use of a typical acidic rust inhibitor, such as EASA, did not increase the emulsion stability of mineral base stocks. At the same treat rate, the use of a basic corrosion inhibitor, such as PA, was found to increase the emulsion stability by acting as an emulsifier. The use of a basic corrosion inhibitor, such as SAA, was found to drastically increase the emulsion stability of 150N mineral base stock by acting as an emulsifier. A further increase in the treat rate of the basic corrosion inhibitors did not increase the emulsion stability any further indicating the saturation at the oil/water interface and formation of micelles.

The use of a typical acidic rust inhibitor, such as EASA, in SWI 6 base stock was found not effective in preventing rust formation but it actually decreased the emulsion separation time by acting as a demulsifier. The use of a basic type corrosion inhibitor, such as PA, in SWI 6 base stock was found not effective in preventing rust formation but it was found to increase the emulsion stability by

TABLE 10.29

Effect of EASA/SAA Blends on Rust Prevention in SWI 6 Base Stock

EASA/SAA Ratio, wt%	Concentration (Total), wt%	ASTM D 665 B Rust Test
100:0	0.03	Fail
95:5	0.03	Fail
90:10	0.03	Fail
80:20	0.03	Fail
70:30	0.03	Fail
60:40	0.03	Pass
50:50	0.03	Pass
40:60	0.03	Pass
30:70	0.03	Pass
20:80	0.03	Pass
10:90	0.03	Pass
0:100	0.03	Fail

Source: From Pillon, L.Z., Reid, L.E., and Asselin, A.E., *Lubricating Oil for Inhibiting Rust Formation*, US Patent 5,397,487, 1995.

acting as an emulsifier. The use of a basic corrosion inhibitor, such as SAA, while effective in preventing rust formation was also found to increase the emulsion stability of SWI 6 base stock by acting as an emulsifier. To prevent degradation in the water separation of base stocks, the use of basic corrosion inhibitors needs to be minimized. The effect of rust inhibitor blends, EASA/SAA, on the rust preventive properties of SWI 6 base stock is shown in Table 10.29.

At the treat rate of 0.03 wt% and higher, the rust inhibitor EASA is not effective in preventing the rust formation in SWI 6 base stock. At the treat rate of 0.03 wt%, SAA is also not effective because its treat rate is too low. Some blends of these rust inhibitors were found effective at a low total treat rate of only 0.03 wt%. The rust inhibitor blends, containing the weight ratio of about 1:1, were found the most effective in preventing the rust formation. A synergism between two different rust inhibitors allows obtaining the rust protection at a lower concentration of the mixture than can be obtained when using a single rust inhibitor. The performance of the EASA/SAA blend, 1:1 weight ratio, on the rust preventive properties of highly saturated base stocks is shown in Table 10.30.

The use of the EASA/SAA blend is effective in preventing the rust formation in highly saturated base stocks and decreases the total treat rate of rust inhibitors, including a decrease in SAA, which causes an increase in emulsion stability. The use of the EASA/SAA blend is also effective in preventing the rust formation in mineral base stocks and decreases the total treat rate of rust inhibitors and the use

TABLE 10.30
Performance of EASA/SAA Blend in Different Highly Saturated Base Stocks

Saturated Base Stocks, ASTM D 665 B Test	Inhibitor, 0.03 wt%	Inhibitor, 0.04 wt%	Inhibitor, 0.05 wt%	Inhibitor, 0.1 wt%	Inhibitor, 0.15 wt%
White oil/EASA	Fail	Fail	Fail	Pass	—
White oil/SAA	Fail	Fail	Fail	Pass	—
White oil/EASA/SAA (1:1)	Fail	Pass	—	—	—
SWI 6/EASA	Fail	Fail	Fail	Fail	Fail
SWI 6/SAA	Fail	Fail	Fail	Pass	—
SWI 6/EASA/SAA (1:1)	Pass	—	—	—	—
PAO 6/EASA	Fail	Fail	Fail	Fail	Fail
PAO 6/SAA	Fail	Fail	Fail	Pass	—
PAO 6/EASA/SAA (1:1)	Pass	—	—	—	—

Source: From Pillon, L.Z., Reid, L.E., and Asselin, A.E., *Lubricating Oil for Inhibiting Rust Formation*, US Patent 5,397,487, 1995.

of SAA which will lead to less degradation in their demulsibility properties. The performance of the EASA/SAA mixture, blended in 1:1 weight ratio, on the rust preventive properties of different viscosity mineral base stocks is shown in Table 10.31.

TABLE 10.31
Performance of EASA/SAA Blend in Different Viscosity Mineral Base Stocks

Mineral Base Stocks, ASTM D 665 B Test	Inhibitor, 0.03 wt%	Inhibitor, 0.04 wt%	Inhibitor, 0.05 wt%	Inhibitor, 0.1 wt%	Inhibitor, 0.15 wt%
150N/EASA	Fail	Pass	—	—	—
150N/SAA	Fail	Fail	Fail	Pass	—
150N/EASA/SAA (1:1)	Pass	—	—	—	—
600N/EASA	Fail	Fail	Pass	—	—
600N/SAA	Fail	Fail	Fail	Pass	—
600N/EASA/SAA (1:1)	Pass	—	—	—	—
1400N/EASA	Fail	Fail	Fail	Fail	Fail
1400N/SAA	Fail	Fail	Fail	Fail	Pass
1400N/EASA/SAA (1:1)	Fail	Fail	Fail	Pass	—

Source: From Pillon, L.Z., Reid, L.E., and Asselin, A.E., *Lubricating Oil for Inhibiting Rust Formation*, US Patent 5,397,487, 1995.

The use of different refining techniques affects the composition and the solvency of base stocks leading to a decrease in the efficiency of typical polar rust inhibitors to prevent rust formation and the use of corrosion inhibitors is required. Many different chemistry rust inhibitors, including partially EASA and PA, and corrosion inhibitors, such as SAA, were found effective in preventing the rust formation in lower viscosity mineral base stocks. While the use of corrosion inhibitors is effective in preventing rust formation in a wide range of different base stocks, their use leads to an increase in emulsion stability. Adsorption of basic type rust and corrosion inhibitors at the oil/water interface will decrease their IFT and delay the wetting of the metal surface. The use of different chemistry rust and corrosion inhibitors is frequently required to assure their solubility and surface activity in different base stocks, having different composition and solvency properties. The understanding of the base stock properties and the surface activity of additives is critical in improving their performance without degrading other properties.

REFERENCES

ASTM D 665-06 Standard Test Method for Rust Preventing Characteristics of Inhibited Mineral Oil in the Presence of Water, *Annual Book of ASTM Standards*, 2006.

Bondi, A., *Physical Chemistry of Lubricating Oils*, Reinhold Publishing Corporation, New York, 1951.

CRC Handbook of Chemistry and Physics, 86th ed., Lide, D.R., Editor-in-Chief, CRC Press, Boca Raton, 2005.

Groszek, A.J., *Nature*, 196, 531, 1962.

Hughes, R.I., *Corrosion Science*, 9, 535, 1969.

Hughes, J.M., Mushrush, G.W., and Hardy, D.R., *Petroleum Science and Technology*, 20 (7&8), 809, 2002.

Khidr, T.T. and Mohamed, M.S., *Petroleum Science and Technology*, 19(5&6), 547, 2001.

Mang, T., Base oils, in *Lubricants and Lubrications*, Mang, T. and Dresel, W., Eds., Wiley-VCH, Weinheim, 2001.

Nelson, A.J., X-ray photoemission spectroscopy, in *Microanalysis of Solids*, Yacobi, B.G., Holt, D.B., and Kazmerski, L.L., Eds., Plenum Press, New York, 1994.

Oyekunle, L.O. and Dosunmu, A.J., *Petroleum Science and Technology*, 21(9&10), 1499, 2003.

Pillon, L.Z., *Petroleum Science and Technology*, 21(9&10), 1453, 2003a.

Pillon, L.Z., *Petroleum Science and Technology*, 21(9&10), 1461, 2003b.

Pillon, L.Z. and Asselin, A.E., *Lubricating Oil Having Improved Rust Inhibition and Demulsibility*, US Patent 5,227,082, 1993.

Pillon, L.Z., Asselin, A.E., and MacAlpine, G.A., *Lubricating Oil Having an Average Ring Number of Less Than 1.5 Per Mole Containing Succinic Anhydride Amine Rust Inhibitor*, US Patent 5,225,094, 1993.

Pillon, L.Z., Reid, L.E., and Asselin, A.E., *Lubricating Oil for Inhibiting Rust Formation*, US Patent 5,397,487, 1995.

Rooney, P.C., DuPart, M.S., and Bacon, T.R., *Hydrocarbon Processing (International Edition)*, 76, 65, 1997.

Rudnick, L.R. and Shubkin, R.L., Poly(alfa-olefins), in *Synthetic Lubricants and High Performance Functional Fluids*, 2nd ed., Rudnick, L.R. and Shubkin, R.L., Eds., Marcel Dekker, Inc., New York, 1999.

Sosa, E. et al., *Journal of Applied Electrochemistry*, 32, 905, 2002.

Von Fuchs, G.H., *Symposium on Steam Turbine Oils*, ASTM Special Technical Publication No 211, Los Angeles, California, 1956.

Vyazovkin, E.S. et al., *Khim. Tekhnol. Topl. Masel*, 3, 7, 1990.

Yacobi, B.G., Holt, D.B., and Kazmerski, L.L., *Microanalysis of Solids*, Plenum Press, New York, 1994.

11 Ion-Exchange Resin Treatment

11.1 USE OF ADSORBENTS

With an increased environmental demand for cleaner burning fuels, a decrease in S and N contents of fuels is required. Burning of N compounds was reported to contribute to acid rain and greenhouse effects. The literature reports on the use of different methods to desulfurize fuels. The literature reviewed the selective removal of sulfur compounds from petroleum products using ionic liquids. Ionic liquids are salts, containing cationic quaternary nitrogen, which melt below room temperature (Aslanov and Anisimov, 2004). The literature reported on the efficiency of different methods, such as hydrochloric acid extraction, liquid–solid adsorption, and complexation, to remove basic nitrogen (BN) compounds from gasoline. Under the optimum conditions over 90% of BN can be removed (Feng, 2004). The literature reported on the desulfurization of middle distillates by an oxidation and extraction process. The oxidation was carried out using hydrogen peroxide and acetic acid as a catalyst. The extraction of sulfones was carried out using different polar solvents, such as acetonitrile, methanol, 2-ethoxyethanol, and γ-butyrolactone (Ramirez-Verduzco et al., 2004). The literature also reported on the use of hydrogen peroxide and sulfuric acid solutions for the removal of sulfur from fuel oil (Karaca and Yildiz, 2005). The optimum oxidation condition, such as a hydrogen peroxide concentration of 15 wt%, a temperature of 40°C, and a time of 60–50 min, was needed to desulfurize fuel oils (Yildiz and Karaca, 2005). The aromatic and heteroatom content of different light oils is shown in Table 11.1.

Desulfurization and denitrogenation of light oils have been investigated using methyl viologen modified alumino-silicate adsorbent. The literature reported that the use of 0.4–0.8 g of methyl viologen modified alumino-silicate adsorbent and 10 mL of different light oils resulted in a decrease in the S content by 24–29 wt% and a decrease in the N content by over 99% (Shiraishi et al., 2004). The results indicated a significantly higher surface activity of N compounds towards alumino-silicate adsorbent. The literature reported on the use of silica gel, containing transition metals, to desulfurize commercial fuels by adsorption of thiophenes (Ma et al., 2001). The literature also reported on the use of cuprous zeolites to desulfurize commercial diesel and jet fuels. An improved performance

TABLE 11.1
Aromatic and Heteroatom Content of Different Light Oils

Light Oil Composition	Commercial Light Oil	Straight-Run Light Gas Oil	Light Cycle Oil
Density, g/mL	0.8271	0.8548	0.8830
Saturates, vol%	78.2	74.4	33.7
1-Ring aromatics, vol%	19.6	14.9	36.4
2-Ring aromatics, vol%	2.2	9.7	29.9
Sulfur, ppm	560	13,800	1,320
Nitrogen, ppm	72	160	243

Source: Reprinted with permission from Shiraishi, Y., Yamada, A., and Hirai, T., *Energy and Fuels*, 18, 1400, 2004. Copyright American Chemical Society.

was observed when adding activated alumina as a guard bed. The use of 1 g of cuprous zeolites with alumina was reported effective in decreasing the S content of 30–50 mL of jet fuel from 364 to below 1 ppm (Hernandez-Maldonado et al., 2004).

The surface activity of middle distillates towards alumina was studied. Different middle distillates were dissolved in chloroform, adsorbed onto neutral alumina, and eluted with different polarity solvents, ranging from nonpolar *n*-hexane to toluene, 10% tetrahydrofuran (THF) in CHCl$_3$, and highly polar 10% ethanol in THF. A wide range of different heteroatom containing molecules, containing S, N, and O, were found to adsorb onto neutral alumina and, in terms of their surface activity, they were divided into four groups (Severin and David, 2001). The chemistry and surface activity of heterocompounds found in different middle distillates towards neutral alumina is shown in Table 11.2.

Paraffins are soluble in nonpolar paraffinic solvents and the least polar group, eluted with *n*-hexane, included only paraffins. Aromatic compounds are soluble in aromatic solvents and the more polar group, eluted with toluene, included all monoaromatic, polycondensed aromatic, and dibenzothiophene. The sulfur compounds identified in middle distillates were mainly dibenzothiophenes. A significantly more polar solvent, such as 10% THF in CHCl$_3$, was needed to remove all basic and nonbasic N compounds from neutral alumina. The nitrogen compounds were found to vary widely from low molecular weight anilines, pyridines, indoles, and quinolines to higher molecular weight compounds, such as carbazoles, acridines, and diphenylpyridines. The most surface active group of heterocompounds required a very polar solvent, such as 10% ethanol in THF, to remove diphenylpyridines and O containing molecules. The oxygen compounds were mainly phenols and dibenzofurane, but also benzaldehyde, 2-ethylhexanol, and *p*-octylanisol, which is a derivative of methyl phenyl ether (Severin and David, 2001). The surface activity of heteroatom molecules towards neutral alumina was found to increase from nonpolar paraffins to aromatics and dibenzothiophenes,

TABLE 11.2
Surface Activity of Heterocompounds Found in Different Middle Distillate

Neutral Alumina, n-Hexane	Neutral Alumina, Toluene	Neutral Alumina, 10% THF in CHCl₃	Neutral Alumina, 10% Ethanol in THF
Paraffins	Benzene	Aniline	Diphenylpyridines
	Indanes	Naphthylamine	—
	Naphthalenes	Diphenylaniline	Benzaldehyde
	Biphenyls	Pyridines	2-Ethylhexanol
	Fluorenes	Diphenylpyridine	p-Octylanisol
	Phenantrenes	Quinolines	—
	Pyrenes	—	—
	Chrysenes	Indoles	—
	—	Carbazoles	—
	Dibenzothiophenes	Acridines	—

Source: From Severin, D. and David, T., *Petroleum Science and Technology*, 19, 469, 2001.

more polar basic and nonbasic N compounds to highly polar O containing molecules which were found to be the most surface active. Crude oils and different petroleum products contain S, N, and O containing molecules; however, O content is difficult to analyze. The effect of solvent refining on the aromatic and heteroatom content of different hydrofined (HF) mineral base stocks is shown in Table 11.3.

TABLE 11.3
Effect of Solvent Refining on Heteroatom Content of Mineral Base Stocks

Mineral Base Stocks, Extraction Solvent	HF 600N, Phenol	HF 600N, NMP	HF 1400N, Phenol
Kin. Viscosity			
At 40°C, cSt	105.91	111.35	301.7
At 100°C, cSt	11.32	11.55	22.01
Viscosity index (VI)	92	89	89
Aromatics, wt%	19.5	19.6	28.2
Sulfur, wt%	0.12	0.12	0.19
Total nitrogen, ppm	30	100	141
Basic nitrogen, ppm	16	88	51

Source: From Pillon, L.Z., Asselin, A.E., and MacAlpine, G.A., *Lubricating Oil Having an Average Ring Number of Less Than 1.5 Per Mole Containing Succinic Anhydride Amine Rust Inhibitor*, US Patent 5,225,094, 1993.

With an increase in viscosity of HF phenol extracted mineral base stocks, their aromatic, S, N, and BN content increases. The use of NMP extraction solvent has no significant effect on aromatic and S content but leads to an increase in N and BN content. The review of patent literature identified the Canadian patent CA 1,266,007 which discusses the use of solid adsorbents, such as silica, alumina, different clays, zeolites, and charcoal, to increase the viscosity index (VI) of base stocks. The molecules of base stocks are completely adsorbed on the solid adsorbent and a selective diluent is used to remove molecules having a higher VI (Butler and Henderson, 1990). The review of the patent literature also mentions the US patent 3,620,969 which discusses the use of crystalline zeolite alumino-silicates for removal of S compounds from petroleum oils and the US patent 3,542,969 which mentions the use of activated carbon for removal of arsenic and its derivatives from petroleum oils (Butler and Henderson, 1990). The main components of the S content of mineral base stocks are dibenzothiophenes. The literature also reported that treating the aviation lubricating oil with activated charcoal decreased the foaming; however, the chemistry of surface active molecules was not identified (Zaki et al., 2002). Silica adsorbents were reported to be the most effective in removing polar molecules from petroleum oils (Basu et al., 1985). The effect of silica gel treatment on kinematic viscosity and composition of HF mineral base stocks is shown in Table 11.4.

The separation of saturate and aromatic fractions from petroleum high boiling oils can be achieved by elution chromatography. A weighted amount of petroleum oil is charged to the top of a glass chromatographic column and n-pentane can be used to elute the saturates. When all saturated hydrocarbons are eluted, polar solvents (diethyl ether, chloroform, and ethyl alcohol) are used to elute the aromatic and polar fraction. The solvents are completely evaporated and the residues are weighted. After the silica adsorbent treatment of HF 600N phenol

TABLE 11.4
Effect of Silica Gel on Viscosity and Composition of Mineral Base Stocks

Mineral Base Stocks, Extraction Solvent	HF 600N, Phenol	HF 600N, NMP	HF 1400N, Phenol
Kin. Viscosity			
At 40°C, cSt	75.4	76.4	155.7
Aromatics, wt%	1.6	0.7	1.7
Sulfur, ppm	—	6	37
Total nitrogen, ppm	0	0	0
Basic nitrogen, ppm	0	0	0

Source: From Pillon, L.Z., Asselin, A.E., and MacAlpine, G.A., *Lubricating Oil Having an Average Ring Number of Less Than 1.5 Per Mole Containing Succinic Anhydride Amine Rust Inhibitor*, US Patent 5,225,094, 1993.

extracted base stock, with a significant decrease in viscosity from 105.91 to 75.4 cSt at 40°C, a decrease in the aromatic content from 19.5 to 1.6 wt% and the total removal of N content were observed. After the silica adsorbent treatment of HF 600N NMP extracted base stock, with a significant decrease in viscosity from 111.35 to 76.4 cSt at 40°C, a decrease in the aromatic content from 19.6 to 0.7 wt%, a decrease in the sulfur content from 0.12 wt% to 6 ppm, and the total removal of N content were observed. After the silica adsorbent treatment of HF 1400N mineral base stock, with a drastic decrease in viscosity from 301.71 to 155.7 cSt at 40°C, a decrease in the aromatic content from 31.7 to 1.7 wt%, a decrease in the sulfur content from 0.19 wt% to 37 ppm, and also the total removal of N content were observed. The use of silica gel treatment confirmed a higher surface activity of N compounds and a lower surface activity of S molecules, similar to aromatic hydrocarbons. The mineral base stocks were diluted to decrease their viscosity and passed through different adsorbents to decrease their surface active content. The effect of different adsorbents, including alumina, on BN content and foaming tendency of diluted mineral base stocks is shown in Table 11.5.

The mineral base stocks were mixed with an equal volume of 1:1 heptane/toluene (v/v) to reduce their viscosity. At the temperature of 21°C, the diluted oils were stirred with selected adsorbents for 3 h, followed by filtration. After the treatment, the patent reported that only 1 wt% of base stocks were adsorbed onto adsorbents (Butler and Henderson, 1990). The ASTM D 892 foam test is used to determine the foaming characteristics of lubricating oils at 24°C (Seq. I) and 93.5°C (Seq. II), and then, after collapsing the foam, at 24°C (Seq. III). The foaming tendency is the volume of foam measured immediately after the cessation of airflow. The use of calcium oxide adsorbent, having basic properties, was found not effective in reducing the BN content; however, a decrease in foaming tendency of 150N mineral base stock is observed. The use of alumina adsorbent was reported effective in reducing the BN content of 150N mineral base stock,

TABLE 11.5
Effect of Adsorbents on BN and Foaming Tendency of Diluted Base Stocks

Diluted Mineral Base Stocks	BN, ppm	Seq. I at 24°C, Foaming, mL	Seq. II at 93.5°C, Foaming, mL	Seq. III at 24°C, Foaming, mL
150N feed	28	355	80	435
Calcium oxide	26	40	20	80
Activated calcium oxide	27	80	20	80
Alumina	9	0	10	0
600N feed	42	460	100	600
Calcium oxide	42	5	25	10
Florisil (MgO_3Si)	0	0	20	0

Source: From Butler, K.D. and Henderson, H.E., *Adsorbent Processing to Reduce Base Stock Foaming*, US Patent 4,600,502, 1990.

from 28 to 9 ppm, and a practically total resistance to foaming was observed (Butler and Henderson, 1990). The use of activated calcium oxide in higher viscosity 600N mineral base stock did not change the BN content; however, a drastic decrease in foaming tendency is observed. The use of fluorosil adsorbent, which is activated magnesium silicate, removed the BN content and a practically total resistance to foaming is observed (Butler and Henderson, 1990). The use of alumina adsorbent can decrease the foaming tendency of mineral base stocks without removing all BN molecules indicating a selective adsorption of more surface active molecules which would be heteroatom compounds containing oxygen. Different petroleum products contain S, N, and O containing molecules; however, usually only S and N contents are analyzed.

11.2   BASIC ION-EXCHANGE RESIN TREATMENT

An ion-exchange resin is one in which either anions or cations are loosely bonded to a larger three-dimensional network structure. The early literature reported that the synthetic resins, prepared by condensing aromatic amines with formaldehyde, have basic binding sites and exhibit anion-exchange or acid adsorbent properties (Daniels et al., 1970). Modern ion-exchange resins are mainly produced from styrene and divinylbenzene. The gel phase porosity can be manipulated by controlling the resin's degree of crosslinking with divinylbenzene (Kim and McNulty, 1997). Most of inorganic and organic molecules having a molecular weight below 200 can diffuse into and out of gelular structure. Other resins might be produced from acrylic acid polymers or by condensation of epichlorohydrin and amines (Kim and McNulty, 1997). Other type resins, known as macroreticular (MR), are available as polymeric adsorbents, having a high surface area, and as ion-exchange resins. Their pore size, porosity, and surface can be controlled (Kim and McNulty, 1997). Different ion-exchange resins and related polymeric adsorbents are commercially available and some are designed for use in nonaqueous applications. The functionality and properties of commercial basic Amberlite and Amberlyst resins are shown in Table 11.6.

While the purification of water is the most important application of ion-exchange resins, other applications include acid or base catalysis, manufacture of high purity solvents, separation of fermentation by-products, deacidification of organic solvents, and metal recovery. The recovery of iron, nickel, copper, vanadium, palladium, tungsten, gold, silver, mercury, and aluminum was reported (Kim and McNulty, 1997). The technical literature reported on the use of weak basic resin, A-21, in catalysis and deacidification of nonaqueous solutions. The strong basic resin, A-26, was reported to be used in catalysis while the strong basic resin, A-27, was reported used in base catalysis (Sigma-Aldrich, 2007). The functionality and properties of commercial acidic Amberlite and Amberlyst resins are shown in Table 11.7.

The technical literature reported on the use of strong acidic resin, A-15 (wet), in aqueous catalysis while the strong acidic resin, A-15, is being used in nonaqueous catalysis. The strong acidic resin, XN-1010, was recommended for systems

TABLE 11.6
Functionality and Properties of Basic Type Amberlite and Amberlyst Resins

Basic Resin Type	A-21, Weak Basic	A-26, Strong Basic	A-27, Strong Basic
Functionality	Dimethylamino	Quaternary Ammonium	Quaternary Ammonium
Max operating temp., °C	100	60(OH)/75(Cl)	60(OH)/75(Cl)
Moisture content, %	57	61	45
Surface area, m²/g	25	28	65
Porosity, %	48	27	51
Ion-exchange capacity, meq/mL	1.3	1.0	0.7
Ion-exchange capacity, meq/g	4.8	4.4	2.6

Source: Modified From Sigma-Aldrich Co., *Technical Information Bulletin*, Number AL-142, 2007.

with high surface area requirements (Sigma-Aldrich, 2007). The ion-exchange resins were reported to be used extensively in hydrometallurgy and the literature reported on increasing interest in applying ion-exchange technology to environmental remediation and waste minimization. The review of patents related to ion-exchange resins manufacture and their use was published (Duyvesteyn, 1998). The patent literature reported that the lube oil distillates were mixed with an equal volume of 1:1 heptane/toluene (v/v) to reduce their viscosity and, at the temperature of 21°C, the diluted oils were stirred with weak base type ion-exchange resin for 3 h, followed by filtration. The weak basic ion-exchange resin treatment

TABLE 11.7
Functionality and Properties of Acidic Type Amberlite and Amberlyst Resins

Acidic Resin Type	A-15 (wet), Strong Acidic	A-15, Strong Acidic	XN-1010, Strong Acidic
Functionality	Sulfonic Acid	Sulfonic Acid	Sulfonic Acid
Max operating temp., °C	120	120	120
Moisture content, %	50–55	<3	<3
Surface area, m²/g	45	45	540
Porosity, %	32	32	47
Ion-exchange capacity, meq/mL	1.8	1.8	1.0
Ion-exchange capacity, meq/g	4.7	4.7	3.3

Source: Modified From Sigma-Aldrich Co., *Technical Information Bulletin*, Number AL-142, 2007.

TABLE 11.8
Effect of Weak Basic Resin on BN Content and Foaming
of Lube Oil Distillates

Batch Treatment, Weak Basic Resin	BN, ppm	Seq. I at 24°C, Foaming, mL	Seq. II at 93.5°C, Foaming, mL	Seq. III at 24°C, Foaming, mL
60N lube distillate feed	128	160	30	120
After treatment	120	70	30	90
150N lube distillate feed	11	535	20	470
After treatment	9	30	15	20

Source: From Butler, K.D. and Henderson, H.E., *Adsorbent Processing to Reduce Base Stock Foaming*, US Patent 4,600,502, 1990.

was reported effective in decreasing the foaming tendency of lube oil distillates (Butler and Henderson, 1990). The effect of weak basic ion-exchange resin treatment on BN content and the foaming tendency of diluted lube oil distillates is shown in Table 11.8.

After the weak basic ion-exchange resin treatment, the BN content 60N lube oil distillate only decreased from 128 to 120 ppm; however, a significant decrease in Seq. I foaming tendency from 160 to 70 mL and Seq. III foaming tendency from 120 to 90 mL was observed. After the basic ion-exchange resin treatment, the BN content of higher viscosity 150N lube oil distillate also only decreased from 11 to 9 ppm; however, a drastic decrease in the Seq. I foaming tendency from 535 to 30 mL and Seq. III foaming tendency from 470 to 20 mL was observed. Under the same experimental conditions, the effectiveness of weak basic ion-exchange resin treatment in decreasing the foaming tendency of diluted lube oil distillates was found to increase with an increase in their viscosity. With an increase in the viscosity of lube oil distillate, their acidic content increases which would indicate a selective removal of acidic surface active contaminants. After solvent extraction and solvent dewaxing, HF mineral base stocks continue to have a high tendency to foam. The effect of weak basic ion-exchange resin on the BN content and the foaming tendency of diluted mineral base stocks is shown in Table 11.9.

The mineral base stocks were mixed with an equal volume of 1:1 heptane/toluene (v/v) and were treated with weak basic ion-exchange resins at 21°C. The patent literature reported that to optimize the processing conditions, the oil/adsorbent ratio for different viscosity base stocks needed to vary (Butler and Henderson, 1990). After the weak basic resin treatment, using the oil/resin ratio of 10.3 wt%/wt%, the low BN content of diluted 100N mineral base stock practically did not change; however, a drastic decrease in the Seq. I foaming tendency from 385 to 40 mL and the Seq. III foaming tendency from 370 to only 80 mL was observed. Using the oil/resin ratio of only 3.6 wt%/wt%, the BN content of diluted 600N base stock did not change but the Seq. I foaming

TABLE 11.9
Effect of Weak Basic Resin on BN Content and Foaming
of Diluted Base Stocks

Batch Treatment, Weak Basic Resin	BN, ppm	Seq. I at 24°C, Foaming, mL	Seq. II at 93.5°C, Foaming, mL	Seq. III at 24°C, Foaming, mL
100N base stock feed	2	385	25	370
After treatment	1	40	25	80
600N base stock feed	42	460	100	600
After treatment	42	5	10	0
Bright stock feed	68	30	220	120
After treatment	62	0	0	0

Source: From Butler, K.D. and Henderson, H.E., *Adsorbent Processing to Reduce Base Stock Foaming*, US Patent 4,600,502, 1990.

tendency decreased from 460 to 5 mL and the Seq. III foaming tendency decreased from 600 to 0 mL. Using a different oil/resin ratio of 10.8 wt%/wt%, a small decrease in the BN content of diluted bright stock from 68 to 62 ppm with a total elimination of foaming was reported (Butler and Henderson, 1990). The effect of dilution and temperature on the weak basic resin treatment of 150N mineral base stock is shown in Table 11.10.

Using the oil/resin ratio of 3.6 wt%/wt%, no significant decrease in the BN content of diluted 150N base stock was observed; however, the Seq. I foaming tendency decreased from 350 to 10–30 mL and the Seq. III foaming tendency decreased from 330 to 10–130 mL. After the weak basic resin treatment at 21°C,

TABLE 11.10
Effect of Weak Basic Resin Treatment on BN Content
and Foaming of 150N Oil

Batch Treatment, Weak Basic Resin	BN, ppm	Seq. I at 24°C, Foaming, mL	Seq. II at 93.5°C, Foaming, mL	Seq. III at 24°C, Foaming, mL
150N base stock feed	28	350	35	330
After treatment				
Diluted at 21°C	27	10	10	10
Diluted at 21°C[a]	26	30	20	130
Nondiluted at 21°C	27	0	30	10
Nondiluted at 40°C	28	0	20	15

Source: From Butler, K.D. and Henderson, H.E., *Adsorbent Processing to Reduce Base Stock Foaming*, US Patent 4,600,502, 1990.

[a] Repeated experiment

no significant decrease in the BN content of nondiluted 150N base stock was observed; however, the Seq. I foaming tendency was found to further decrease from 350 to 0 mL and the Seq. III foaming tendency was found to decrease from 330 to 10 mL. The use of nondiluted 150N base stock was reported to increase the efficiency of weak basic resin in decreasing the foaming tendency without any change in the BN content. After the weak basic resin treatment at 40°C, no significant decrease in the BN content of nondiluted 150N base stock was observed; however, the Seq. I foaming tendency was found to decrease from 350 to 0 mL and the Seq. III foaming tendency was found to decrease from 330 to 15 mL indicating no significant effect of an increased temperature. The use of the strong basic resin treatment and different viscosity nondiluted mineral base stocks also indicated no change in the BN content and a decrease in the foaming tendency with a decrease in the water content from a typical of 20–25 to less than 10 ppm. The use of the strong basic resin treatment and different viscosity nondiluted mineral base stocks also confirmed the viscosity effect. With an increase in the base stock viscosity, the efficiency of the strong basic resin treatment to decrease foaming was found to increase.

11.3 ACIDIC ION-EXCHANGE RESIN TREATMENT

The patent literature reported that the use of a strong acidic ion-exchange resin treatment was found effective in improving the demulsibility properties of HF mineral base stocks (Pillon and Asselin, 1994). After the strong acidic ion-exchange resin treatment of different viscosity nondiluted mineral base stocks, no presence of BN content was found and their water content was found to decrease to below 10 ppm. After the strong acidic ion-exchange resin treatment of nondiluted 150N mineral base stock, the Seq. I foaming tendency was reduced from 330 to 140 mL, while the foaming tendency of nondiluted 600N phenol extracted base stock was drastically reduced from 570 to 5 mL, and no foam stability was observed. With an increase in the viscosity of mineral base stocks, the efficiency of strong acidic resin treatment to decrease their foaming tendency was found to increase. The strong acidic ion-exchange resin treatment was found to be effective in removing the surface active molecules, having an affinity for water, and having emulsifying properties. The effect of hydrofining, silica gel, and acidic ion-exchange resin treatments on composition, IFT, and demulsibility properties of 150N NMP extracted base stock is shown in Table 11.11.

HF 150N NMP extracted base stock, containing 14.0 wt% of aromatics, 0.06 wt% of S, 36 ppm of N, and 33 ppm of BN, was found to have a low IFT of 33.4 mN/m and a positive spreading coefficient (S) of 5.6 mN/m on the water surface. After mixing with water, it required over 60 min to separate the emulsion indicating poor demulsibility properties. After the silica gel treatment, with a drastic decrease in the viscosity, the aromatic content decreased to 0.7 wt%, practically no heteroatom and no BN contents were present. The IFT of base stock increased from 33.4 to 42 mN/m and the emulsion separation time decreased from over 60 to 5 min. After the strong acidic ion-exchange resin

TABLE 11.11
Effect of Strong Acidic Resin Treatment on Properties of 150N Base Stock

150N Mineral Base Stock	Finishing Step, Hydrofining	Finishing Step, Silica Gel	Finishing Step, Strong Acidic Resin
Aromatics, wt%	14.0	0.7	14.7
Sulfur, wt%	0.06	2 ppm	0.06
Nitrogen, ppm	36	0	2
Basic N, ppm	33	0	0
IFT, mN/m	33.4	42.0	42.0
Spreading (S), mN/m	5.6	—	−3
ASTM D 1401 (54°C)			
O/W/E, mL	41/25/14	40/40/0	40/41/0
Separation time, min	60	5	5

Source: From Pillon, L.Z. and Asselin, A.E., *Method for Improving the Demulsibility of Base Oils*, US Patent 5,282,960, 1994.

treatment, no significant change in the aromatic and the sulfur contents of 150N mineral base stock was observed. Only the N content decreased to 2 ppm and no presence of BN content was found. The IFT of 150N NMP extracted base stock increased from 33.4 to 42 mN/m and the spreading coefficient (S) on the water surface decreased to a negative −3 mN/m. The time required to separate the emulsion also decreased to 5 min. The effect of strong acidic ion-exchange resin treatment on composition, IFT, and interfacial properties of 600N NMP extracted base stock is shown in Table 11.12.

The HF 600N NMP extracted base stock, containing 19.6 wt% aromatics, 0.12 wt% of S, 100 ppm of N, and 88 ppm of BN, was found to have an IFT of 39.4 mN/m and a negative (S) of −1.4 mN/m. After mixing with water, HF 600N base stock was found to require only 5 min to separate the emulsion at 54°C indicating excellent demulsibility properties. However, after mixing with air, HF 600N base stock was found to foam. After the strong acidic ion-exchange resin treatment, no change in the aromatic and the S contents but a significant decrease in N from 100 to 10 ppm and the total removal of BN content were observed. With an increase in the IFT from 39.4 to 42.4 mN/m, a further decrease in spreading coefficient (S) from −1.4 to −4.4 mN/m was observed. While HF 600N was found to have good demulsibility properties and no further decrease in the emulsion separation time was observed, the Seq. I foaming tendency was drastically decreased from 570 to 0 mL measured at 24°C. The results indicate the presence of different surface active contaminants adsorbing at the oil/air and oil/water interfaces of HF mineral base stocks. The use of the strong basic ion-exchange resin, while effective in decreasing the foaming tendency of base stocks, did not decrease the BN content and no increase in the IFT was observed. The effect of strong acidic ion-exchange resin treatments on composition,

TABLE 11.12
Effect of Strong Acidic Resin on Properties of 600N NMP Base Stock

600N Mineral Base Stock	Finishing Step, Hydrofining	Finishing Step, Strong Acidic Resin
Aromatics, wt%	19.6	19.6
Sulfur, wt%	0.12	0.12
Nitrogen, ppm	100	10
Basic N, ppm	88	0
IFT, mN/m	39.4	42.4
Spreading (S), mN/m	−1.4	−4.4
ASTM D 1401 (54°C)		
O/W/E, mL	39/40/1	40/40/0
Separation time, min	5	5
Foaming tendency	Yes	No

Source: From Pillon, L.Z. and Asselin, A.E., *Method for Improving the Demulsibility of Base Oils*, US Patent 5,282,960, 1994.

IFT, and demulsibility properties of 1400N phenol extracted base stock is shown in Table 11.13.

High viscosity mineral base stocks require more time to separate the emulsion. HF 1400N phenol extracted base stock, containing 28.2 wt% aromatics, 0.19 wt% of S, 141 ppm of N, and 51 ppm of BN, was found to have an IFT of

TABLE 11.13
Effect of Strong Acidic Resin Treatment on Properties of 1400N Base Stock

1400N Mineral Base Stock	Finishing Step, Hydrofining	Finishing Step, Silica Gel	Finishing Step, Strong Acidic Resin
Aromatics, wt%	28.2	1.7	28.7
Sulfur, wt%	0.19	37 ppm	0.18
Nitrogen, ppm	141	0	75
Basic N, ppm	51	0	0
IFT, mN/m	40.8	44.4	44.3
ASTM D 1401 (54°C)			
O/W/E, mL	44/36/2	40/40/0	41/38/2
Separation time, min	45	5	15

Source: From Pillon, L.Z. and Asselin, A.E., *Method for Improving the Demulsibility of Base Oils*, US Patent 5,282,960, 1994.

TABLE 11.14

Surface Active Content of Mineral Base Stocks Separated by Acidic Resin

Strong Acidic Resin Treatment	Isolated Surface Active Contaminants
1400N phenol extracted base stock	Mostly esters
150N NMP extracted base stock	Acids, pyridines, amines
600N NMP extracted base stock	Acids, pyridines, amines

Source: From Pillon, L.Z., *Petroleum Science and Technology*, 21, 1469, 2003.

40.8 mN/m and it required 45 min to separate the emulsion. After the silica gel treatment, with a drastic decrease in viscosity, the aromatic content of 1400N phenol extracted base stock decreased to 1.7 wt%, the sulfur content decreased to 37 ppm, and no nitrogen and BN content were present. The IFT of 1400N phenol extracted base stock increased from 40.8 to 44.4 mN/m and the time required to separate the emulsion decreased from 45 to 5 min. After the strong acidic ion-exchange resin treatment, no significant decrease in the aromatics and the S content was observed but the N content decreased from 141 to 75 ppm and no presence of BN was found. The IFT increased from 40.8 to 44.3 mN/m and the emulsion separation time decreased to 15 min. The FTIR analysis of the surface active content of mineral base stocks, separated by the strong acidic ion-exchange resin, is shown in Table 11.14.

The FTIR analysis of the surface active content separated from 1400N phenol extracted base stock, having poor demulsibility properties, indicated mostly the presence of esters. Heteroatom compounds, containing O, are known to be highly surface active and adsorb at the oil/water interface. The FTIR analysis of the surface active content separated from 150N NMP extracted base stock, having poor demulsibility properties, indicated the presence of acids, pyridines, and amines. The FTIR analysis of the surface active content separated from 600N NMP extracted base stock, having good demulsibility properties, also indicated the presence of acids, pyridines, and amines. The early literature reported that the long chain aliphatic amines can easily oxidize to form oxides which are surface active and used as foam stabilizers in liquid dishwashing formulations. In the presence of acids, amine oxides become protonated to form salts which further increases their surface activity and foam stabilizing properties (Lake and Hoh, 1963). The literature reported that the BN compounds found in crude oils are usually pyridines and quinolines and, if they have a low molecular weight, can be extracted with dilute mineral acids and analyzed (Speight, 1999). The presence of acids, pyridines, and amines in HF mineral oils indicates the presence of surface active salts which might be resistant to hydrofining.

11.4 THERMAL STABILITY

Most petroleum oils, when exposed to air over time, react with oxygen and a long-term storage of base stocks can lead to their discoloration. The color of mineral base stocks can vary from pale to red, and heavy base stocks are darker in color. In finished lubricants, color has little significance except in the case of medicinal and industrial white oils (Pirro and Wessol, 2001). According to the literature, poor color is related to the presence of nitrogen containing molecules and a darker color might indicate the presence of oxidation by-products. The literature reported on 900N base stock, produced at Yumen oil refinery, which contained 574 ppm of sulfur and 598 ppm of nitrogen. To improve its oxidation stability, a new method of denitrogenation by solid acid was implemented (Wang and Li, 2000). The polymerization and polycondensation of oxidation by-products can lead to sludge and deposit formation (Bardasz and Lamb, 2003). The early literature reported that aliphatic amines are readily oxidized to form amine oxides while aromatic and heterocyclic amines require significantly more severe conditions (Lake and Hoh, 1963). The chemistry and the content of N molecules can affect the thermal stability of mineral base stocks and their storage stability. Alumina is used to make acidic catalysts for the catalytic cracking process and the literature reports on the catalyst deactivation caused by high surface activity of nitrogen compounds leading to sludge and deposit formation. Petroleum feedstocks and products are mixtures of hydrocarbons and heterocompounds, and molecules which are surface active towards one interface are usually also surface active towards another surface. The effect of molecular weight and chemistry of different hydrocarbons and heterocompounds on their spreading coefficient (S) on the water surface, at 20°C, is shown in Table 11.15.

TABLE 11.15

Effect of Different Compounds on Their Spreading Coefficient on Water Surface

Organic Liquids at 20°C	Formula	ST, mN/m	IFT, mN/m	(S), mN/m
n-Hexane	C_6H_{14}	18.4	51.1	3.3
n-Octane	C_8H_{18}	21.8	50.8	−0.1
Dodecane	$C_{12}H_{26}$	24.4	52.9	−4.5
Hexadecane	$C_{16}H_{34}$	27.5	53.8	−8.5
Benzene	C_6H_6	28.9	35.1	8.8
Aniline	$C_6H_5NH_2$	42.9	5.9	24
Diethylether	$C_4H_{10}O$	17	10.7	45.1
Butanol	$C_4H_{10}O$	24.6	1.6	46.6
Octanol	$C_8H_{18}O$	27.5	8.5	36.8
Heptanoic acid	$C_7H_{14}O_2$	27.8	7.0	38.8
Ethyl acetate	$C_4H_8O_2$	23.9	3	45.9

The saturated hydrocarbons, such as n-octane, dodecane, and hexadecane, have a relatively low ST of 21.7–27.5 mN/m but a high liquid/water tension of 51.7–53.8 mN/m and do not spread on the water surface. The paraffins are the least polar fraction of crude oils. The monoaromatic hydrocarbon, such as benzene, has increased ST and a lower IFT which leads to a positive spreading coefficient (S) of 8.8 mN/m. The aromatic hydrocarbons, along with dibenzothiophenes, were found to be a more polar fraction of crude oils and petroleum products. Aromatic amine, such as aniline, has a significantly higher surface tension of 42.9 mN/m and a drastically lower liquid/water tension of only 5.9 mN/m. Aminobenzene has a significantly higher spreading coefficient of 23.2 mN/m indicating an increased surface activity on the water surface.

However, the most polar and surface active fraction of crude oils and petroleum products are heterocompounds containing O. Diethyl ether has a low surface tension of 17 mN/m and a low liquid/water tension of 10.7 mN/m which leads to a very high spreading coefficient of 45.1 mN/m. Butanol and octanol have an increased ST of 24.6–27.5 mN/m but significantly lower liquid/water tension of only 1.6–8.5 mN/m which also leads to a high spreading coefficient of 36.8–46.6 mN/m. Heptanoic acid has an increased ST of 27.8 mN/m but a low IFT of 7.0 mN/m and a high spreading coefficient of 38.8 mN/m on the water surface. Ethyl ethanoate was reported to have an ST of 23.9 mN/m and a very low IFT of only 3 mN/m leading to a very high spreading coefficient of 45.9 mN/m on the water surface (Jaycock and Parfitt, 1981). The use of hydrotreated (HT) slack wax feed, followed by hydroisomerization, leads to lube oil base stocks having practically no aromatic, S, and N contents. The effect of strong acidic ion-exchange resin treatment on composition, IFT, and demulsibility properties of solvent dewaxed SWI 6 base stock is shown in Table 11.16.

SWI 6 base stock, containing 1.4 wt% of aromatic content, 6 ppm of S, and no N and BN contents, was found to have a low IFT of only 24.4 mN/m and a high spreading coefficient (S) of 15.7 mN/m. Solvent dewaxed SWI 6 base stock required 4 min to separate the emulsion. After the solvent dewaxing and the strong acidic ion-exchange resin treatment, only a small decrease in the aromatic content from 1.4 to 1.3 wt%, with no decrease in the S content, is observed, thus confirming a low surface activity of aromatic and S compounds. However, the IFT of SWI 6 base stock drastically increased from 24.4 to 46.8 mN/m indicating the removal of surface active O containing molecules. With a decrease in spreading coefficient (S) from 15.7 to −6.7 mN/m, the emulsion separation time decreased from 4 to 1 min indicating a decrease in O containing molecules having emulsifying properties. The effect of strong acidic ion-exchange resin treatment on composition, IFT, and demulsibility properties of HF SWI 6 base stock is shown in Table 11.17.

After hydrofining, the aromatic content of SWI 6 base stock decreased from 1.4 wt% to less than 0.5 wt% and the S content was eliminated indicating that some saturation was taking place. The IFT of SWI 6 also drastically increased from 24.4 to 45.6 mN/m leading to a decrease in the (S) to −5.6 mN/m. After hydrofining, the time required to separate the emulsion decreased to only 2 min

TABLE 11.16
Effect of Strong Acidic Resin on IFT of Solvent Dewaxed
SWI 6 Base Stock

Hydroprocessing SWI 6 Base Stock	After Solvent Dewaxing	After Strong Acidic Resin Treatment
Aromatics, wt%	1.4	1.3
Sulfur, ppm	6	6
Nitrogen, ppm	0	0
Basic N (BN), ppm	0	0
Equilibrium Time at 24°C		
IFT, mN/m (30 min)	24.4	46.8
Spreading (S), mN/m	15.7	−6.7
ASTM D 1401 (54°C)		
O/W/E, mL	40/38/2	40/40/0
Separation time, min	4	1

Source: From Pillon, L.Z. and Asselin, A.E., *Method for Improving the Demulsibility of Base Oils*, US Patent 5,282,960, 1994.

TABLE 11.17
Effect of Strong Acidic Resin on IFT of Hydrofined
SWI 6 Base Stock

Hydroprocessing SWI 6 Base Stock	After Hydrofining	After Strong Acidic Resin Treatment
Aromatics, wt%	<0.5	<0.5
Sulfur, ppm	0	0
Nitrogen, ppm	0	0
Basic N (BN), ppm	0	0
Equilibrium Time at 24°C		
IFT, mN/m (30 min)	45.6	48.1
Spreading (S), mN/m	−5.6	−8.1
ASTM D 1401 (54°C)		
O/W/E, mL	40/40/0	40/40/0
Separation time, min	2	1

Source: From Pillon, L.Z. and Asselin, A.E., *Method for Improving the Demulsibility of Base Oils*, US Patent 5,282,960, 1994.

TABLE 11.18
Effect of Hydrotreatment on Heteroatom Content of VGO

HT VGO from Light Arabian Crude Oil	Properties and Composition
Viscosity at 75°C, cSt	10.55
Viscosity at 100°C, cSt	5.87
Sulfur, wt%	0.08
Nitrogen, wt%	0.03
Carbon residue (CCR), wt%	0.12

Source: From Okuhara, T. et al., *Petroleum Science and Technology*, 19, 685, 2001. Reproduced by permission of Taylor & Francis Group, LLC, http://www.taylorandfrancis.com.

indicating the presence of some residual surface active molecules. After hydrofining and the strong acidic ion-exchange resin treatment, a further increase in the IFT from 45.6 to 48.1 mN/m is observed. With a further decrease in the (S) to −8.1 mN/m, the emulsion separation time further decreased from 2 to 1 min indicating a further decrease in the surface active contaminants adsorbing at the oil/water interface and having emulsifying properties. The FTIR analysis of HF SWI 6 base stock, before and after the strong acidic ion-exchange resin treatment, indicated a decrease in the carboxylate acid content. After the use of HT slack wax feed, hydroisomerization process, and hydrofinishing step, the presence of highly surface active carboxylate acids, adsorbing on strong acidic resin and having emulsifying properties, was found. A severe hydrotreatment is used to desulfurize fuels and reduce their N content and the literature reported on the hydrotreatment of VGO, from light Arabian crude oil, used as feed to an FCC unit (Okuhara et al., 2001). The effect of hydrotreatment on the S and N contents of VGO, from light Arabian crude oil, is shown in Table 11.18.

After hydrotreatment, VGO was reported to contain 0.08 wt% of S and 0.03 wt% of N and have a carbon residue of 0.12 wt%. The nonbasic N compounds found in crude oils are usually the pyrrole, indole, and the carbazole types. The basic type N molecules, such as pyridines and quinolines, have a tendency to exist in the higher boiling fractions of crude oils and residue (Speight, 1999). After high temperature processing conditions, the chemistry of hydrocarbons and the residual heteroatom content, including N compounds, might be different from the ones present in the VGO feed. Also, the use of basic corrosion inhibitors, which are aliphatic type N containing molecules, might affect the BN content of the VGO feed and many other petroleum products. The basic type heterocyclic N compound, such as 2,4,6-trimethylpyridine, has a surface tension of 33.7 mN/m and does not

TABLE 11.19

Effect of Temperature on Stability of Heterocyclic and Aliphatic
N Compounds

Thermal Stability of SWI 6 Base Stock	ST, mN/m	IFT, mN/m	(S), mN/m
Neat SWI 6	33.1	46.7	−7.0
SWI 6 (2,4,6-trimethylpyridine)			
At room temperature	No change	No change	No change
At 150°C/air/>3 h	No change	Decrease	Increase
SWI 6 (N,N-dimethyldecylamine)			
At room temperature	No change	Decrease	Increase
At 150°C/air/>3 h[a]	No change	Decrease	Increase

[a] White sediment formed.

adsorb at the oil/water interface of base stocks, indicating a low surface activity; however, it can form salts. The basic type aliphatic amine, such as *N,N*-dimethyldecylamine having a formula $CH_3(CH_2)_9N(CH_3)_2$, has a lower ST of 27.2 mN/m and adsorbs at the oil/water interface indicating a significant increase in the surface activity and can also form salts. The effect of temperature on stability of 2,4,6-trimethylpyridine and *N,N*-dimethyldecylamine in SWI 6 is shown in Table 11.19.

The presence of oxygen and sunlight during storage was reported to accelerate the deterioration of mineral base stocks and the literature reported on the oxidation stability of base stocks (Pillon, 2003). At room temperature, the addition of 2,4,6-trimethylpyridine had no effect on the ST and the IFT of SWI 6 base stock. At 150°C and the presence of air, with an increase in time there was no change in the ST but a gradual decrease in the IFT of SWI 6 base stock was observed indicating the formation of oxidation by-products adsorbing at the oil/water interface. At room temperature, the addition of *N,N*-dimethyldecylamine had no effect on the ST but decreased the IFT of SWI 6 base stock leading to an increase in the spreading coefficient on the water surface. At 150°C and the presence of air, with an increase in time there was no change in the ST but a further decrease in the IFT of SWI 6 base stock was observed indicating the formation of oxidation by-products adsorbing at the oil/water interface. After 3 h of heating, a white sediment was formed. At room temperature, the addition of *N,N*-dimethyldecylamine had no effect on 150N mineral base stock but with an increase in time at 150°C and the presence of air, a brown sludge was formed. Solvent refined NMP extracted base stocks, having higher nitrogen and BN contents, are known to be more susceptible to oxidation and sludge formation. The basic type N molecules found in petroleum products might include aromatic, heterocyclic, and aliphatic molecules and their surface activity will be different. Aliphatic amines were found to have a higher surface activity and, at increased temperatures, leading to sediments and sludge.

11.5 FISCHER–TROPSCH PROCESS

Fuels and lube oil base stocks can be produced without the S and N contents using the Fischer–Tropsch technology which can use natural gas and coal to produce a carbon monoxide (CO) called syngas. According to the literature, the methane from natural gas, oxygen from air and water vapor can turn syngas (CO and H_2) into fluid and solid hydrocarbons waxes (Mang, 2001). There is a high research interest in the Fischer–Tropsch synthesis because it can use stranded natural gas and directly yield hydrocarbon-based transportation fuels. The fuel products are low in the S content and other impurities and are classified as clean fuels (Mahajan et al., 2003). The literature reported on Fisher–Tropsch cobalt catalysts used for syngas conversion to diesel fuel. A high temperature of 340°C produces alkenes, gasolines, and methanes while a low temperature, below 250°C, produces synthetic wax (Dunn et al., 2004). The Fischer–Tropsch process has been known for over 70 years and the recent literature reported on the commercial status and the challenges of converting natural gas to liquid transportation fuels (Pavone, 2007). The products from Fischer–Tropsch synthesis have a carbon range from methane to heavy paraffins, such as C_{50}–C_{200} (Henderson, 2003). The Fischer–Tropsch synthesis products are shown in Table 11.20.

The literature reported that the gas to liquid conversion, known as GTL, can be used to produce naphtha, kerosene, jet fuel, fuel oils, and other specialty products, such as petroleum waxes, solvents, polymers, and alcohols. The commercialization of the Fischer–Tropsch technology has a potential to produce high volume, low cost, and high quality base stocks (Kline & Company, 2002). The Fischer–Tropsch wax was reported to be over 99% paraffinic and can be hydrocracked, hydroisomerized, and iso-dewaxed to make high quality base stocks (Mang, 2001). The Fischer–Tropsch base stocks, also called GTL base stocks, were reported to have a high VI, 100% saturate content, and low volatility similar to synthetic products (Henderson, 2003). The properties and composition of

TABLE 11.20
Fischer–Tropsch Synthesis Products

Carbon Range	Products
C_1–C_2	Methane
C_3–C_4	NGL
C_5–C_{10}	Gasoline
C_{11}–C_{19}	Diesel
C_{20}–C_{25}	Low viscosity base stocks
C_{26}–C_{35}	Medium viscosity base stocks
C_{36+}	High viscosity base stocks

Source: From Henderson, H.E., *ACCN Canadian Chemical News*, 55, 17, 2003.

TABLE 11.21

Properties and Composition of GTL Base Stocks

Properties	GTL 2	GTL 3	GTL 5	GTL 7
Kin. Viscosity				
At 40°C, cSt	5.0	9.6	20.1	38.4
At 100°C, cSt	1.7	2.7	4.5	7.0
VI	—	117	144	147
Depressed pour point, °C	−59	−57	−39	−39
Iso-paraffins, wt%	100	100	100	100
Flash point, °C	159	199	238	260
Noack volatility, % off	—	34	8	2

Source: From Henderson, H.E., *Synthetics, Mineral Oils, and Bio-Based Lubricants: Chemistry and Technology*, Rudnick, L.R., Ed., CRC Press, Boca Raton, 2006. Reproduced by permission of Routledge/Taylor & Francis Group, LLC.

different viscosity GTL base stocks, containing 0.1 wt% of pour point depressant, are shown in Table 11.21.

The early patent literature reported that slack wax and Fischer–Tropsch synthetic waxes can be isomerized to produce lubricating oils, having a boiling point above 370°C, in the presence of hydrogen and catalysts comprising a noble group VIII metal on a small particle size, fluorided, refractory metal oxide (Cody and Brown, 1990). The Fischer–Tropsch process, used to produce synthetic wax followed by hydroisomerization, was reported to produce extra high VI base stocks, having a high saturate content and excellent oxidation stability (Phillipps, 1999). According to the literature, their only shortfall was reported to be their pour point. While the MS analysis indicated the presence of only iso-paraffins and no presence of *n*-paraffins, the use of pour point depressants is required to meet the low temperature requirements of lubricating oils (Henderson, 2003). Nonconventional base stocks, produced by the Fischer–Tropsch synthesis and converting gas to liquids, were reported to have high VI, above 140, and no S and N contents; however, their O content and interfacial properties were not reported.

REFERENCES

Aslanov, L.A. and Anisimov, A.V., *Petroleum Chemistry*, 44(2), 65, 2004.
Bardasz, E.A. and Lamb, G.D., *Lubricant Additives Chemistry and Applications*, Rudnick, L.R., Ed., Marcel Dekker, Inc., New York, 2003.
Basu, B. et al., *ASLE Transactions*, 28, 313, 1985.
Butler, K.D. and Henderson, H.E., *Adsorbent Processing to Reduce Base Stock Foaming*, US Patent 4,600,502, 1990.

Cody, I.A. and Brown, D.L., *Wax Isomerization Using Small Particle Low Fluoride Content Catalysts*, US Patent 4,923,588, 1990.

Daniels, F. et al., *Experimental Physical Chemistry*, 7th ed., McGraw-Hill Book Company, New York, 1970.

Dunn, B.C. et al., *Energy and Fuels*, 18, 1519, 2004.

Duyvesteyn, S., *JOM*, 62, October 1998.

Feng, Y., *Petroleum Science and Technology*, 22(11&12), 1517, 2004.

Henderson, H.E., *ACCN Canadian Chemical News*, 55(8), 17, 2003.

Henderson, H.E., Chemically modified mineral oils, in *Synthetics, Mineral Oils, and Bio-Based Lubricants: Chemistry and Technology*, Rudnick, L.R., Ed., CRC Press, Boca Raton, 2006.

Hernandez-Maldonado, A.J., Yang, R.T., and Cannella, W., *Industrial and Engineering Chemistry Research*, 43, 6142, 2004.

Jaycock, M.J. and Parfitt, G.D., *Chemistry of Interfaces*, Ellis Horwood Ltd., Chichester, 1981.

Karaca, H. and Yildiz, Z., *Petroleum Science and Technology*, 23(3&4), 285, 2005.

Kim, I. and McNulty, J.T., *Chemical Engineering*, 104(6), 94, 1997.

Kline & Company, Inc., *Report on GLT Specialties: High Value Opportunity or Threat*, 2002.

Lake, D.B. and Hoh, G.L.K., *Journal of American Oil Chemists' Society*, 40, 628, 1963.

Ma, X. et al., *Pre-prints division of petroleum chemistry*, *ACS*, 46, 648, 2001.

Mahajan, D. et al., *Energy and Fuels*, 17, 1210, 2003.

Mang, T., Base oils, in *Lubricants and Lubrications*, Mang, T. and Dresel, W., Eds., Wiley-VCH, Weinheim, 2001.

Okuhara, T. et al., *Petroleum Science and Technology*, 19(5&6), 685, 2001.

Pavone, T., *PTQ Gas*, 25, 2007.

Phillipps, R.A., Highly refined mineral oils, in *Synthetic Lubricants and High-Performance Functional Fluids*, Rudnick, L.R. and Shubkin, R.L., Eds., 2nd ed., Marcel Dekker, Inc., New York, 1999.

Pillon, L.Z., *Petroleum Science and Technology*, 21(9&10), 1469, 2003.

Pillon, L.Z. and Asselin, A.E., *Method for Improving the Demulsibility of Base Oils*, US Patent 5,282,960, 1994.

Pillon, L.Z., Asselin, A.E., and MacAlpine, G.A., *Lubricating Oil Having an Average Ring Number of Less Than 1.5 Per Mole Containing Succinic Anhydride Amine Rust Inhibitor*, US Patent 5,225,094, 1993.

Pirro, D.M. and Wessol, A.A., *Lubrication Fundamentals*, 2nd ed., Marcel Dekker, Inc., New York, 2001.

Ramirez-Verduzco, L.F. et al., *Petroleum Science and Technology*, 22(1&2), 129, 2004.

Severin, D. and David, T., *Petroleum Science and Technology*, 19(5&6), 469, 2001.

Shiraishi, Y., Yamada, A., and Hirai, T., *Energy and Fuels*, 18, 1400, 2004.

Sigma-Aldrich Co., *Technical Information Bulletin*, Ion-Exchange Resins and Related Polymeric Adsorbents, Number AL-142, http://www.sigmaaldrich.com (accessed February 2007).

Speight, J.G., *The Chemistry and Technology of Petroleum*, 3rd ed., Marcel Dekker, Inc., New York, 1999.

Wang, Y. and Li, R., *Petroleum Science and Technology*, 18(7&8), 965, 2000.

Yildiz, Z. and Karaca, H., *Petroleum Science and Technology*, 23(3&4), 371, 2005.

Zaki, N.N., Poindexter, M.K., and Kilpatrick, P.K., *Energy and Fuels*, 16, 711, 2002.

Index

Printed in the United States
by Baker & Taylor Publisher Services